MODERN ASPECTS OF ELECTROCHEMISTRY

No. 52

Series Editors:
Ralph E. White
Department of Chemical Engineering
University of South Carolina
Columbia, SC 29208

Constantinos G. Vayenas
Department of Chemical Engineering
University of Patras
Patras 265 00
Greece

Managing Editor:
Maria E. Gamboa-Aldeco
1107 Raymer Lane
Superior, CO 80027

For further volumes:
http://www.springer.com/series/6251

Noam Eliaz
Editor

Applications of Electrochemistry and Nanotechnology in Biology and Medicine I

Springer

Editor
Prof. Noam Eliaz
Faculty of Engineering
School of Mechanical Engineering
Tel-Aviv University
Ramat Aviv
69978 Tel-Aviv
Israel
neliaz@eng.tau.ac.il

ISSN 0076-9924
ISBN 978-1-4614-0346-3 e-ISBN 978-1-4614-0347-0
DOI 10.1007/978-1-4614-0347-0
Springer New York Dordrecht Heidelberg London

Library of Congress Control Number: 2011934050

Printed on acid-free paper

Springer is part of Springer Science+Business Media (www.springer.com)

Preface

The emergence of nanoscience and nanotechnology has led to new developments in and applications of electrochemistry. These two volumes of Modern Aspects of Electrochemistry, entitled: "Applications of Electrochemistry and Nanotechnology in Biology and Medicine, I and II" address both fundamental and practical aspects of several emerging key technologies. All Chapters were written by internationally renowned experts who are leaders in their area.

The chapter by A. Heiskanen and J. Emnéus provides a lucid and authoritative overview of electrochemical detection techniques for real-time monitoring of the dynamics of different cellular processes. First, biological phenomena such as the cellular redox environment, release of neurotransmitters and other signaling substances based on exocytosis, and cellular adhesion, are discussed thoroughly. Next, the capabilities of electrochemical amperometric and impedance spectroscopic techniques in monitoring cellular dynamics are highlighted, in comparison to optical and other techniques. The applications of such techniques already include biosensors and microchip-based biological systems for cell biological research, medical research and drug development. Finally, the state-of-the-art and future developments, e.g. miniaturization of planar interdigitated electrodes in order to achieve a gap/width size regime on the nanometer scale and thus considerable signal amplification, are summarized.

Electron transfer by thermally activated hopping through localized centers is an essential element for a broad variety of vital biological and technological processes. The use of electrode/self-assembled monolayer (SAM) assemblies to explore fundamental aspects of long- and short-range electron exchange between electrodes and redox active molecules, such as proteins, is reviewed comprehensively in a chapter by D.H. Waldeck and D.E. Khoshtariya. The authors, who are pioneers in this area, nicely demonstrate that such bioelectrochemical devices with nanoscopically tunable physical properties provide a uniquely powerful system for fundamental electron transfer studies and nanotechnological applications. Studies on protein systems also reveal how the binding motif of the protein to the electrode can be changed to manipulate its behavior, thus offering many promising opportunities for creating arrays of redox active biomolecules.

A microbial fuel cell (MFC) is a bio-electrochemical transducer that converts microbial biochemical energy directly to electrical energy. In their authoritative chapter, J. Greenman, I.A. Ieropoulos and C. Melhuish overview lucidly the principles of biofilms, biofilm electrodes, conventional fuel cells, and MFCs. Potential applications of both biofilm electrodes and MFCs are suggested, including sensing, wastewater treatment, denitrification, power packs, and robots with full energy autonomy. The symbiotic association between microbial life-forms and mechatronic systems is discussed in detail by the authors, who are internationally renowned experts in this field.

The last three chapters in Volume I deal with surface modification of implants, namely surface biofunctionalization or coating. First, R. Guslitzer-Okner and D. Mandler provide concise survey of different electrochemical processes (electrodeposition, electrophoretic deposition, microarc deposition, electropolymerization, and electrografting) to form different coatings (conducting polymers, non-conducting polymers, sol-gel inorganic-organic polymer materials, oxides, ceramics, bioglass, hydroxyapatite and other calcium phosphates) on different substrates (titanium and its alloys, stainless steels, cobalt-chrome alloys, nitinol, and magnesium alloys). The authors who are highly experienced in this field demonstrate the applicability of these coatings for medical devices such as drug eluting stents and orthopedic implants.

Different electrochemical processes to render metal implants more biofunctional and various electrochemical techniques to characterize the corrosion resistance of implants or the adsorption of biomolecules on the surface are reviewed by T. Hanawa in his authoritative chapter. Electrodeposition of calcium phosphates or polyethylene glycol (PEG), as well as anodizing and micro-arc oxidation processes to obtain TiO_2 nanotube-type oxide film on Ti substrate, or electrochemical treatment to obtain nickel-free oxide layer on nitinol alloys, are described. The effects of different surfaces on phenomena such as cell adhesion, bacterial attachment and calcification are presented.

The last chapter in Volume I, by T. Kokubo and S. Yamaguchi, lucidly summarizes the pioneering work and inventions of these authors in the field of bone-bonding bioactive metals for orthopedic and dental implants. The metals include titanium, zirconium, niobium, tantalum and their alloys. The main surface modification technique presented in this chapter is chemical, followed by heat treatment, although other techniques such as ion implantation, micro-arc treatment, hydrothermal treatment and sputtering are also described. The bone-bonding ability of metals with modified surfaces is attributable to the formation of apatite on their surface in the body environment, which can be interpreted in terms of the electrostatic interaction of the metal surface with the calcium or phosphate ions in a body fluid. These findings open numerous opportunities for future work.

Volume II begins with a chapter by P.S. Singh, E.D. Goluch, H.A. Heering and S.G. Lemay which provides a lucid overview of the fundamentals and applications of nanoelectrochemistry in biology and medicine. First, some key concepts related to the double layer, mass transport and electrode kinetics and their dependence on the dimension and geometry of the electrode are discussed. Next, various fabrication schemes utilized in making nano-sized electrodes are reviewed, along with the inherent challenges in characterizing them accurately. Then, the "mesoscopic" regime is discussed, with emphasis on what happens when the Debye length becomes comparable to the size of the electrode and the diffusion region. Quantum-dot electrodes and charging and finite-size effects seen in such systems are also described. Then, recent advances in the electrochemistry of freely-diffusing single molecules as well as electrochemical scanning probe techniques used in the investigations of immobilized biomolecules are presented by the authors, who have pioneered several of the developments in this area. Finally, a brief survey of the applications of nanoelectrodes in biosensors and biological systems is provided.

During the last decade, nanowire-based electronic devices emerged as a powerful and universal platform for ultra-sensitive, rapid, direct electrical detection and quantification of biological and chemical species in solution. In their authoritative chapter, M. Kwiat and F. Patolsky describe examples where these novel electrical devices can be used for sensing of proteins, DNA, viruses and cells, down to the ultimate level of a single molecule. Additionally, nanowire-based field-effect sensor devices are discussed as promising building blocks for nanoscale bioelectronic interfaces with living cells and tissues, since they have the potential to form strongly coupled interfaces with cell membranes. The examples described in this chapter demonstrate nicely the potential of these novel devices to significantly impact disease diagnosis, drug discovery and neurosciences, as well as to serve as powerful new tools for research in many areas of biology and medicine.

The Human Genome Project has altered the mindset and approach in biomedical research and medicine. Currently, a wide selection of DNA microarrays offers researchers a high throughput method for simultaneously evaluating large numbers of genes. It is anticipated that electrochemical detection-based DNA arrays will provide many advantages over radioisotope- or fluorophore-based detection systems. Due to the high spatial resolution of the scanning electrochemical microscope (SECM), this technology has been suggested as a readout method for locally immobilized, micrometer-sized biological recognition elements, including a variety of DNA arrays with different formats and detection modes. In his concise review, K. Nakano explains the underlying electrochemistry facets of SECM and examines how it can facilitate DNA array analysis. Some recent achievements of Nakano and his colleagues in SECM imaging of DNA microdots that respond toward the target DNA through hybridization are presented.

Biological membranes are the most important electrified interfaces in living systems. They consist of a lipid bilayer incorporating integral proteins. In view of the complexity and diversity of the functions performed by the different integral proteins, it has been found convenient to incorporate single integral proteins or smaller lipophilic biomolecules into experimental models of biological membranes (i.e. biomimetic membranes), so as to isolate and investigate their functions. Biomimetic membranes are common in pharmaceuticals, as well as for the investigation of phase stability, protein-membrane interactions, and membrane-membrane processes. They are also relevant to the design of membrane-based biosensors and devices, and to analytical platforms for assaying membrane-based processes. The last two chapters in Volume II are dedicated to these systems. In their thorough chapter, R. Guidelli and L. Becucci review the principles and types of biomimetic membranes, the advantages and disadvantages of these systems, their applications, their fabrication methodologies, and their investigation by electrochemical techniques – mainly electrochemical impedance spectroscopy (EIS). This definitie chapter was written by two authors who are among the leaders leaders in the field of bioelectrochemistry worldwide.

Ion channels represent a class of membrane spanning protein pores that mediate the flux of ions in a variety of cell types. They reside virtually in all the cell membranes in mammals, insects and fungi, and are essential for life, serving as key components in inter- and intracellular communication. The last chapter in Volume II, by E.K. Schmitt and C. Steinem, provides a lucid overview of the potential of pore-suspending membranes for electrical monitoring of ion channel and transporter activities. The authors, who are internationally acclaimed experts in this area, have developed two different methods to prepare pore-suspending membranes, which both exhibit a high long-term stability, while they are accessible from both aqueous sides. The first system, nowadays known as nano black lipid membrane (nano-BLM), allows for ion channel recordings on the single channel level. The second system – pore-suspending membranes obtained from fusing unilamellar vesicles on a functionalized porous alumina substrate – makes it possible to generate membranes with high protein content. The electrochemical analysis of these systems is described thoroughly in this chapter, and is largely based on EIS.

I believe that the two volumes will be of interest to electrochemists, chemists, materials, biomedical and electrochemical engineers, surface scientists, biologists and medical doctors. I hope that they become reference source for scientists, engineers, graduate students, college and university professors, and research professionals working both in academia and industry.

I wish to thank Professor Eliezer Gileadi who was the driving force making me edit these two volumes. I dedicate this project to my wife Billie and our two daughters, Ofri and Shahaf, for their infinite love and support.

N. Eliaz
Tel-Aviv University
Tel-Aviv, Israel

Professsor Noam E Eliaz

Noam Eliaz is an Associate Professor at Tel-Aviv University, Israel, where he serves as the Head of The Biomaterials and Corrosion Laboratory and as the first Head of the multi faculty Materials and Nanotechnologies Program. He also serves as a Chief Editor of the journal *Corrosion Reviews* (jointly with Professor Ron Latanision). He received his B.Sc. and Ph.D. (direct track) in Materials Enngineering, both *cum laude*, from Ben-Gurion University. After completing his doctorate, he became the first ever materials scientist to receive, simultaneously, a Fulbright postdoctoral award and a Rothschild postdoctoral fellowship. He then worked for two years in the H. H. Uhlig Corrosion Laboratory at M.I.T. To-date, he has contributed more than 220 journal and conference publications, including 28 invited talks, as well as 4 book chapters. He is currently editing a book on *Degradation of Implant Materials*, to be published by Springer during 2011. He has garnered numerous accolades, including the T.P. Hoar Award for the best paper published in *Corrosion Science* during 2001 (with co-authors), and the 2010 Herbert H. Uhlig Award granted by NACE International in recognition of outstanding effectiveness in postsecondary corrosion education. His main research interests include corrosion, electrodeposition, biomaterials and bio-ferrography.

Contents

Chapter 1

MONITORING OF CELLULAR DYNAMICS WITH ELECTROCHEMICAL DETECTION TECHNIQUES

A. Heiskanen and J. Emnéus

Chapter 2

FUNDAMENTAL STUDIES OF LONG- AND SHORT-RANGE ELECTRON EXCHANGE MECHANISMS BETWEEN ELECTRODES AND PROTEINS

David H. Waldeck and Dimitri E. Khoshtariya

Chapter 3

MICROBIAL FUEL CELLS – SCALABILITY AND THEIR USE IN ROBOTICS

John Greenman, Ioannis Ieropoulos, and Chris Melhuish

Chapter 4

ELECTROCHEMICAL COATING OF MEDICAL IMPLANTS

Regina Guslitzer-Okner and Daniel Mandler

Chapter 5

ELECTROCHEMICAL TECHNIQUES TO OBTAIN BIOFUNCTIONAL MATERIALS

Takao Hanawa

Chapter 6

BIOACTIVE METALS PREPARED BY SURFACE MODIFICATION: PREPARATION AND PROPERTIES

Tadashi Kokubo and Seiji Yamaguchi

List of Contributors, MAE 52

Professor Jenny Emnéus
Department of Micro- and Nanotechnology, Technical University of Denmark, DK-2800 Kgs. Lyngby, Denmark
jenny.emneus@nanotech.dtu.dk
http://www.nanotech.dtu.dk/English/Medarbejdere/Staff/Emneus_Jenny.aspx.

Dr. Regina Guslitzer-Okner
Department of Inorganic and Analytical Chemistry, The Hebrew University of Jerusalem, Edmond J. Safra Campus, Givat Ram, Jerusalem 91904, Israel
reginaokner@gmail.com.

Professor John Greenman
School of Life Sciences, Faculty of Health and Life Sciences, Frenchay Campus, University of the West of England, Coldharbour Lane, BS16 1QY, Bristol, UK
John.Greenman@uwe.ac.uk,
http://sls.uwe.ac.uk/staff/StaffPage.asp?StaffID=J-GREENMAN.

Professor Takao Hanawa
Institute of Biomaterials and Bioengineering, Tokyo Medical and Dental University, 2-3-10 Kanda-Surugadai, Chiyoda-ku Tokyo 101-0062, Japan
Telefax: +81-3-5280-8006
hanawa.met@tmd.ac.jp,
http://www.tmd.ac.jp/i-mde/www/index-e.html.

Dr. Arto Heiskanen
Department of Micro- and Nanotechnology, Technical University of Denmark, DK-2800 Kgs. Lyngby, Denmark
Arto.Heiskanen@nanotech.dtu.dk.

Dr. Ioannis A. Ieropoulos
School of Life Sciences, Faculty of Health and Life Sciences, Frenchay Campus, University of the West of England, Coldharbour Lane, BS16 1QY, Bristol, UK
Ioannis.Ieropoulos@brl.ac.uk

Professor Dimitri E. Khoshtariya
Department of Chemistry, University of Pittsburgh, Chevron Science Center, 219 Parkman Avenue, Pittsburgh, PA 15260, USA
dimitri.k@joker.ge.

Professor Tadashi Kokubo
Department of Biomedical Sciences, College of Life and Health Sciences, Chubu University, 1200 Matsumoto, Kasugai, Aichi 487-8501, Japan
kokubo@isc.chubu.ac.jp,
http://www.chubu.ac.jp/english/faculty/profile/f12576b5c3fc3b10a9eeccd7a0c8aeb67e99ff85.html.

Professor Daniel Mandler
Department of Inorganic and Analytical Chemistry, The Hebrew University of Jerusalem, Edmond J. Safra Campus, Givat Ram, Jerusalem 91904, Israel
mandler@vms.huji.ac.il,
http://chem.ch.huji.ac.il/employee/mandler/imandler.htm.

Chris Melhuish
Bristol Robotics Laboratory, University of the West of England, Bristol Business Park, Coldharbour Lane, Bristol, BS16 1QD, UK

Professor David H. Waldeck
Department of Chemistry, University of Pittsburgh, Chevron Science Center, 219 Parkman Avenue, Pittsburgh, PA 15260, USA
dave@pitt.edu
http://www.pitt.edu/~dave/About.html.

Dr. Seiji Yamaguchi
Department of Biomedical Sciences, College of Life and Health Sciences, Chubu University, 1200 Matsumoto, Kasugai, Aichi 487-8501, Japan

Modern Aspects of Electrochemistry

Topics in Number 50 include:

- Investigation of alloy cathode Electrocatalysts
- A model Hamiltonian that incorporates the solvent effect to gas-phase density functional theory (DFT) calculations
- DFT-based theoretical analysis of ORR mechanisms
- Structure of the polymer electrolyte membranes (PEM)
- ORR investigated through a DFT-Green function analysis of small clusters
- Electrocatalytic oxidation and hydrogenation of chemisorbed aromatic compounds on palladium
  Electrodes
- New models that connect the continuum descriptions with atomistic Monte Carlo simulations
- ORR reaction in acid revisited through DFT studies that address the complexity of Pt-based alloys in electrocatalytic processes
- Use of surface science methods and electrochemical techniques to elucidate reaction mechanisms in electrocatalytic processes
- In-situ synchrotron spectroscopy to analyze electrocatalysts dispersed on nanomaterials

Topics in Number 51 include:

- Temperature effects on platinum single-crystal and aqueous solution interphases
- Surface thermodynamics of the metal and solution interface
- XAS investigations of PEM fuel cells
- Palladium-Based electrocatalysts required for alcohol oxidation in direct alcohol fuel cells
- Structure and Reactivity of transition metal chalcogenides used for molecular oxygen reduction reactions
- Proton conductivity and electrocatalysis in high temperature PEM fuel cells

1

Monitoring of Cellular Dynamics with Electrochemical Detection Techniques

A. Heiskanen and J. Emnéus

Department of Micro- and Nanotechnology, Technical University of Denmark, Ørsteds Plads 345 East, DK-2800 Kgs. Lyngby, Denmark

I. Introduction

The elucidation of the human and *Saccharomyces cerevisiae* genomes has opened new possibilities in biology, medicine and drug discovery. This knowledge has increased the number of targets and ways to treat diseases; clinical studies of disease manifestation can effectively be translated to the level of certain genes and their products, proteins. The traditional way of finding active compounds to treat different diseases has relied on biochemical assays to modulate the function of a certain target, such as an enzyme, a cell surface receptor or an ion channel. The targets have been isolated from the natural environment to be used in high-throughput screening (HTS) assays to test huge libraries of possible active compounds. At the same time has come the realization that cells and organisms are more than just the sum of their functional units. More needs to be known in order to fully understand the living organisms in terms of their functional and constructional units, the cells and proteins. This has further led to the development of ana-

N. Eliaz (ed.), *Applications of Electrochemistry and Nanotechnology in Biology and Medicine I*, Modern Aspects of Electrochemistry 52,
DOI 10.1007/978-1-4614-0347-0_1, © Springer Science+Business Media, LLC 2011

lytical systems, screening platforms, which rely on cell-based assays instead of dissecting cells to the component level. Techniques that have been applied for assaying biological parameters in intact cells range from intracellular monitoring of temperature[1], pH[2] and oxygen consumption[3] as well as extracellular monitoring of oxygen consumption[4] and acidification[4,5] to cellular adhesion[6] and wound healing,[7] G-protein coupled receptor (GPCR) activation,[8, 9] monitoring of enzyme activity,[10-13] cofactor availability,[14-17] cellularly released secondary metabolites,[18-21] gene expression detected as appearance of mRNA[22] and ion channel activity.[23]

A new development in treating diseases has also emerged; instead of solely relying on medication, cells that die due to the pathological manifestation of a disease could potentially be replaced by new cells that can restore the functions impaired by the disease. Such development is on the way to treat Type 1 diabetes mellitus (T1DM)[24] and Parkinson's disease (PD).[25] In both cases, a possible treatment could rely on embryonic stem cells (ESCs), which can differentiate into the desired type of cells to replace the destroyed pancreatic insulin secreting β-cells of T1DM patients or dopaminergic neurons in striatum of PD patients. Despite the great possibilities, these treatments are still far from being realized in clinical trials. The major obstacles are insufficient knowledge on how to control the differentiation process of ESCs and their integration into the host tissue to restore the impaired functions. Miniaturization of analytical platforms has emerged as a possibility to address these challenges encountered in biology and medicine. Developments in microfluidic technology can be adopted to alleviate problems in liquid handling at the same time increasing capabilities of parallelization.[26,27] However, no matter how much improvement microfluidics can provide to handling and cultivation of cells, the development has to be conducted hand-in-hand with development of analytical techniques to more effectively discover the dynamics of cellular processes. This emphasizes the significance of research on finding and understanding ways of detecting biologically relevant cellular parameters and then integrating the protocols into analytical systems to achieve a new generation of tools for biology, medicine and drug discovery.

At present, the primary detection technique in systems that are used to study the behavior of living cells is optical or fluorescence detection that can be automated in X-Y direction to monitor cellular responses and behavior at multiple positions.[28] Due to, for instance, photodamage on the studied cells, these techniques do not always provide well-developed capabilities to conduct real-time measurements that could provide more information than just an endpoint. Capability for real-time measurements that facilitate monitoring of cellular dynamics under conditions, where an observed cell or cell population can function as its own control, is crucial for effective realization of systems applied for cellular studies in biology, medicine and drug discovery. Electrochemical detection can provide the necessary capability for real-time monitoring and has been applied in different formats to study the dynamic processes of living cells based on, for instance, amperometric,[12,16,17,20,29-31] potentiometric[14] and impedance spectroscopic[6,9] measurements. However, despite ongoing research, the full potentials of electrochemical techniques have not yet been utilized. Micro- and nanofabrication of electrode systems[32] with the inherent capability for scaling of electrode dimensions can provide possibilities for parallel detection of different cellular parameters to suit equally well single-cell and cell population monitoring. Additionally, systems built for the application of electrochemical techniques can also be integrated with capabilities for simultaneous microscopic observations, facilitating more effective multiparameter monitoring of cellular dynamics.

This chapter illustrates the potentials of amperometric and impedance spectroscopic monitoring of cellular dynamics. In order to provide a complete understanding of what biological and medical aspects can be studied using these electrochemical techniques, the following sections contain a thorough description of biological phenomena, such as cellular redox environment (CRE), release of neurotransmitters and other signaling substances based on Ca^{2+} ion triggered mechanism (exocytosis), as well as cellular adhesion with emphasis on cell-cell and cell-growth substrate interactions. The discussion of the biological phenomena is then lifted to the context of electrochemical measurements to highlight the capabilities of amperometry in monitoring of the dynamic changes in CRE and exocytotic behavior of cells as well as impedance spectroscopy in monitoring of factors influencing cellular adhesion. Additionally, the capabilities of other, especially optical, techniques are discussed to provide a perspective to what benefits and possible limitations electrochemical techniques possess.

II. MONITORING OF CELLULAR REDOX ENVIRONMENT

1. Biological Significance of Cellular Redox Environment

(*i*) *Cellular Redox Processes*

Cellular functions, such as locomotion, contraction and biosynthesis of cellular building blocks require energy, which organisms acquire by digesting the main constituents of food, carbohydrates, proteins and fats, to form the biopolymeric structures from monomers. Digestion of carbohydrates, proteins and fats results in formation of hexoses (glucose, fructose, galactose, etc.), amino acids, glycerol and fatty acids that are taken up by the cell, in which they undergo further degradation in enzymatic processes referred to as catabolism. Ultimately, the catabolic processes provide precursors for the synthesis of new biomolecules (anabolic processes) needed for the maintenance of cellular structures and growth as well as energy. Based on purely thermodynamic considerations, the energy contents stored in biomolecules is too large to be released in a single process without being detrimental to the cells. Instead, all types of cells, from prokaryotes to eukaryotes, store the released energy in the form of catabolic intermediates that facilitate energy release in a stepwise manner. Examples of such catabolic intermediates are the reduced cofactors nicotinamide adenine dinucleotide (NADH), nicotinamide adenine dinucleotide phosphate (NADPH), flavin adenine dinucleotide ($FADH_2$) as well as acetyl coenzyme A (Acetyl-CoA). NADH, NADPH and $FADH_2$ as well as their corresponding oxidized forms, NAD^+, $NADP^+$ and FAD, respectively, are known as cellular redox couples. An examples of other redox couples, aside from those that are involved in metabolic processes, is glutathione (GSSG) and its oxidized form (GSSG), involved in cellular detoxification processes to alleviate, for instance, oxidative stress [33]. The different cellular redox couples participate in enzymatic processes that catalyze oxidation or reduction of biomolecules. In these processes, the reduced form of a redox couple functions as an electron donor and, in an analogous manner, the oxidized form functions as an electron acceptor.

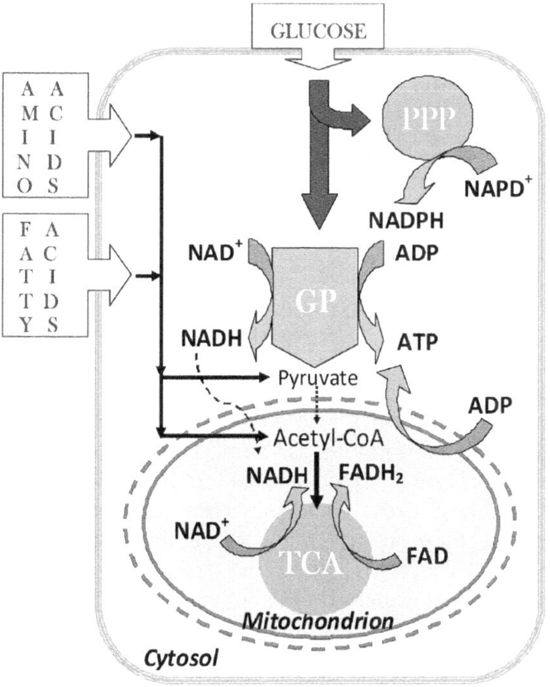

Figure 1. Glucose catabolism proceeds through the cytosolic pentose phosphate pathway (PPP) and glycolytic pathway (GP) generating NADPH and NADH, respectively. GP also produces ATP in product level phosphorylation. The GP endproduct, pyruvate, enters the mitochondria and upon conversion into acetyl coenzymeA (Acetyl-CoA) contributes to formation of NADH and $FADH_2$ in the tricarboxylic (TCA) cycle and further ATP through oxidative phosphorylation. Electrons from cytosolic NADH are shuttled into the mitochondria. Degradation of amino acids and fatty acids primarily leads to formation of pyruvate or Acetyl-CoA, and further processing in the mitochondria.

Figure 1 shows a schematic overview of the main catabolic pathways. In respiratory eukaryotic cells, the metabolism of different biomolecules leads to a common intermediate, the mitochondrially formed Acetyl-CoA, which is catabolized further in the tricarboxylic acid (TCA) cycle, resulting in the formation of, for

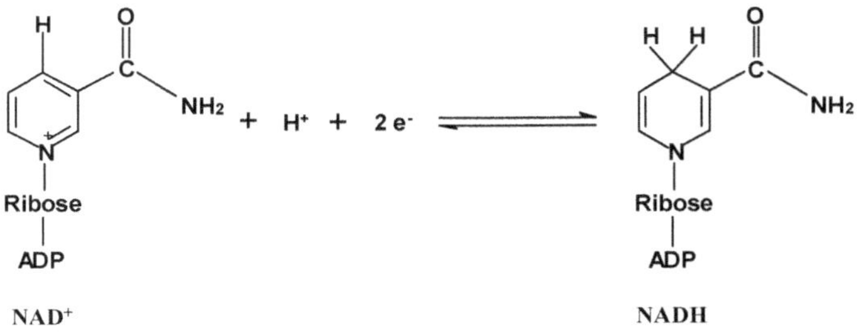

Scheme 1. The reduction of NAD^+ to NADH as part of enzyme-catalyzed reactions proceeds as a transfer of two electrons and one proton, i.e., a hydride.

instance, three molecules of NADH and one molecule of $FADH_2$ per each cycle. Amino acids form, however, a heterogeneous group, some of which are catabolized to other TCA cycle intermediates than Acetyl-CoA.[34] The enzymatic reactions resulting in the formation of these compounds are oxidation-reduction (redox) reactions, in which NAD^+ gains two electrons and one proton (the combination is also referred to as a hydride) (Scheme 1) and FAD gains two electrons and two protons. These compounds can then deliver their electrons in subsequent redox processes as part of the mitochondrial electron transport chain (ETC) (also referred to as the respiratory chain). NADH is oxidized by NADH-quinone oxidoreductase (Complex I) and $FADH_2$ by Succinate-coenzyme Q reductase (Complex II), resulting in reduction of ubiquinone (UQ) to ubiquinole (UQH_2) in the UQ-pool of the mitochondrial membrane. Further activity of the ETC, i.e.,

(*a*) oxidation of UQH_2 by Coenzyme Q: cytochrome *c*-oxidoreductase (Complex III)
(*b*) upon reduction of cytochrome c (Cyt c), and
(*c*) the subsequent oxidation of Cyt c by Cytochrome c oxidase (Complex IV), finally results in reduction of molecular oxygen (O_2) to water.

The gradually lost potential energy of the electrons of NADH and $FADH_2$ is stored in the form of a proton gradient (ΔH^+) that is built up and maintained by Complexes I, III and IV, which are able to expel protons from the mitochondrial matrix into the mitochondrial intermembrane space (IMS).[35] ATP synthase utilizes the formed ΔH^+ to synthesize ATP that serves as the universal cellular energy, predominantly formed in the mitochondria but transported into the cytosol,[35] where it is utilized for biosynthetic processes and other energy-requiring functions.

The cytosolic catabolism of glucose may follow two different pathways depending on cellular needs. The glycolytic pathway (GP) provides two molecules of NADH, ATP and pyruvate for each catabolized glucose molecule. The formed NADH can, to some extent, be utilized either in cytosolic redox processes, such as lactic (in muscle cells) and alcoholic fermentation (in *Saccharomyces cerevisiae*), or the electrons of NADH may be shuttled into the mitochondria to be processed in the ETC.[35] The formed pyruvate is also transported into the mitochondria, where pyruvate dehydrogenase converts it into Acetyl-CoA followed by processing in the TCA cycle. The enzymatic reactions of the oxidative part of the pentose phosphate pathway (PPP) form two molecules of NADPH per glucose molecule. These serve as redox equivalents in biosynthetic processes, such as synthesis of fatty acids. The non-oxidative part of the PPP continues the catabolism of glucose, resulting in the formation of ribulose-5-phosphate, which is further converted into ribose-5-phosphate, the precursor of nucleotide synthesis.[34]

(ii) Definition of Cellular Redox Environment

The oxidized and reduced cofactors, such as NAD^+-NADH and $NADP^+$-NADPH (collectively denoted as $NAD(P)^+$-NAD(P)H), form redox couples, the reduction potential (E) of which is defined by the Nernst equation (Eq. 1) for the half reaction shown in Scheme 1, where E° is the standard reduction potential under standard conditions, where T = 298 K, activity $a_{NA(P)D^+} = a_{NAD(P)H} = a_{H^+} = 1$ ($[NAD(P)^+] = [NAD(P)H] = 1$ molal and pH = 0), R is the gas constant (8.314 J K^{-1} mol^{-1}), F is the Faraday's constant (96485 C mol^{-1}) and n is the number of transferred electrons, n = 2,

$$E_{NAD(P)^+/NAD(P)H} = E^\circ_{NAD(P)^+/NAD(P)H} + \frac{RT}{nF} \ln \frac{\alpha_{NAD(P)^+}\alpha_{H^+}}{\alpha_{NAD(P)H}} \quad (1)$$

The standard conditions are not, however, applicable to biological systems. If concentrations are used instead of activities, these can be replaced by $\gamma_i[i]$, where γ_i is the activity coefficient and [i] the molar concentration of the species i. In literature, tabulated values of reduction potential for biological systems are usually valid at pH = 7 and defined at an equimolar composition with respect to the redox couple. Under such conditions the standard reduction potential, E°, is replaced by the formal potential, $E^{\circ'}$, which implicitly comprises the contribution of pH and the activity coefficients. Equation (2) shows this relation for the $NAD(P)^+$/NAD(P)H redox couple,

$$E_{NAD(P)^+/NAD(P)H} = E^{\circ'}_{NAD(P)^+/NAD(P)H} + \frac{RT}{nF}\ln\frac{[NAD(P)^+]}{[NAD(P)H]} \qquad (2)$$

where only the concentrations of the components of the redox couple affect the formal potential.

In redox reactions catalyzed by enzymes, generally referred to as redox enzymes, electrons are shuttled from a donor compound that is oxidized to an acceptor compound that is reduced. According to some estimations, about one fourth of all known proteins belong to the category of redox enzymes.[36] These can be classified based on the employed natural electron donating and accepting compounds into the following classes: dehydrogenases, oxidases, peroxidases and oxygenases. Dehydrogenases catalyze oxidation of a substrate (donor) through hydride transfer to a cofactor (acceptor), which can be a prosthetic group permanently bound to the enzyme, such as flavin mononucleotide (FMN) and FAD, or a soluble co-substrate transiently associated with the enzyme, such as $NAD(P)^+$.[37] In a dehydrogenase catalyzed reduction, however, the cofactor functions as the donor, while the substrate to be reduced is the acceptor. In a redox reaction, the substrate forms the other half reaction needed for the complete process. Equations (3a) and (3b) show the two half reactions for a dehydrogenase (DHG) catalyzed oxidation of an arbitrary substrate using $NADP^+$ as the cofactor,

$$NADP^+ + H^+ + 2e^- \xrightarrow{DHG} NADPH \qquad (3a)$$

$$SH_{red} \xrightarrow{DHG} S_{ox} + H^{+} + 2\,e^{-} \tag{3b}$$

where S_{red} and S_{ox} are the reduced and oxidized forms of the substrate, respectively. In an analogous way as was shown above for the $NAD(P)^{+}$-NAD(P)H redox couple, the substrate forms another redox couple, the reduction potential of which is also determined by the Nernst equation. Equation (4) shows the corresponding form of the Nernst equation for the substrate with the earlier described convention, using the formal potential, $E^{\circ'}$. Although Eq. (3b) indicates the involvement of a proton, its contribution to the reduction potential is included in the formal potential. Equation (5) shows the overall oxidation of the substrate,

$$E_{S_{ox}SH_{red}} = E^{\circ'}_{S_{ox}SH_{red}} + \frac{RT}{nF}\ln\frac{[S_{ox}]}{[SH_{red}]} \tag{4}$$

$$SH_{red} + NADP^{+} \xrightarrow{DHG} S_{ox} + NADPH \tag{5}$$

The overall electromotive force (ΔE) of the enzymatic reduction is obtained as the difference between the reduction potentials of the half reactions as shown by Eq. (6):

$$\Delta E = \left(E^{\circ'}_{NAD(P)^{+}/NAD(P)H} + \frac{RT}{nF}\ln\frac{[NAD(P)^{+}]}{[NAD(P)H]} \right) - \left(E^{\circ'}_{S_{ox}SH_{red}} + \frac{RT}{nF}\ln\frac{[S_{ox}]}{[SH_{red}]} \right) \tag{6}$$

The validity of Eq. (2) implies that inside living cells the actual reduction potential is primarily dependent on the value of the ratio $[NAD(P)^{+}]/[NAD(P)H]$. At pH 7 and equimolar concentration of both components, the reduction potential of the $NAD^{+}/NADH$ and $NADP^{+}/NADPH$ redox couple is –320 mV vs. the normal hydrogen electrode (NHE).[38] However, if the pH of interest for a redox reaction differs from pH 7, which is the reference value for tabulated biological reduction potentials, the actual reduction potential ($E_{pH=x}$) may be calculated based on Eq. (7) below,[33]

$$E_{NAD(P)^+/NAD(P)H\,(pH=X)} = E^{o'}_{NAD(P)^+/NAD(P)H} + \left[(pH - 7.0) \times \left(\frac{\Delta E}{\Delta pH}\right)\right] mV \tag{7}$$

which indicates that for $NAD(P)^+/NAD(P)H$, involving a transfer of two electrons ($n = 2$), the actual reduction potential is decreased by 30.8 mV per unit increase in pH at 37°C. Although the reduction potential of $NAD^+/NADH$ and $NADP^+/NADPH$ is the same, their participation in different metabolic processes is rigorously controlled by the existence of enzymes with strict dependence on either one of the cofactors as well as compartmentalisation of the metabolic processes.[39,40]

Living cells exert a rigid control of the catabolic and anabolic processes, which maintain the $NADP^+/NADPH$ and $NAD^+/NADH$ ratios as well as pH in different sub-cellular compartments at a constant level.[41] Additionally, the involvement of the GP and the PPP in the catabolism of glucose varies between different type of cells, depending on whether they have a greater need for NADH and Acetyl-CoA to be used for production of ATP in the mitochondria or NADPH and intermediates of the PPP for biosynthesis.[42] The $NADP^+/NADPH$ and $NAD^+/NADH$ ratios in different type of mammalian cells have been shown to be of the order of magnitude of 0.01[43] and 10–1000,[41] respectively. These values implicate that the reduction potential of the $NADP^+$-NADPH redox couple is kept on average about 60 mV more negative than the formal potential, whereas that of the NAD^+-NADH redox couple is kept 30–90 mV more positive than the formal potential (the estimated values are valid at 37°C). The general consequence is that the $NADP^+$-NADPH redox couple maintains the intracellular environment reductive as is necessary in reductive biosynthesis, whereas the NAD^+-NADH redox couple effectively serves as a sink of electrons as needed in oxidative catabolism.

Based on the description above, the formal potential of the cellular redox couples, such as $NADP^+$-NADPH and NAD^+-NADH, is determined by the concentration ratio of the individual components, indicating the direction of cellular processes, reductive or oxidative. Furthermore, the actual concentrations of the reduced components are also significant, since these determine the cellular reducing capacity. The overall cellular redox environment (CRE) is hence a combined contribution of the reduction potential of different cellular redox couples and their reducing capacity. Equation (8) shows the mathematical formulation of CRE, defined by Schafer and Buettner as the sum of products of reduction potential and reducing capacity of each cellular redox couple:[33]

$$CRE = \sum_{i=1}^{n\,(\text{redox couple})} E_i \times [Red]_i \qquad (8)$$

(*iii*) *Perturbations in CRE*

CRE and its cellular adjustment have significance in controlling different cellular functions, e.g., signaling,[44,45] DNA and RNA synthesis,[46] protein synthesis,[47] enzyme activity,[44] cell cycle,[47] ligand binding[48] as well as DNA binding and nuclear translocation,[49] all of which are ultimately involved in controlling cellular proliferation and differentiation. Too drastic or uncontrolled perturbations in CRE may lead to either programmed cell death (apoptosis) or necrosis. Schafer and Buetner have proposed a model, viewing changes in formal potential, controlled or noncontrolled, as nano-switches that trigger different cellular functions or ultimately death.[33] The model is devised based on alterations in the formal potential of the redox couple glutathione-oxidized glutathione during growth, differentiation, apoptosis and necrosis. Figure 2 shows a schematic representation of the model.

Perturbations in CRE result from cellular functions and pathological disorders as well as environmental factors. However, in many cases, the causative factors of the perturbations and their consequences are interrelated, which makes it difficult to categorize the causative relations. Some pathological disorders, e.g., cancer, are caused by perturbations in CRE, and upon reaching the full pathogenic state of the disorder, in this may cause a perturbation in CRE. A perturbation in CRE resulting from one pathological disorder may also trigger another pathological disorder, which is the case in mitochondrial disorders and neurodegenerative diseases.[50]

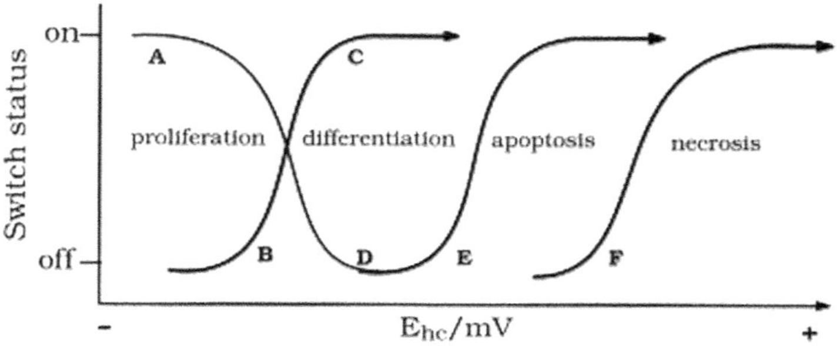

Figure 2. A schematic view of a nano-switch proposed by Schafer and Buettner to describe what biological events are triggered upon change of glutathione reduction potential, and hence the redox environment. The most reductive (negative) environment is required for growth. Differentiation is triggered by a moderate increase in the redox potential, whereas greater positive shifts lead to the onset of apoptosis or if sufficiently drastic to uncontrolled cell death, necrosis. (The figure is used with permission from Ref. 33, © Elsevier Science, Inc.)

In microbial strain engineering, perturbations in CRE are desired results of genetic modifications and able to improve the strain properties. In the following sections, the diversity of perturbations in CRE and significance of studying them are illustrated with examples.

(*a*) *Oxidative stress caused by mitochondrial functions*

In the mitochondrial ETC, oxygen is reduced to water by Complex IV through four consecutive one-electron transfer steps. Estimations have shown, however, that 1–2 % of the oxygen that is taken up by cells yields H_2O_2 instead of water.[51] The formed H_2O_2 originates from the superoxide radical ($O_2^{\cdot-}$) generated by Complex I[52] and III.[53-55] $O_2^{\cdot-}$ formation in Complex I is mediated by the fully reduced flavine mononucleotide ($FMNH_2$) facing the mitochondrial matrix.[56] In Complex III, on the other hand, $O_2^{\cdot-}$ formation is mediated by the ubisemiquinone[55] in the ubiquinole oxidation site facing the intermembrane space.[57] Formation of $O_2^{\cdot-}$ is a nonenzymatic process,[58] whereas H_2O_2 is a result of enzymatic activity acting on the formed $O_2^{\cdot-}$.[55] Through interaction with transition metal ions, such as Fe^{2+}, H_2O_2 can be further converted into the highly reactive hydroxyl radical (·OH).[59]

Due to the generation of $O_2^{\cdot -}$, mitochondria considerably contribute to cellular oxidative stress, which has been implicated as a causal factor for neurodegenerative diseases, such as Parkinson's disease,[60] Alzheimer's disease[61] and amyotrophic lateral sclerosis,[62] as well as cancer[63] and aging.[50,64] Oxidative stress contributes to the development of different pathological disorders by lipid peroxidation, protein modification and DNA fragmentation.[62] To counteract the detrimental effects of oxidative stress, cells utilize different defense mechanisms, e.g., conversion of $O_2^{\cdot -}$ into H_2O_2 by superoxide dismutase and further into water and O_2 by catalase or into water by glutathione peroxidase and peroxiredoxin.[64] The reactions with glutathione peroxidase and peroxiredoxin involve GSH and thioredoxin ($Trx(SH)_2$), respectively. Upon reduction of H_2O_2, these are oxidized to GSSG and $TrxS_2$. GSSG-GSH and $TrxS_2$-$Trx(SH)_2$ are cellular redox couples, which significantly contribute to the overall CRE. In order to maintain the reducing capacity of these redox couples, and hence an effective protection against oxidative stress, cells utilize enzymatic reactions for reducing GSSG and $TrxS_2$. These reactions, catalyzed by glutathione reductase and thioredoxin reductase, respectively, are dependent on NADPH as cofactor.[38] This shows, inspite of varying functions, that the pools of different cellular redox couples are interconnected. The rigorous cellular defense against oxidative stress is capable of normalizing the perturbations of CRE caused by normal activity of the ETC. However, the effect of mitochondrially caused oxidative stress may be potentiated in pathological disorders causing, e.g., abnormal function of the ETC and superoxide dismutase[50] as well as glucose-6-phosphate dehydrogenase deficiency.[65]

(*b*) *Oxidative stress caused by environmental factors*

Organisms are exposed to a multitude of toxic chemicals that induce oxidative stress (pro-oxidants), e.g., quinines,[66] reactive aldehydes[67] and dithiocarbamates,[68] by increasing the intracellular concentration of reactive oxygen species (ROS) including $O_2^{\cdot -}$. Cells have enzymes, e.g., cytochrome P450 (cyt P450) and NAD(P)H quinone oxidoreductase1 (NQO1; also referred to as DT-diaphorase), that function as a defense against oxidative stress caused by external factors. The redox reactions catalyzed by NQO1 utilize either NADH or NADPH as cofactor, whereas those catalyzed by cyt P450 are strictly NADPH dependent. GSH forms an additional defense mechanism against toxic chemicals by scavenging the formed ROS or directly conjugating with the chemicals, resulting in their expulsion from the cells.[69] In light of these examples, it is clear that there is a direct connection between the toxicity of many chemicals and CRE, which involves the pools of cellular redox couples and cellular catabolism.

The defense mechanisms provided by Cyt P450 and NQO1 have a fundamental difference; Cyt P450 catalyzes one-electron reduction, while NQO1 catalyzes two-electron reduction. In the former case, for instance quinones are reduced to semiquinone free radicals, which are short lived and tend to oxidize biomolecules, such as proteins, DNA, RNA or oxygen resulting in the formation of $O_2^{\cdot -}$.[70] In the latter case, reduction of quinones yields the fully reduced form, hydroquinone.[71] The activity of NQO1 has been shown to function as a strong defense against the effect of harmful chemicals without directly contributing to oxidative stress.[72]

In certain types of cancer, e.g., liver, lung, colon, breast[73] and pancreatic[74] cancer, NQO1 gene expression is up-regulated. This forms the basis for selectively utilizing certain quinones in chemotherapy, while the harmful effects on normal cells are minimized.[75]. By fine tuning the chemical structure of the quinones, these can be made more prone to autoxidation after having been reduced by NQO1.[76] Upon autoxidation of these compounds, the resulting oxidative stress selectively causes apoptosis in cancer cells. Determination of NQO1 activity in general,[77] the screening of NQO1 substrates suitable as chemotherapeutic drugs,[74] or the screening of compounds inducing expression of NQO1, have become significant in pharmacology. Aside from up-regulated NQO1 gene expression, cancer cells have also shown an up-regulated function of the GP[78] as a consequence of depressed vascularization, and hence lack of oxygen. This results in increased ATP production through the GP instead of ATP synthesis in the mitochondria. Furthermore, the increased GP activity results in an increased NADH availability, which is significant for NQO1 activity.

(*c*) *Oxidative stress caused by strain engineering*

Microbial strain engineering, with applications ranging from fundamental research of cellular functions to industrial applications using microbes, is characterized by modification of metabolic pathways by gene deletion or over-expression as well as cloning of a gene from another organism.

The gene *PGI1,* encoding for phosphoglucose isomerase (PGI), the branch point between the GP and the PPP, has been deleted as part of the research for the elucidation of mechanisms that control the function of catabolic pathways in *Saccharomyces cerevisiae* (baker's yeast). *S. cerevisiae* strains, having undergone such a genetic modification, are unable to grow only on glucose.[79] Results showed that the consequence of *PGI1* deletion was diversion of glucose catabolism into the PPP, causing a rapid depletion of the $NADP^+$ pool,[80] and hence inability to grow on glucose. In other words, the growth inhibition in the presence of glucose was caused by an initial increase in the concentration of NADPH, i.e., the reducing capacity of the $NADP^+$-NADPH redox couple, to the maximal value at the expense of the $NADP^+$ concentration. Also the reduction potential of the redox couple ($E_{NADP^+/NADPH}$) was affected. According to Eq. (2), it was shifted towards more negative values. Since the increase in the NADPH concentration and increase in the absolute value of $E_{NADP^+/NADPH}$ causes an increase in the CRE, it can be expected that the genetic modification could render the cells more effective in redox reactions relying on NADPH as the cofactor. This example demonstrates that NADPH-dependent cellular functions require a CRE that promotes reduction, but that extreme shifts without a functional mechanism for recycling between the reduced and oxidized component of the redox couple, growth is not possible.

S. cerevisiae is used for fuel ethanol production by fermenting lignocellulosic hydrolysates, which contain pentose sugars, such as xylose, which however cannot naturally be metabolized, by *S. cerevisiae*. Through genetic modifications of the metabolic pathways in *S. cerevisiae*, the organism can be made able to ferment xylose resulting in a high yield of ethanol. Ethanol production from lignocellulosic hydrolysates imposes, however, a problem: the acidic pretreatment of the raw material yields toxic compounds that inhibit fermentation and growth,[81] e.g., 5-hydroxymethyl furfural (HMF). Gene expression analysis of *S. cerevisiae* has revealed that increased tolerance to HMF is caused by up-regulation of the *ADH6* gene encoding for the NADPH-dependent alcohol dehydrogenase6 (ADH6), rendering *S. cerevisiae* with aldehyde reducing activity.[82] Based on this finding, an originally non-tolerant *S. cerevisiae* strain can be made able to overcome the fermentation/growth inhibition by over-expressing the *ADH6* gene. This results in a concomitant increase in the consumption of NADPH. In spite of the subsequent modulation of CRE in terms of the $NADP^+$-NADPH redox couple, the cells can grow in the presence of HMF.

(*iv*) *A Functional Definition of CRE*

A multitude of diverse cellular functions that are involved in the responses to and defense mechanisms against oxidative stress as well as strain engineering necessitate a modification of the definition of CRE presented by Schafer and Buettner.[33] By only considering the reduction potential, i.e., ratio of the oxidized and reduced component of cellular redox couples, and the concentration of the reduced component, i.e., the reducing capacity, an instantaneous view of CRE can be obtained. Time-dependent dynamics of the reduction potential and reducing capacity of a certain redox couple is dependent on the enzymes/metabolic pathways involved in its metabolism. Hence, the cellular activity of different redox enzymes, e.g., cyt P450, NQO1 and the complexes of the ETC, should be determined in order to obtain a more functional definition of CRE.

2. Techniques to Monitor Cellular Redox Environment

(*i*) *Techniques Utilizing Cell Lysates*

The increased knowledge regarding the different metabolic pathways and their relation to the cellular redox couples has raised the interest to determine the real amounts of the individual components or the ratios between the oxidized and reduced components. The obtained information has then been correlated with the effect of different nutritional and pathological conditions on the cellular redox environment. Accurate determination of the concentrations of the individual components of redox couples requires utilization of techniques which are not applicable in the context of live cells. Methods that still are being used were predominantly developed already during the 1960's. The determination of NAD^+/NADH, $NADP^+$/NADPH and GSSG/GSH ratios comprises two main tasks, extraction and detection, which have been separately developed by different workers. Despite the development of methods to accurately determine the parameters of CRE, the methods based on disintegration of cells lack a fundamental aspect. They cannot yield any information of the cellular dynamics with respect to factors affecting CRE. Each measurement only gives an endpoint result, i.e., if a concentration-dependent response is needed an individual determination in a different aliquot of the cell lysate has to be made. Furthermore, when using cell lysates, each obtained result is an isolated piece of information lifted away from its original context, i.e., each determined factor of CRE is independent of the cellular machinery that controls the metabolic fluxes and cellular signaling. These disadvantages in performing assays using cell lysates have led to development towards methods capable of real-time monitoring of cellular dynamics.

(ii) Techniques for Live Cells

(*a*) *Fluorescence techniques*

The reduced forms, NADH and NADPH, possess characteristic fluorescence spectra, which can be used for their detection in living cells. However, due to the fact that their spectra are equal, they can only be detected collectively without distinguishing between the phosphorylated and non-phosphorylated forms.[83] The employed wavelengths are usually 360 nm (excitation) and 400–500 nm (emission). In the case of proteins, containing a flavin prosthetic group, such as FAD in Complex II, the oxidized form possesses a characteristic fluorescence spectrum with peak excitation at 450 nm and peak emission at 540 nm.[84] Different metabolic processes result in changes in the ratio between the oxidized and reduced forms, $NAD(P)^+$/NAD(P)H and FAD/$FADH_2$, causing an increase or decrease in cellular autofluorescence, when using the characteristic excitation and emission wavelengths. These observations can be interpreted as temporal changes in CRE.

Since cellular autofluorescence, based on for instance NADH, is weak and background fluorescence in a suspension would be high enough to mask the minute changes, relevant applications are primarily based on fluorescence microscopy instead of fluorimetry in suspensions,[85] although recently high-throughput single-cell detection has also been demonstrated using a flow cytometer with fluorescence detection.[86] Microscopic detection of cellular autofluorescence ranges from general mapping of glucose metabolism,[83,87] which is reflected in increased NADPH fluorescence by the activity of the PPP and increased NADH fluorescence by the activity of the GP as well as the mitochondrial production of NADH, to applications involved in pharmacology,[88] toxicology[89,90] and neuroscience.[91,92] Different modes of microscopic detection are, however, affected by a common problem, photodamage, to the studied biological material,[83] caused by the necessarily high light intensity when exciting the weak fluorophores, NAD(P)H and $FADH_2$. A solution to the problem has been found in *two-photon excitation microscopy*[93,94] which applies the energy of two photons to reach the energy needed for excitation of a certain fluorophore. The energy of each of the two photons is exactly half of the needed energy and hence twice the wavelength of a single photon used in other modes of microscopic detection. An additional feature that eliminates photodamage is that only photons in the focus plane have the sufficient energy to generate two-photon excitation.[95]

Microscopic detection of cellular autofluorescence is generally considered as a non-invasive method due to elimination of exogenously added fluorophores, facilitated by the utilization of cellular fluorophores, which are directly associated with the target of study, CRE. However, complete non-invasiveness cannot be obtained, unless two-photon excitation microscopy is used. Aside from the problems associated with photodamage, autofluorescence detection does not facilitate high throughput and easy automation, which are the prerequisites for applications in areas, such as high-throughput screening in drug discovery. The lack in high-throughput can be overcome by using flow cytometry, a technique that is also amenable to microfabricated systems capable of providing fast and parallelized detection at single-cell level. Flow cytometry brings, however, another fundamental problem; each observation represents an individual response instead of providing a real-time response of a single cell functioning as its own control. This emphasizes the significance of further development of techniques capable of determining both the cellular availability of relevant redox couples, such as $NAD(P)^+$-NAD(P)H, and the activity of certain key enzymes directly in living cells.

(*b*) *Electrochemical techniques*

Bioelectrocatalysis – from Enzymes to Cells. Enzyme catalyzed reactions have been coupled to electrochemical reactions in order to construct enzyme-based electrochemical biosensors. In the approach, generally referred to as bioelectrocatalysis,[96] the flow of electrons from a donor via the enzyme to an acceptor must reach the electrode for the corresponding current to be detected. When an enzyme is immobilized on an electrode, the prevailing question is whether the electrons flow directly from the active site of the enzyme to the electrode or whether some artificial arrangement is needed to enable the flow. When the flow of electrons takes place directly from the active site of an enzyme to an electrode, e.g., heme-containing enzymes, which have been employed in biosensor applications,[97-99] the phenomenon is referred to as direct electron transfer (DET). A schematic view of DET is shown in Fig. 3 (upper left panel). DET is not possible, if the distance from the active site of the enzyme to the electrode is too long or if there are moieties that cover the active site so that the electrons cannot be directly transferred to the electrode. In such a case, a small molecular redox active species, such as hydrophobic ferrocene, meldola blue, menadione or hydrophilic ferricyanide, is needed as an electron transfer mediator. The electrons from the active site of the enzyme are delivered to the mediator molecule, which in turn can diffuse to the electrode where it undergoes oxidation. This mediated electron transfer (MET) has been widely applied in biosensor applications [100, 101] and is depicted in Fig. 3 (upper right panel). The coupling of an enzymatic redox reaction to an electrode process at a constant potential using mediator molecules is referred to as mediated bioelectrocatalysis or mediated amperometry.

In cell-based systems, e.g., anodes of microbial fuel cells, DET has been demonstrated using both prokaryotic[102] and eukaryotic[103] cells. In order for DET to function, the utilized cells need to possess membrane-bound enzymes having the active site toward the extracellular environment and be sterically oriented toward the electrode at a distance that facilitates tunneling of electrons. These applications are, however, still quite few. On the other hand, coupling of the small-molecular electron acceptor, ferricyanide, to the microbe-catalyzed oxidation of glucose was reported already in the 1960's.[104] Since then, numerous applications of cell-based MET have been published, demonstrating biosensors for, e.g., lactate,[105] glucose,[106,107] ethanol[108] and phenol.[109] The utilized mediators range from the hydrophilic ferricyanide and hydrophobic *p*-benzoquinone, 2,6-dichlorophenolindophenol and 2-methyl-5,6-dimethoxy benzoquinone to redox hydrogels with redoxactive osmium center [110]. The common feature of all these biosensors is that only one analyte-sensing enzyme, located in the plasma membrane, communicates with the mediator participating in the bioelectrocatalytic process (Fig. 3).

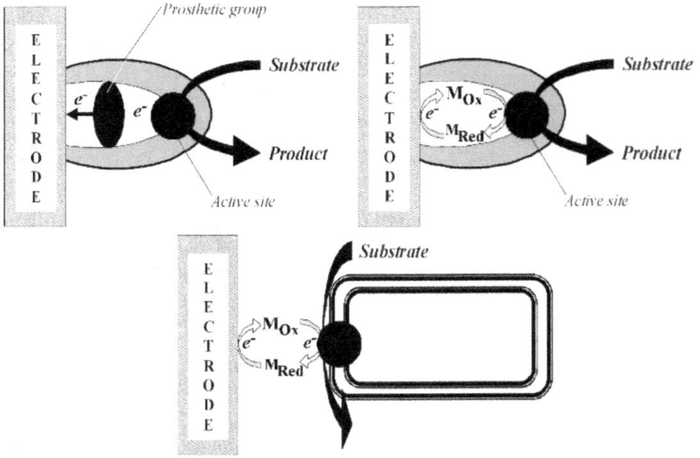

Figure 3. (Upper left panel) Direct electron transfer (DET) from a prosthetic group of a redox enzyme. An electron from the active site is transferred to the prosthetic group. (Upper right panel) Enzymes without a prosthetic group or with too long a distance to the electrode can be connected to the electrode using a small-molecular mediator that reaches the active site to receive an electron, which it then delivers to the electrode. The process is referred to as mediated electron transfer (MET). (Lower panel) Cells with membrane-bound enzymes can communicated with an electrode via a small-molecular mediator that delivers the electrons to the electrode.

lower panel). The ability of plasma membrane-bound enzymes of bacteria to communicate with ferricyanide has also been utilized in whole-cell biosensors for environmental monitoring. In these cases, instead of sensing a certain analyte, the mediator communicates with, e.g., the cyanobacterial plasma membrane-bound NADPH dehydrogenase[111] or *E. coli* plasma membrane-bound respiratory chain.[112] In the cyanobacterial biosensor, NADPH is generated intracellularly by the photosynthetic electron transport chain, which is inhibited in a concentration dependent manner by herbicides. The *E. coli*-based biosensor was developed to replace the conventional biochemical oxygen demand (BOD) test. The presence of organic material in the incubation medium increases the respiratory activity that increases the reduction of ferricyanide, and hence the bioelectrocatalytic current.

Monitoring of the intracellular redox activity in eukaryotic cells strictly requires the utilization of a hydrophobic mediator that can readily cross the plasma membrane into the intracellular environment. This approach has been applied in an environmental whole-cell biosensor based on fish cells and the hydrophobic mediator 2,6-dimethylbenzoquinone (DMBQ).[113] The cofactors that are needed for intracellular reduction of DMBQ are generated in metabolic processes. The presence of toxic chemicals in the incubation medium perturbs the metabolism, being reflected in the cellular reducing capacity and consequently in the recorded bioelectrocatalytic current. Scanning Electrochemical Microscopy (SECM) has been applied to study intracellular redox activity of human breast epithelial cells (MCF-10A), genetically engineered breast epithelial cells with over-expression of protein kinase Cα (11α) and metastatic breast cancer cells (MDA-MB-231).[114] Menadione in the incubation medium was first reduced to menadiol at an ultramicroelectrode (UME) of the SECM instrument, which subsequently diffused into the intracellular environment, where it underwent reoxidation. The intracellularly generated menadione was detected at the UME positioned adjacent to the plasma membrane. The technique was capable of mapping differences in the CRE of the different cell lines, even being able to distinguish individual cancer cells in a field of normal epithelial cells.[115]

Real-Time Monitoring of Cellular Bioelectrocatalysis. The double mediator system menadione-ferricyanide has been utilized for real-time monitoring of dynamic changes in CRE. A lightaddressable potentiometric system was used to study responses of different mammalian cell lines.[14] The responses were obtained as ferricyanide reduction rate expressed in μV/s. The study covered the effects of different inhibitors, such as dicoumarol, epiandrosterone, iodoacetate, oxamate, rotenone, antimycin A and azide. The results showed that dicoumarol caused a decrease in the reduction rate due to cytosolic inhibition of DT-diaphorase. In this study, the utilized concentrations did not give any indication of mitochondrial effects. Inhibition of the PPP with epiandrosterone strongly decreased the reduction rate in comparison with the effect of iodoacetate on the GP. This led to the conclusion that NADPH is the preferred cofactor for menadione reduction. Administered oxamate increased the reduction rate by blocking the flow of electrons from NADH to pyruvate and further to lactate, resulting in increased availability of NADH for menadione reduction. The ETC inhibitors, antimycin A and azide, increased the reduction rate by blocking reoxidation of menadiol by cytochrome *b* of Complex III and transfer of electrons to O_2 by Complex IV, respectively. Inhibition of Complex IV hence increased the availability of electrons for menadione reduction. Inhibition of Complex I, on the other hand, decreased the menadione reduction rate, since this inhibition also blocks the transfer of electrons from NADH to menadione. Work with *S. cerevisiae* cells immobilized in a Ca^{2+}-alginate gel on microelectrodes and amperometric detection has given further insight into the function of mediated bioelectrocatalysis in probing the dynamics of CRE in real-time. The results revealed the significance of a fully hydrophobic mediator when probing eukaryotic cells with target enzymes in the cytosol or mitochondria.[116] Monitoring of the responses of a deletion mutant strain of *S. cerevisiae*, EBY44, lacking the enzyme PGI[117] that functions as a branch point between the PPP and the GP, showed in accordance with mammalian cells[14] that NADPH is the preferred cofactor for menadione reduction in the cytosol.[16] The study further revealed that the mediator system causes a shift in the metabolic fluxes from ethanol fermentation to increased formation of acetate with concomitant generation of NADPH. EBY44 was also used together with its parental strain to demonstrate the possibility of employing multiple microelectrode arrays in conjunction with multichannel amperometric detection to monitor simultaneously

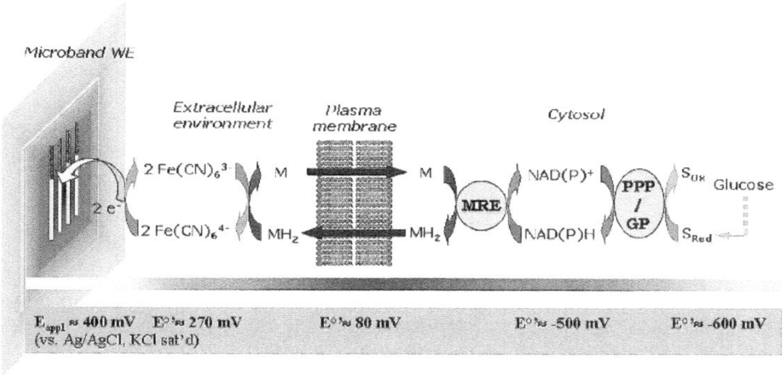

Figure 4. The functional principle of the ferricyanide-menadione double mediator system. Menadione (M) can freely diffuse through the plasma membrane into the intracellular environment where it undergoes reduction by NAD(P)H-dependent menadione reducing enzymes (MREs). The reduced form, menadiol (MH_2), is also lilpophilic and able to diffuse from the cell to the extracellular environment, where it delivers its electrons to ferricyanide ($Fe(CN)_6^{3-}$). Oxidation of the formed ferrocyanide ($Fe(CN)_6^{4-}$) at the electrode poised at 400 mV vs. Ag/AgCl reference electrode gives a current response that is directly proportional to the intracellular menadione reduction. The technique facilitates the probing of metabolic pathways through the involvement of NAD(P)H in, e.g., the pentose phosphate pathway (PPP) and glycolytic pathway (GP), which utilize glucose and other carbon sources to generate NAD(P)H. The driving force for electron flow from the highly reduced substrates to the mediators and finally to the utilized electrodes is the increasing reduction potential of the components involved in the processes along the shown potential gradient.

the responses of cells representing two genotypes and two phenotypes, i.e., respiratory and fermentative.[17] Furthermore, a genetically engineered strain with overexpressed *ADH6* gene[81] was used together with its parental strain to demonstrate that increased ADH6 activity resulted in an increased reduction of HMF and consequently in a decreased availability of NADPH.[12]

The Function of the Double-Mediator System. Figure 4 depicts the functional principle of the double mediator system menadione (M)-ferricyanide ($[Fe(CN)_6]^{3-}$) in probing CRE. The hydrophobic menadione can diffuse through the plasma membrane into the intracellular environment, where it undergoes reduction to menadiol (MH_2) (Scheme 2) by NAD(P)H-dependent menadione reducing enzymes (MREs). In virtue of its hydrophobicity, menadione is

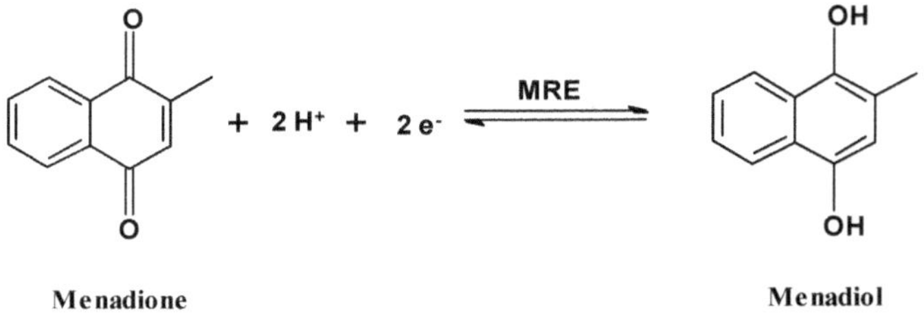

Scheme 2. Intracellularly, menadione undergoes reduction by NAD(P)H dependent menadione reducing enzymes (MREs) that in many cases catalyze a two-electron transfer in combination with the transfer of two protons.

distributed between different subcellular compartments, e.g., cytosol and mitochondria. The Supplementary Material of[13] gives a short review of MREs in *S. cerevisiae* as well as the different metabolic pathways, e.g., the cytosolic PPP and GP as well as the mitochondrial TCA, that provide the reduced cofactors, NAD(P)H, for menadione reduction. MH_2, being equally hydrophobic as menadione, can traverse through the plasma membrane into the extracellular environment, where it delivers its electrons to $[Fe(CN)_6]^{3-}$, resulting in reduction of $[Fe(CN)_6]^{3-}$ to ferrocyanide ($[Fe(CN)_6]^{4-}$) and concomitant reoxidation of MH_2. The formed M is capable of diffusing back into the intracellular environment and continuing redox cycling. The formed $[Fe(CN)_6]^{4-}$ diffuses to the electrode, where it is oxidized at the poised potential of 400 mV vs. a Ag/AgCl electrode, completing the bioelectrocatalytic process that couples the intracellular metabolic pathways to the final electrode process. For each intracellularly reduced M, two electrons are delivered to the electrode.

The utilized hydrophobic mediator, menadione, induces a controlled metabolic perturbation to which the cells respond based on their metabolic status determined by the available energy source (e.g., glucose or fructose), as well as the genetic and phenotypic characteristics of the studied cells. Menadione and other quinones are known to induce oxidative stress, which the cells tend to counteract by shifting their metabolism toward, for instance, cytosolic formation of acetate with the concomitant generation of NADPH[16] that has significance in the utilization of GSH as a means to overcome oxidative stress. The utilized concentration of menadione, 100 μM, can be regarded, however, as non-invasive [118] when short-lived measurements using *S. cerevisiae* cells are conducted. An example of the function exerted by GSH is the conjugation of menadione. Although this means that a certain fraction of the introduced menadione is consumed in the conjugation process, the technique functions as an analytical tool since the same effect is exerted on all the studied cellular samples facilitating comparison between different strains, one of which may be a control strain without any genetic modification. The hydrophilic nature of $[Fe(CN)_6]^{3-}$ strictly retains it in the extracellular environment, where it facilitates a fast regeneration of menadione and enhances the amplitude of the amperometric signal as well as the kinetics of the response.[16] Due to continued reoxidation of $[Fe(CN)_6]^{4-}$ at the electrode, only 2 mM concentration is needed to obtain a sufficiently enhanced response during continued measurements to monitor the dynamic changes in CRE.

Menadione competes for the available NAD(P)H in the entire intracellular environment, and hence, depending on the monitored cellular processes that influence the CRE, these either produce or consume NAD(P)H, subsequently increasing or decreasing, respectively, the amperometric response generated by a certain cell type with respect to the utilized control. Figure 5 shows an example of the current response generated by *S. cerevisiae* cells to introduced menadione based on the CRE represented by the basal availability of NAD(P)H as well as the increased level of NAD(P)H as the consequence of introduced glucose that is metabolized through the PPP and the GP. Depending on how an experiment is designed, double mediator-assisted amperometry can provide information of different factors influencing the CRE as represented by changes in the availability of NADH and NADPH. Generally, the application of this technique can be divided into two categories, i.e., study of the effect of

- genetic modifications, e.g., gene deletion[16,17] and overexpression,[12] and
- changes in the cellular phenotype, e.g., between respiratory and fermentative metabolism.[17]

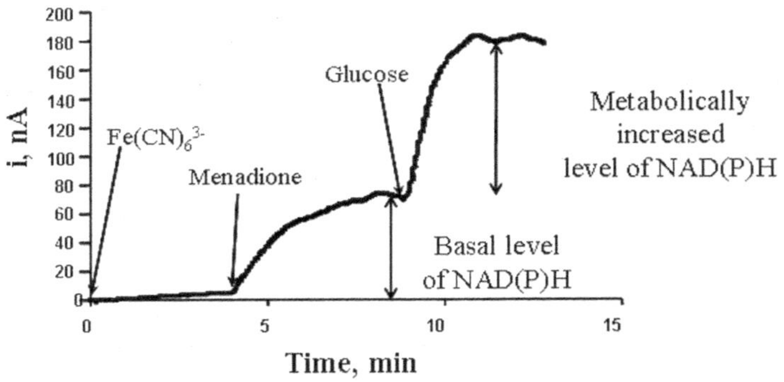

Figure 5. A current-time trace obtained as the response of *S. cerevisiae* cells to the introduction of $Fe(CN)_6^{3-}$, menadione and glucose. Due to its hydrophilicity, $Fe(CN)_6^{3-}$ cannot enter the cells, and hence no appreciable response is obtained. Introduced menadione is reduced by MREs based on the basal level of NADH and NADPH. Upon introduction of glucose, the NAD(P)H availability strongly increases. This is observed as a proportional increase in the recorded current.

Both genetic modifications and changes in the cellular phenotype can be induced by either mutations, exemplified by different pathogenic disorders, or strain engineering in industrial applications of microbes. Furthermore, in both of the aforementioned categories, the technique can yield information on the influence of cellular effectors, such as inhibitors,[17] and the activity of cellular redox enzymes[12] as well as simultaneous interaction between different subcellular compartments, e.g., cytosol and mitochondria.[17] Through utilization of microfabricated multiple electrode arrays (Fig. 6) and multichannel electrochemical detectors, the technique can be parallelized facilitating, for instance, simultaneous monitoring of dynamic changes in the CRE of different geno- and phenotypes.[17]

Study of genetic modifications. The effect of gene deletion has been studied using a *S. cerevisiae* deletion mutant strain lacking the gene *PGI1*[16] (EBY44)[117] and as the control a laboratory strain, CEN.PK, having a non-modified expression of *PGI1*. When cells take up glucose and fructose, these are converted to glucose-6-phosphate (G-6-P) and fructose-6-phosphate (F-6-P), respectively. In cells having a normal function of PGI, the interconversion be-

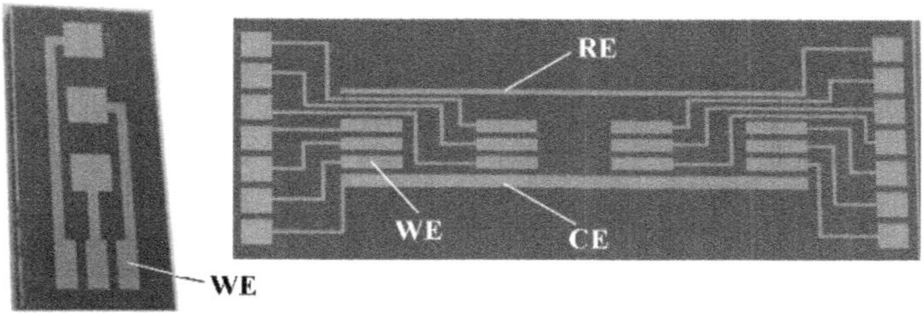

Figure 6. Examples of microfabricated electrode arrays that facilitate parallel measurements when monitoring the dynamic changes in CRE. Both arrays have working electrodes (WE) with the dimension 25 μm × 1000 μm. The array in the right panel has integrated counter electrodes (CE) and reference electrodes (RE).

tween G-6-P and F-6-P facilitates channeling of both glucose and fructose into the two major cytosolic catabolic pathways, the PPP and the GP (Fig. 7A, right panel). In cells lacking functional PGI, glucose is predominantly catabolized through the PPP and fructose through the GP (Fig. 7B, right panel). Hence, introduction of glucose primarily results in an increased availability of NADPH, whereas introduction of fructose increases the availability of NADH. The application of double mediator-assisted amperometry in monitoring the effect of the gene deletion on the cofactor availability and, hence, the change in the cellular CRE is directly observable: In the presence of PGI, both glucose and fructose yield the same response (Fig. 7A, left panel), i.e., both carbon sources result in the same availability of NADPH and NADH. In the case of *PGI1* deletion, the obtained response generated by the introduction of glucose is considerably higher than the one obtained upon the introduction of fructose (Fig. 7B, left panel), attributed to the predominant formation of NADPH and NADH, respectively. In both cases, the responses are presented as relative responses with respect to the baseline current obtained upon introduction of menadione. Although the technique yields data that is analyzed with respect to the utilized control strain, the results facilitate the conclusion that NADPH is the preferable cofactor for cytosolic MREs. This illustrates the versatility of the technique in fast screening of the effect of genetic modifications. The validity of the conclusion made based on double mediator-assisted amperometry was con-

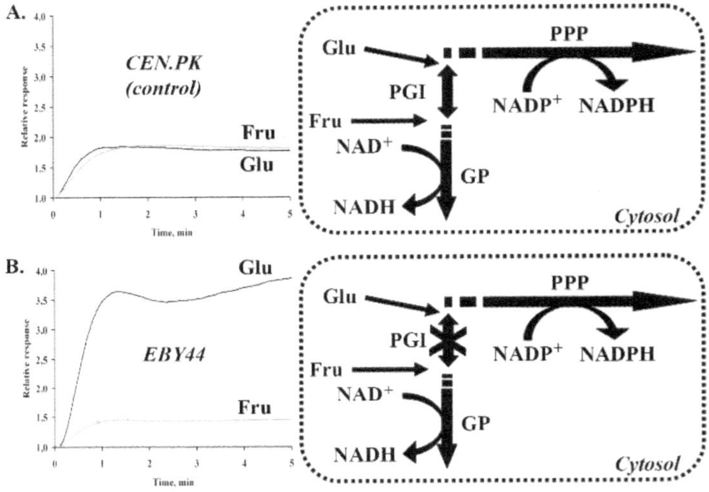

Figure 7. Relative response of A) CEN.PK and B) EBY44 strain of *S.cerevisiae* to introduced glucose and fructose. (Reprinted with permission from [13], © 2007 American Chemical Society.) CEN.PK has normal activity of phosphoglucose isomerase (PGI), which leads to channeling of both glucose and fructose into the PPP and GP (right panel in A). In EBY44, the gene encoding for PGI has been deleted (right panel in B).

firmed by performing complementary studies through metabolic flux analysis.[16]

The effect of gene overexpression was illustrated using a *S. cerevisiae* strain overexpressing the *ADH6* gene (ADH6 strain),[12] capable of reducing HMF, the toxic byproduct formed during chemical pretreatment of lignocellulose. The obtained responses were correlated to those generated by the utilized control strain having an empty plasmid. HMF reduction by ADH6 results in an increased consumption of NADPH consequently causing a competition between ADH6 and MREs with respect to the available NADPH pool (either the basal level or an increased level upon introduction of glucose) (Fig. 8C). Figure 8A shows the response of ADH6 strain and the control strain in the absence of HMF, indicating that both strains had a similar NADPH availability with or without added glucose. Figure 8B shows the corresponding re-

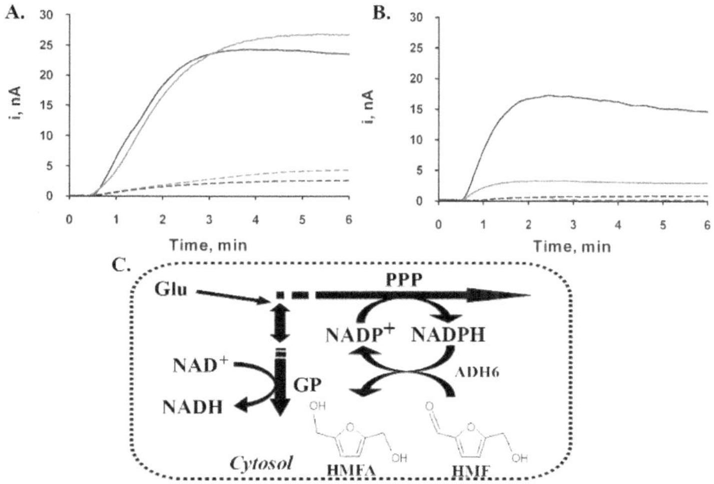

Figure 8. A) Current-time traces recorded as the response of a *S. cerevisiae* strain overepressing the gene encoding for alcoholdehydrogenase6 (ADH6 strain) (grey lines) and its parental strain (control strain) (black lines) to the introduction of 2 mM $Fe(CN)_6^{3-}$ and 100 μM menadione after 15-min incubation with 10 mM glucose (solid lines) and without glucose (dashed lines). B) The same as A) with the exception that the cells were first incubated for 15 min in the presence of 20 mM 5-hydroxymethyl furfural (HMF). Reprinted with permission from Ref. 13, Copyright (2009) American Chemical Society. C) The reduction of HMF to 5-hydrowymethyl furfuryl alcohol (HMFA) by ADH6 is NADPH dependent causing a competition between ADH6 and MREs.

sponses in the presence of HMF, revealing that the ADH6 strain had a several fold increase in the effect of HMF reduction indicated by the considerably lowered NADPH availability.

The ADH6 strain and its control strain were also utilized to illustrate the capability of double mediator-assisted amperometry in yielding concentration response curves, which facilitate the determination of enzyme kinetic parameters.[12] Using this technique, enzyme kinetic data are obtained by titrating with a substrate or a cellular effector, e.g., an inhibitor or a competing substrate. Figure 9A shows a titration curve (current-time trace) for the ADH6 strain and its control strain recorded upon consecutive addition of the substrate HMF. The curves represent the response of MREs, which is decreasing due to the decreasing availability of NADPH as a

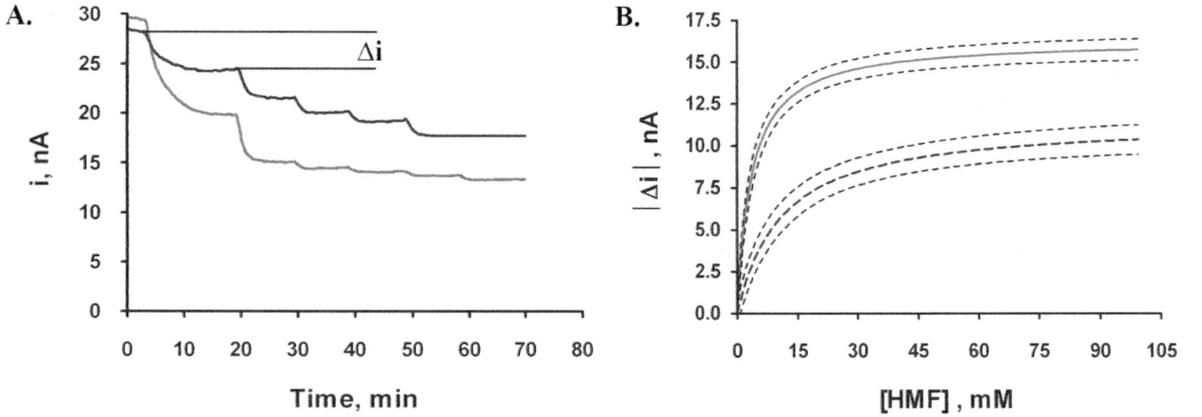

Figure 9. A) Recorded current-time traces of *S. cerevisiae* ADH6 strain (gray line) and control strain (black line) upon titration with HMF. The decreasing current response is caused by the competition between MREs and ADH6. B) Concentration response curves were obtained by extracting data from the titration curves shown in A). Reprinted with permission from Ref. 13, Copyright (2009) American Chemical Society.

consequence of the competition with respect to the common NADPH pool. The concentration response curves (Fig. 9B) are obtained by plotting the absolute value of the difference between two consecutive steady-states, generated upon addition of HMF (in units of current) (indicated in Fig. 9A) as a function of the total concentration of added HMF. Enzyme kinetic parameters are obtained upon fitting a concentration response curve to the four-parameter logistic equation (Eq. 9),

$$|\Delta i| = \text{Bottom} + \frac{\text{Top - Bottom}}{1+\left(\dfrac{10^{\log(XC_{50})}}{10^{\log[S]}}\right)^{(\text{Hill slope})}} \tag{9}$$

where the bottom and top are the response for the minimal and maximal curve asymptote, respectively, XC_{50} is either the IC_{50} or EC_{50} and the Hill slope is the midpoint slope. The concentration response curve depicted in Fig. 9B does not show, however, the characteristic sigmoidal shape expected based on a four-parameter logistic equation. In the case when the Hill slope is equal to 1 and the initial response in the absence of an added substrate is zero, i.e., the bottom is zero, the obtained curve is a hyperbolic function, analogous to the one known as the Michaelis-Menten equation. Generally, concentration response curves describe the cellular behavior in a setting, where either several enzymes or other factors, such as transport of a substrate or inhibitor into cells, are simultaneously active. Hence, even if the mathematical form of the curve is the same as the Michaelis-Menten equation, the four-parameter logistic equation is the preferred mathematical representation. Since the concentration response curves are expressed in the form of current vs. concentration, the maximal curve asymptote can be denoted by I_{MAX} and the XC_{50} by $K^{app}_{M,cell}$, which comprises the contribution of different factors simultaneously active in the cellular context. The hyperbolic equation obtained based on amperometric measurements can be represented by Eq. (10),

$$|\Delta i| = \frac{I_{MAX} \times [S]}{K^{app}_{M,cell} + [S]} \tag{10}$$

In order to convert I_{MAX} into a more informative form expressed as the maximal rate of enzyme activity (V_{MAX}) in, for instance, μmol substrate/min, Faraday's law of electrolysis (Eq. 11) can be applied to calculate the number of moles of substrate (n_S) that corresponds to a certain value of current,

$$V_{MAX} = \frac{n_S}{\tau} = \frac{60 \times I_{MAX}}{nF} \tag{11}$$

The number of coulombs corresponding to the number of moles per unit time can be obtained by integrating the current over the chosen time base (τ) of V_{MAX}, such as one minute. In this case, the necessary integration is simple because each value of the current is related to a steady-state, and hence constant. F and n are as previously defined. Since oxidation of NADPH, which in this case is the redox equivalent, involves transfer of two electrons, n is equal to 2, other enzyme kinetic parameters, such as k_{cat} and k_{cat}/K_M, can be derived based on the determined values of V_{MAX} and $K^{app}_{M,cell}$ by using the number of cells involved in an assay.[17] Although the determined kinetic parameters should purely reflect the intracellular electron transfer, they cannot be separated from other competing contributions. Hence, generally, the obtained kinetic parameters also reflect the influence of mass transport of a substrate or a cellular effector across the plasma membrane into the cells and out of the cells, the rate of menadiol oxidation by ferricyanide, mass transfer of ferrocyanide to the electrode and oxidation of ferrocyanide at the electrode surface.

Study of changes in cellular phenotype. The ability of double mediator-assisted amperometry in distinguishing between cells having a different phenotype has been illustrated using *S. cerevisiae* cells that are fermentative (without fully functional mitochondria) and respiratory (with fully functional mitochondria).[17] In respiratory cells, the CRE is characterized by the function of cytosolic metabolic pathways, the PPP and the GP, which adjust the level of NADPH and NADH, respectively. In respiratory cells, the functional mitochondria contribute to the CRE with an additional pool of NADH originating from the TCA as well as through oxidation of NADH in the ETC. Aside from influencing the availability of NAD(P)H, the two phenotypes also determine what MREs are active:[13] In fermentative cells, only cytosolic dehydrogenases contribute to menadione reduction, whereas in respiratory cells, also the mitochondrial complexes of the ETC function as MREs. The study was conducted using fermentative and respiratory cells of the *S. cerevisiae* strain EBY44 with the properties described in the previous section and a control strain (ENY.WA) having an empty plasmid. Since EBY44 represents a strain with a genetic modification, the entire study comprised simultaneous monitoring of dynamic changes in CRE due to change in cellular phenotype and genetic modification. Additionally, the study illustrated the possibility to monitor changes in CRE as a consequence of an inhibitor as well as simultaneous interaction between the cytosol and mitochondria. The utilized inhibitor, dicoumarol (at the concentration of 30 μM), has a dual effect, i.e., it is capable of inhibiting cytosolic dehydrogenases, also comprising MREs, as well as functioning as an uncoupler of oxidative phosphorylation.

The obtained results when using glucose as the energy source indicate that the current response generated by fermentative EBY44 and ENY.WA (control strain) cells in the absence of dicoumarol (Fig. 10A) is significantly higher than that obtained in the presence of dicoumarol (Fig. 10B). This is the consequence of cytosolic inhibition of dehydrogenases. In the case of respiratory cells, the current response of EBY44 cells is also lower in the presence of dicoumarol, whereas the corresponding response of ENY.WA cells is significantly higher in the presence of dicoumarol. In the case of EBY44 cells, glucose is predominantly catabolized in the PPP resulting in the formation of NADPH, which does not influence the mitochondrial functions, hence primarily revealing the effect of cytosolic inhibition of dehydrogenases. ENY.WA cells, on the other hand, catabolize glucose in both the PPP and GP, the latter of which results in formation of cytosolic NADH. Electrons from cytosolic NADH as well as metabolic intermediates from the downstream GP are shuttled into the mitochondria, being utilized in the TCA and hence able to influence the function

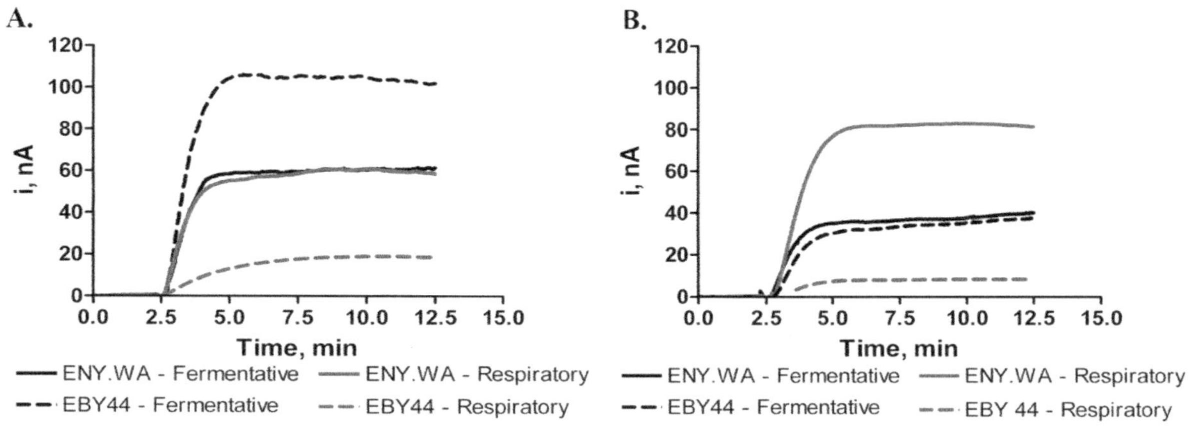

Figure 10. A) Recorded current-time traces of fermentative and respiratory *S. cerevisiae* cells of EBY44 strain and ENY.WA (control) strain after 15-min incubation with 10 mM glucose A) without and B) with 15-min incubation with 30 μM dicoumarol. Reprinted with permission from Ref. 16, © (2009) Elsevier B.V.

of the ETC. The increased current response in the presence of dicoumarol is caused by the uncoupling of the oxidative phosphorylation, which functionally enhances the complexes of the ETC to counteract the disruption of the proton gradient between the mitochondrial matrix and the intermembrane space. Although an increase in the current response is observed, it also comprises the simultaneous effect of cytosolic inhibition that tends to decrease the current response. However, the effect of the mitochondrial functions prevails over the cytosolic effect. In this study, corresponding results were obtained when using fructose as the energy source. In this case, even respiratory EBY44 cells showed a significantly increased current response due to NADH generation in the GP, having influence on the mitochondrial functions.

Aside from evaluating the capability of double mediator-assisted amperometry in revealing different biological functions, this study also illustrated the possibility of utilizing microelectrode arrays (Fig. 6) facilitating parallel measurements with a multichannel potentiostat. Each cell type can be immobilized on multiple electrodes of a dedicated electrode array. This facilitates simultaneous recording of multiple current responses, yielding current-time curves for each cell type representing different phenotypes and genotypes during one measurement. In this work, the presented results represent the average of two individual current responses for each phenotype- genotype combination.

III. MONITORING OF EXOCYTOSIS

1. Biological Function of Exocytosis

Neuronal signal propagation comprises both an electrical and a chemical mode, the former of which propagates as action potentials generated by opening and closing of Na^+ and K^+ ion channels along an axon increasing the conductance to Na^+ and K^+ ions. Upon reaching the gap between two neurons, the electrical mode of signal propagation represented by ionic current is converted into the chemical mode, which is functional in the finite gap, the synaptic cleft, of about 100 nm[119] between the two neurons. When a propagating action potential reaches the synapse, it triggers opening of Ca^{2+} ion channels, increasing the Ca^{2+} ion concentration in the microenvironment near the Ca^{2+} ion channels of the presynaptic neuron to concentrations that may be 1000 times higher than the normal level of about 100 nM. The entering Ca^{2+} ions, facilitated by a cascade of biochemical processes, trigger a fast and regulated excretion, exocytosis, of a signaling molecule, neurotransmitter that propagates the neuronal signal across the synaptic cleft. In some literature sources, only acetylcholine, glutamate, glycine and γ-aminobutyric acid (GABA) are categorized as neurotransmitters, whereas others are included in the category of peptides/neuromodulators [120]. In this text, based on the general convention, all the substances that chemically propagate a signal between two neurons or a neuron and another target cell are classified as neurotransmitters.[121]

The Ca^{2+} dependent mode of triggering the secretion of different regulatory substances, including neurotransmitters, is not, however, only limited to neurons. Instead, it is conserved from lower organisms, such as yeasts, to humans,[122] including other types of cells, such as insulin-secreting pancreatic β-cells and catecholamine-secreting chromaffin cells of the adrenal gland (a mixture of dopamine (DA), epinephrine (E) and norepinephrine (NE)) and pheochromocytoma cells of adrenal gland cancer (PC12) (DA) [120]. In order for a compound to be classified as a neurotransmitter, it has to be secreted in the Ca^{2+} ion dependent manner from the presynaptic cells, and the postsynaptic cell must have specialized receptors, to which this compound functions as a ligand. In neurons, neurotransmitters are stored in small synaptic vesicles (SVs) of 20–50 nm in diameter, whereas in non-synaptic cells (e.g., chromaffin and PC12 cells) the signaling substances are stored in larger dense-core vesicles (DCVs) of 0.25-1 μm in diameter.[18] Two classes of postsynaptic receptors are found in neurons:

(*a*) ion channels that facilitate fast responses, such as excitatory and inhibitory postsynaptic potentials, and
(*b*) GPCRs that facilitate slower signals with longer duration.

Neurotransmitters, such as acetylcholine, glutamate, glycin and GABA may bind to and activate both types of receptors,[123] whereas DA, NE and E, exclusively bind and activate GPCRs,[121] which then via a cascade of intracellular reactions modulate the function of a certain type of ion channel.

Studies on chromaffin and PC12 cells have provided to a great extent the information describing the overall process of neurotransmitter exocytosis from the storage vesicles. Despite the fact that these cells are not neurons and hence, the catecholamines they release do not function as neurotransmitters, the release mechanism is the same as the one involved in exocytotic release of neurotransmitters by presynaptic neurons. The overall process, schematically illustrated in Fig. 11, can be divided into four stages:[120]

(*a*) docking,
(*b*) priming,
(*c*) triggering, and
(*d*) fusion/exocytosis.

The first stage, docking, involves the movement of the vesicles, loaded with neurotransmitters or other signaling molecules, to the plasma membrane and their subsequent tethering at the active zone where exocytosis takes place. Over 30% of the vesicles in chromaffin cells are located within 50 nm of the plasma membrane and are considered to be in the docking state. During priming, the binding of the vesicles to the plasma membrane mediated by complex formation between certain plasma membrane proteins and vesicle membrane proteins is matured making them exocytotically competent. The influx of Ca^{2+} ions as the result of the opening of ion channels, mediated by plasma membrane polarization, which in neurons is represented by the propagating axon potential, serves as the triggering signal that leads to the final fusion of the plasma membrane and the vesicle membrane and further to opening of a fusion pore. Although the formed fusion pores allow diffusion of molecules out of the vesicles, during an exocytotic event the fusion pores expand, facilitating an effective expulsion of the contents. Hence, in exocytosis, the signaling molecules are released as packages from individual vesicles, referred to as quantal release. In neuronal synapses, the response of a postsynaptic neuron is the sum of the individual responses caused by each individual quantum. Even without a propagating presynaptic action potential, individual quanta can be released from neurons into the synaptic cleft, however, only resulting in miniature postsynaptic action potentials. Only when Ca^{2+} ions enter a presynaptic cell as the consequence of an action potential, multiple quanta are released in a synchronized manner, resulting in a postsynaptic action potential that propagates the incoming signal.

The process by which the emptied vesicles are recycled is not fully known. The proposed mechanisms are:

(*a*) *kiss-and-run*,
(*b*) endocytosis, and
(*c*) endocytosis/endosomal fusion.[119]

According to the *kiss and run* mechanism, the fusion pore, through which the neurotransmitter molecules are released, is directly re-closed followed by formation of a vesicle. The endocytosis mechanism, on the other hand, entails a complete fusion of a vesicle membrane with the plasma membrane, resulting in a flattening of the vesicle membrane on the surface of the plasma membrane. This process is then followed by endocytosis mediated by a protein-assisted guided process to form a closed vesicular structure. The endocytosis/endosomal fusion mechanism is related to the endocytosis mechanism with the exception that the endocytotically formed new vesicles are fused in an endosome, from which mature vesicles are formed by budding. These proposed mechanisms are, in fact, not exclusively related to recycling of vesicles that have undergone exocytosis; the mechanisms involving endocytosis are derived from other areas of cell biology,[124] microbiology[125] and virology.[126] In terms of exocytosis, the *kiss-and-run* recycling mechanism involves a transient incomplete fusion, resulting in incomplete emptying of the vesicular contents, whereas the endocytotic mechanisms, generally termed as *all-or-none* exocytosis,[127] results in complete emptying of the neurotransmitters. Figure 11 illustrates the distinction between the *kiss and run* and *all-or-none* mechanism.

The *all-or-none* mechanism has generally been considered as the primary mechanism in cells containing large vesicles, such as chromaffin,[29,128,129] mast,[29,130] PC12 and pancreatic β-cells.[29] *All-or-none* exocytosis has also been shown to be the primary mechanism at hippocampal synapses,[131,132] pituitary nerve terminals[133] and neuromuscular junctions.[134] However, in some cellular systems, a combination of *all-or-none* and *kiss-and-run* mechanisms have been observed,[135] such as hippocampal synapses,[131,136] pituitary nerve terminals,[133] and mast cells.[130,137] Generally, the *kiss-and-run* mechanism is associated with synapses that have low re-

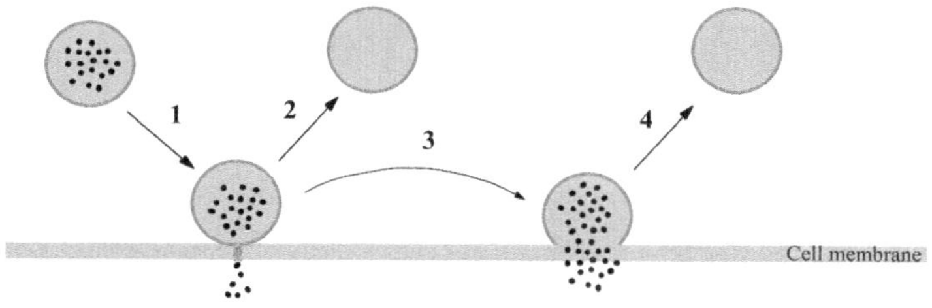

Figure 11. A schematic view of exocytosis. (1) A vesicle filled with neurotransmitter molecules approaches the plasma membrane in the active zone. Ca^{2+} influx triggers the biochemical cascade that results in the tethering of the vesicle to the plasma membrane. A fusion pore is opened in the tethered vesicle allowing the neurotransmitter molecules to be partially released. (2) According to the *kiss and run* mechanism, the still partially filled vesicle is directly retrieved from the membrane and returned to the pool of reusable vesicles. (3) According to the *all-or-none* mechanism, the partially emptied vesicle may undergo a total fusion with the plasma membrane, during which the contents are fully released. (4) The fully emptied fused membrane section is then retrieved through endocytosis.

lease probability,[131] small number of functional vesicles coupled with high rate of release,[136] or small synaptic vesicles.[127] These three cellular conditions clearly require a high efficiency in utilizing vesicles, which is the advantage of the *kiss-and-run* mechanism. A special case of the *kiss-and-run* mechanism, a flickering fusion pore, has been observed in connection with DA exocytosis from small synaptic vesicles of rat ventral midbrain neurons.[138] This mechanism further increases the efficiency of utilizing vesicles as well as controlling the amount of released neurotransmitters. The size of initial fusion pores of small vesicles undergoing exocytosis by the *kiss-and-run* mechanism has been determined to be of the order of magnitude of one ion channel as well as that the opening has to be a protein structure that is similar to an ion channel.[119,139] Figure 12 shows a schematic overview of exocytosis based on the *kiss-and-run* mechanism with flickering fusion pore. In the case of the *all-or-none* mechanism, the number of released neurotransmitter molecules per quantum is equal to that of the vesicular content, which can be several millions in large vesicles. In small neuronal vesicles the number of neurotransmitter molecules

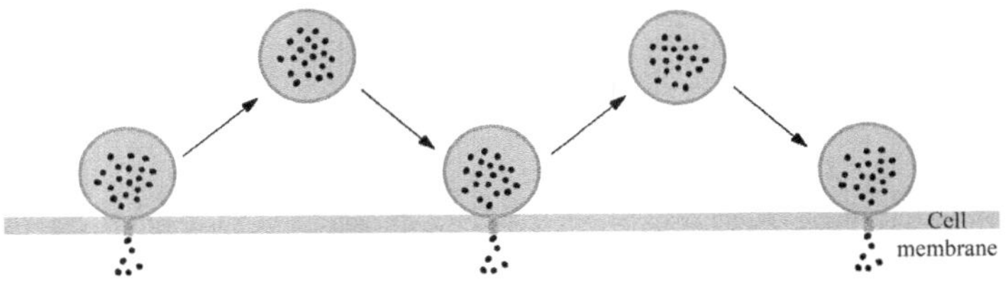

Figure 12. In an exocytotic event based the mechanism *kiss and run* with a flickering fusion pore, multiple transient fusion pores are formed. Each fusion pore formation and neurotransmitter release event is interrupted by a temporary retrieval of the partially emptied vesicle.

can be as low as 3,000–30,000.[127] If exocytosis follows the *kiss-and-run* mechanism with a flickering fusion pore, the vesicular content is not fully released in one exocytotic event; instead, only 25–30% is released.[138] One vesicle can undergo, however, multiple exocytotic events.

2. Techniques to Monitor Exocytosis

Instrumental methods for monitoring exocytosis have been the primary source of information that describes the behavior of the biological phenomenon in general and specifically for different cellular model systems. Although electron microscopy has qualitatively revealed many features of the exocytotic machinery,[140] such as location and distribution of docked vesicles, only instrumental methods of monitoring exocytosis upon electrical, mechanical or chemical stimulation have revealed the fundamental behavior of, not only the mechanism involved in release of the neurotransmitters or other signaling substances, but also the fusion of storage vesicles. Exocytosis has been monitored from a wide range of cell lines, such as chromaffin cells,[141-145] PC12 cells[146-149] and pancreatic β-cells.[150-153] Additional significance for monitoring exocytosis has been found by *in vitro* elucidation of the mechanism, by which pathological[154,155] and biological[147,156] states, pharmacological substances,[157] toxicants[156,158-161] or drugs of abuse[162,163] affect the exocytotic function and machinery as well as how such factors influence the *in vivo* function of exocytosis.[164-168] The different techniques that have been utilized for real-time monitoring of exocytosis can roughly be divided into two categories, techniques applying (i) indirect and (ii) direct detection of the released compound.

(*i*) *Indirect Monitoring Techniques*

(*a*) *Impedance based techniques*

The patch-clamp technique was developed for recording of ion channel activity in terms of ionic currents.[169] Modifications of the principle gave possibility to utilize the method for monitoring changes in plasma membrane capacitance brought about by biological activity, such as exocytosis.[128] The capacitance and conductance changes are measured as components of the overall impedance of the cellular system. The technique utilizes a glass pipette, with a bore in the micrometer regime, which houses internal electrodes and electrolyte. The two possible configurations of the impedance-based technique are the whole-cell and the cell-attached configuration. In the whole-cell configuration, the pipette is inserted into the studied cell, facilitating a direct contact between the electrolyte inside the pipette and the cytoplasm. The measured impedance during exocytosis comprises the contribution of the plasma membrane capacitance, vesicle capacitance and fusion pore conductance. Exocytotic events are detected as a minute increase in capacitance brought about by the increased membrane area as the result of fusion of the vesicle and the plasma membrane. The quantal nature of exocytosis can be detected as stepwise increments in the measured capacitance. Generally, this approach is applicable for cells with a single membrane-bound cytosolic compartment, such as chromaffin cells,[170] which can be modeled using a simple equivalent circuit,[171] although it has also been successfully applied for monitoring of exocytosis in hippocampal mossy fiber terminals.[172] The disadvantage of this approach is that the small increase in capacitance due to vesicle fusion is easily obscured by the larger plasma membrane capacitance. Instead of perforating the cell, the method was modified by applying suction that facilitates the attachment of the pipette on the studied cell. When using this cell-attached configuration, the equivalent circuit only comprises the capacitance of the vesicle and the conductance of the fusion pore. This modification improved the sensitivity, making it possible to detect the capacitance changes caused by fusion of small vesicles in nerve cells and revealed the presence of the *kiss and run* mechanism[133] as well as the conductance change associated with the initial opening of a fusion pore in human neutrophils.[139]

Since this technique only measures the capacitance and conductance changes associated with fusion of a vesicle and opening of a fusion pore, respectively, the measurements do not give any indication regarding the quantal size or nature of the released neurotransmitter or other signaling substance. In the ultimate case, fusion of an empty vesicle would yield the same response as a filled vesicle. Additionally, recycling of the emptied vesicles through endocytosis (*all-or-none* mechanism) overlaps with exocytosis, which especially when using the whole-cell configuration causes problems in resolving the two events.[171] Ion channel activity can also affect the measured capacitance without being associated with exocytosis[170] and changes in membrane conductance due to other biological activity[173] can mask the fusion pore conductance.

(*b*) *Fluorescence microscopic techniques*

Both epifluorescence[131,136,174] and total internal reflection fluorescence (TIRF) (this technique is described more detailed in Section IV)[129,175-177] microscopy have been applied for monitoring exocytotic events. Generally, TIRF microscopy facilitates accurate observations in virtue of the thin volume element mapped by the evanescent wave. This has made it possible to monitor exocytotic events from single vesicles and monitor their behavior, revealing the presence of *kiss-and-run* mechanism.[177] When using epifluorescence microscopy, the major problem is caused by accurately detecting and distinguishing the minute fluorescence changes associated with exocytosis from the background noise due to the overall fluorescence. This problem can be avoided by using a technique termed pre-photobleaching, which allows attenuation of the background fluorescence up to 90%. Using this technique it has been possible to resolve even minute fluorescence changes capable of revealing *kiss-and-run* mechanism.[131]

Aside from the two employed instrumental techniques in fluorescence microscopy, both of them comprise different methods depending on which fluorescence label is utilized. One of the methods is based on a pH-dependent mutant of green fluorescent protein (GFP), termed pHluorin.[178] By applying genetic engineering, pHluorin can be fused with a vesicular membrane protein, yielding vesicles with pHluorin inside the lumen.[131,176,177] Prior to fusion, the acidic contents of the lumen quench the fluorescence, and only upon increasing the pH due to mixing of the vesicular contents with the solution in the extracellular environment, green fluorescence can be observed. Using this method, it was possible to resolve the formation of an initial fusion pore followed by complete fusion of the vesicle.[176] Fusion of pH-insensitive GFP to exocytosed peptides and their expression in a host cell line capable of exocytosis, such as expression of GFP-neuropeptide Y fusion complex in PC12 cells,[175] is another possible method of utilizing GFP. Although the fused GFP is released as a component of the vesicular contents, the approach is indirect since the fluorescence in this case is only visible in vesicles that are in the active zone but have not fused. When exocytosis has been triggered and the vesicle opens, the fluorescence disappears. The third method for fluorescence based monitoring is based on loading of the vesicles with a fluorescent dye, such as acridine orange, which is co-exocytosed with the neurotransmitter or other signaling substance.[129,136,174] In this case, the fluorescence appears as the consequence of exocytosis. These two latter techniques have been successfully used for determining the kinetics of exocytotic events.

The advantage of the method utilizing pH-sensitive GFP in comparison with impedance-based methods is that fusion and recycling of a vesicle can be temporally resolved since fluorescence is visible only when a fusion pore is open or the vesicle is fully fused with the plasma membrane. Upon closing the fusion pore (*kiss-and-run*) or endocytotically retrieving the vesicular material (*all-or-none*), the fluorescence disappears. The general disadvantage with the microscopic methods is that no correlation between the appearing or disappearing fluorescence and the amount or nature of the released neurotransmitter or other signaling substance is obtained. In the case of GFP fused with a peptide, the method is limited to exocytosis of peptides without possessing any general applicability. The necessity of utilizing genetic engineering in order to be able to apply GFP limits the applicability of the GFP-based methods to engineered cell lines.

(*ii*) *Direct Monitoring Techniques—Electrochemistry*

Based on published applications since the initial presentation of the technique by Wightman and his co-workers,[141] amperometry has become the most widely used electrochemical technique to monitor of exocytosis of cateholamines, especially DA. In comparison with impedance-based and microscopic techniques, amperometry provides the advantage of directly detecting the release of many electroactive neurotransmitters and other signaling substances, such as catecholamines, serotonin and histamine,[171] based on direct oxidation on the utilized working electrode (WE). The technique is based on the utilization of carbon-fiber microelectrodes (CFMEs) as WEs (the diameters usually range from 1 to 10 μm), the position of which can be adjusted under microscopic observation using a micromanipulator[18] or a scanning electron microscope.[179,180] The electrode surface is placed adjacent to the plasma membrane of the monitored cell; the distance is usually of the order of magnitude corresponding to the diffusion layer.[143] This setup ensures that a sufficient temporal resolution is achieved to monitor single-vesicle exocytotic events. Aside from being a very fast process (in the time regime $\leq$ 10 ms requiring a potentiostat capable of data sampling at the rate of 5–10 kHz.) that requires a good temporal resolution, monitoring of exocytosis of neurotransmitters and other signaling substances requires an extreme sensitivity due to the small number of molecules as mentioned above. Amperometry has been proven to posses the necessary sensitivity to quantify, for instance, 31 zmol (~18700 molecules) of catecholamines[146] and 7.8 zmol (~4700 molecules) of serotonin[181] (1 zmol $\equiv 1 \cdot 10^{-21}$ mol). The utilization of CFMEs with continuous improvements in fabrication techniques has also influenced the capability to accurately determine the position of an exocytotically active zone reaching a spatial resolution of a single catecholamine vesicle (~ 100 nm).[182]

Amperometric monitoring of exocytotic events from single cells provides a range of information that is useful in characterizing the biological phenomenon. Figure 13A presents current-time traces in the form of spikes, each of which represents a single-vesicle exocytotic event [20]. The spikes arise from the nature of the events, where a certain number of molecules, representing a portion of or the complete contents of a vesicle, are fast expelled into the finite gap between the cell membrane and the electrode surface. A current-time trace, showing a succession of exocytotic events, does not reveal, however, the details needed for obtaining the relevant biological information. By utilizing, for instance, a software package for technical graphing, image processing and data analysis, such as IGOR Pro from WaveMetrics, Inc., it is possible to create macros that zoom into the individual spikes and retrieve detailed data of the recorded events. Figure 13B shows a magnified spike of an exocytotic event among those comprised in the entire recording of Fig. 13A together with the quantitative and kinetic parameters that can be amperometrically determined.

The area under a spike corresponds to the charge (Q) that is directly proportional to the number of released neurotransmitter or signaling molecules (N) from one vesicle during the opening of the fusion pore or complete fusion of the vesicle membrane with the cell membrane. The relationship is given by Faraday's law of electrolysis (Eq. 12b) upon integration of the current-time trace under each spike (Eq. 12a),

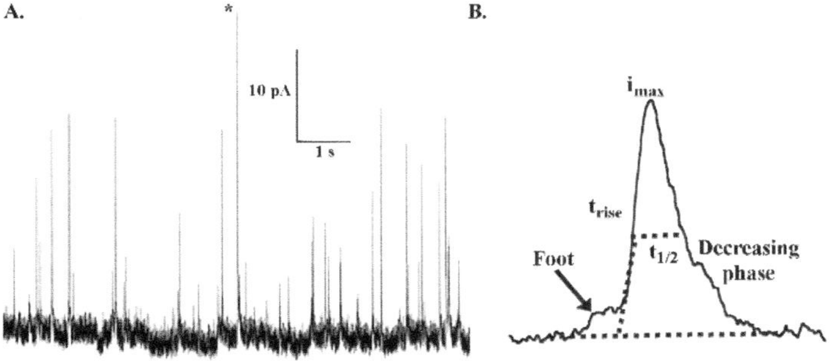

Figure 13 A) A 10-second long amperometric recording from PC12 cells. Each spike corresponds to a single-vesicle exocytotic event. B) A magnification of a spike (indicated by an asterisk in A). The kinetic parameters are described in the text. The spike shows the presence of initial neurotransmitter leakage (foot signal) thorough a fusion pore. Ref. 20, Copyright (2007) WILEY-VCH Verlag GmbH & Co. KGaA. Reproduced with permission.

$$Q = \int_0^t I(\tau)d\tau \tag{12a}$$

$$N = \frac{Q}{nF} N_A \tag{12b}$$

where 0 and t refer to the time at the beginning and end of a spike, respectively, and N_A is the Avogadro's number (6.022×10^{23} molecules/mole). The other parameters are as previously defined. The height of a spike corresponds to the maximum oxidation current (i_{max}) that is directly proportional to the flux of released neurotransmitters. The peak width at half of i_{max} ($t_{1/2}$) is a robust measure of the duration of an exocytotic event. Due to irregularities at the foot of amperometric spikes, the total width is not readily definable. An additional parameter to describe spikes is the rise-time (t_{rise}), the delay between $i_{max(20\%)}$ and $i_{max(90\%)}$,[145] $i_{max(25\%)}$ and $i_{max(75\%)}$[183] or $i_{max(35\%)}$ and $i_{max(60\%)}$,[184] defining the slope of the rising phase of a spike, which serves as a kinetic parameter of the exocytotic event. Chow et al. have pointed out that t_{rise} cannot be explained, however, by an instantaneous release of neurotransmitters from a fused vesicle[185] due to the fact that a varying proportion (in average 34% for PC12 cells[186] and 20–29% for chromaffin cells [185,187]) of the spikes of all monitored exocytotic events have a *foot* (Fig. 13B), indicates an initial release of neurotransmitters through a stable but short-lived fusion pore prior to the onset of the full exocytotic event. The frequency for observing a foot in recorded amperometric spikes has been found to increase with the size of the vesicles.[145] Considering that foot signals, in the case of dopamine exocytosis, have a time duration of only a few milliseconds and are based on the release of 30–40 zmol of dopamine, depending on the type of studied cells and the size of the vesicles, is an additional indication of the performance of amperometric monitoring.

When monitoring exocytosis from single cells, a large number of spikes are recorded and, upon analyzing them, a large number of values for Q, i_{max}, $t_{½}$ and t_{rise} are obtained. In order to facilitate characterization of the cellular behavior, a statistical presentation of the parameters is oftentimes performed to reveal, for instance, variation in a population as reflected by the behavior of single cells. A common approach is to prepare histogram plots presenting the percentage of the total number of exocytotic events as a function of the values of the presented parameter divided into classes with a certain interval (Fig. 14).[20] Since, for instance, the number of released molecules in an exocytotic event can depend on different cellular factors, such as modification of the proteins involved in the exocytotic machinery,[188] as well as the influence of, e.g., drugs of abuse, such as amphetamine,[162,163] statistical comparison of the results from single cell experiments can yield significant information that is relevant for pharmacological studies and activities of drug discovery.

Aside from the statistical presentations showing the distribution of Q, the distribution of $Q^{1/3}$ has also been found to give information of the vesicular radius ($r_{vesicle}$). Upon the assumption that the vesicular shape is spherical and the concentration ($c_{vesicle}$) is constant, although the number of released molecules (directly proportional to Q) varies based on the duration of the fusion pore

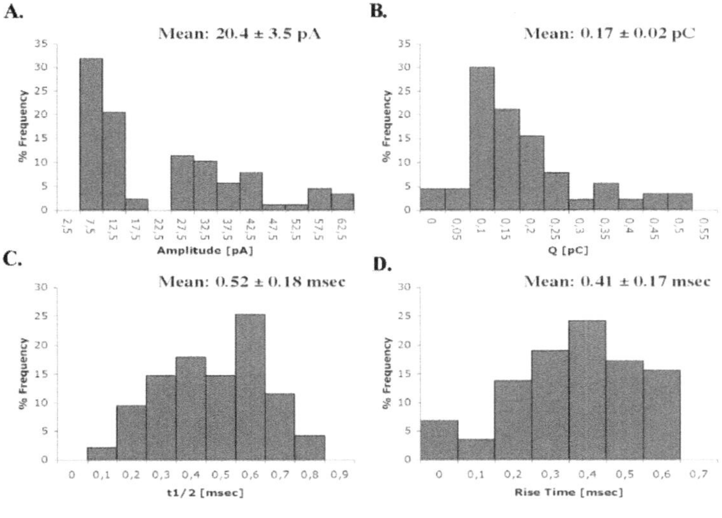

Figure 14. Histogram plots presented as percentage of the total number of exocytotic events for A) i_{max}, B) Q, C) $t_{½}$ and D) t_{rise}. From Ref. 20, Copyright (2007) WILEY-VCH Verlag GmbH & Co. Reproduced with permission.

opening, by applying Faraday's law of electrolysis (Eq. 13a) and the formula giving the volume of a sphere (V_{sphere}) (Eq. 13b), it is possible to obtain an equation relating $r_{vesicle}$ to $Q^{1/3}$ [141] (Eq. 13c):

$$n_{released} = \frac{Q}{nF} \tag{13a}$$

$$V_{sphere} = \frac{4}{3}\pi r^3 \tag{13b}$$

$$r_{vesicle} = \left(\frac{3Q}{4\pi n F c_{vesicle}} \right)^{\frac{1}{3}} \tag{13c}$$

The derivation of this equation comprises the additional assumption that the amount of released substance ($n_{released}$) is dependent on $r_{vesicle}$. Figure 15 shows the statistical distribution of $Q^{1/3}$ based on exocytosis measurements using chromaffin cells and illustrates the significance of determining the statistical distribution of $Q^{1/3}$.[189]

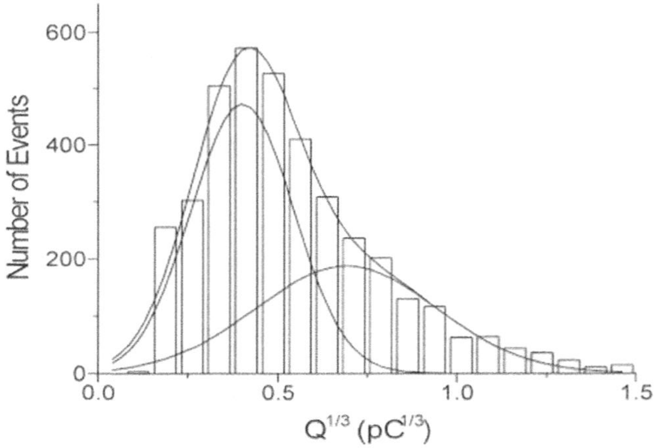

Figure 15. A histogram plot presented for $Q^{1/3}$ indicating the presence of two overlapping distributions. Reproduced with permission from Ref. 189. Copyright (2005), The American Physiological Society.

The presented histogram plot indicates that upon fitting the data to Gaussian distributions, the skewed distribution was found to be the sum of two overlapping Gaussian distributions. The result led to the conclusion that chromaffin cells have two different populations of vesicles differing in size. The finding was confirmed by scanning electron microscopic (SEM) imaging.

Amperometry can yield accurate quantitative and high-resolution spatial-temporal information regarding single-vesicle exocytotic events. However, it cannot distinguish between different released substances, such as DA, NE and E. Since, for instance, chromaffin cells may release a mixture of DA, NE and E, it is of interest to determine what substance is released in a certain exocytotic event. This has been achieved by using cyclic voltammetry (CV) with a fast scan rate (up to 800 V/s) (FSCV) to maintain sufficient temporal resolution for distinguishing between neurotransmitters in a single-vesicle exocytotic event.[143] The method has been applied, e.g., for monitoring the release of NE and E from chromaffin cells[190] as well as histamine and 5-hydroxytryptamine from mast cells.[191] Since the fast scan rate increases the presence of capacitive currents, voltammograms for exocytotic events are corrected by subtracting the background in order to resolve the current that corresponds to the oxidation of the neurotransmitters (Faradaic current).[192] The identification of different neurotransmitters is based on their characteristic redox mechanisms. In the case of NE and E, one oxidation peak is obtained due to oxidation of the catechol moiety into the corresponding quinone, however, E gives rise to two reduction peaks, whereas NE only shows one reduction peak. Both compounds yield a common reduction peak for reduction of the quinone formed during the anodic scan. The additional reduction peak of E at a more negative potential originates from an intermediate, adrenochrome, which is formed as a result of intracyclization of the secondary amine-containing tail of E.[193] In the case of the primary amine-containing tail of NE, such an intermediate is not formed.

The advantage of amperometry is that it only gives a response to a substance released from a fusing vesicle, and hence, not if a fusion involves an empty vesicle. However, it cannot detect membrane-related phenomena involved in exocytosis, such as fusion pore formation, fusion and retrieval of vesicles. In order to merge the quantitative capabilities of amperometry and the ability of cell-attached impedance-based measurements to resolve the fusion and retrieval of single vesicles, a new technique, patch amperometry, has been developed.[194] It has been successfully applied for simultaneous quantitative monitoring of catecholamine release and detection of capacitance and conductance changes involved in fusion pore formation in chromaffin cells[130] and mast cells.[130,195] The technique utilizes a special glass pipette with an inserted carbon fiber WE for amperometric detection.[130]

Microfabricated planar electrode systems. Although the amperometric investigations on exocytosis have been conducted to a great extent using CFMEs, a new approach has emerged based on lithographic techniques that facilitate fabrication of microelectrode chips for monitoring of exocytosis. Microfabricated systems have been used for monitoring of catecholamine exocytosis from chromaffin cells,[196-201] PC12 cells[20,21,30,31,202] and MN9D cells,[203,204] glutamate exocytosis from nerve cells[19,205-210] as well as histamine from RBL-2H3 leukemia cells.[211-213] A more detailed discussion of microchip-based systems for monitoring of exocytosis can be found in a recent review by Spégel et al.[32] These systems introduce a fundamental difference to monitoring of exocytosis; the detection involves vesicles in the active zones at the basal cell membrane that is adjacent to a planar electrode surface. In such a configuration, no micromanipulator under microscopic observation is needed as in the case of CFMEs. An optimal separation between the electrode and the cell under investigation is automatically obtained since, for instance, in the case of adherent cells, the distance between the cell and the underlying electrode is determined by the focal adhesions and close contacts (discussed in detail in Section IV), which form a microenvironment under the cell having the same dimension as the distance between a presynaptic and postsynaptic neuron *in vivo*. This arrangement is schematically illustrated in Fig. 16A, depicting a neuron on a planar electrode and the release of dopamine molecules that are oxidized (Scheme 3) at the applied potential of e.g., 700 mV vs. a Ag/AgCl reference electrode. Figure 16B shows an Au microelectrode with sedimented PC12 cells on top. Hence, systems applying microelectrode chips with planar electrodes facilitate monitoring of exocytosis at a sufficient temporal resolution and sensitivity to reveal the kinetics of a single-vesicle exocytotic event with a foot signal (Fig. 13).

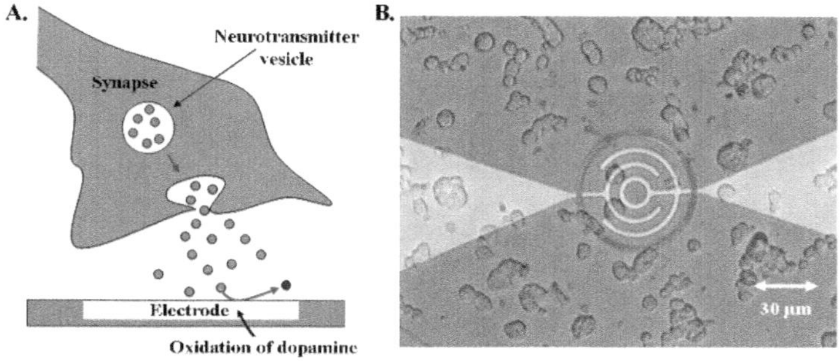

Figure 16. A) A schematic view of a cell's active zone adjacent to a planar electrode surface, where dopamine is oxidized. B) A microscopic view of Sedimented PC12 cells on ring electrodes. From Ref. 20, Copyright (2007) WILEY-VCH Verlag GmbH & Co.KGaA. Reproduced with permission.

However, in such a configuration, the utilized electrodes cannot be positioned to a certain part of the monitored cell. As the consequence of this, the spatial resolution to some extent decreases. To overcome this limitation, microelectrode chips with different systems for guiding and positioning single cells have been devised with an ability to achieve a sufficient spatial resolution to reveal the location of an active zone involved in single-vesicle exocytotic events.

Figure 17 illustrates examples of microfabricated systems that facilitate, for instance, positioning of single cells on planar electrodes to achieve an improved spatial resolution for revealing the site of a single vesicle exocytotic event.[31,198] The microelectrode chip in Fig. 17A (upper panel) features an array of four platinum microelectrodes around a central opening, an *electrochemical*

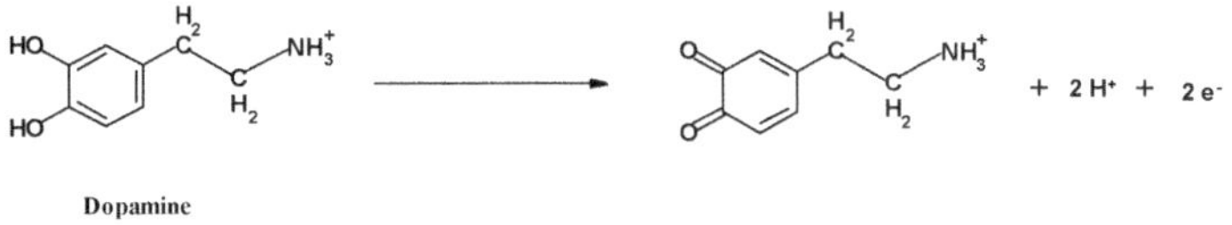

Scheme 3. In oxidation reactions on an electrode, dopamine looses two electrons and protons to form the fully oxidized quinoid product.

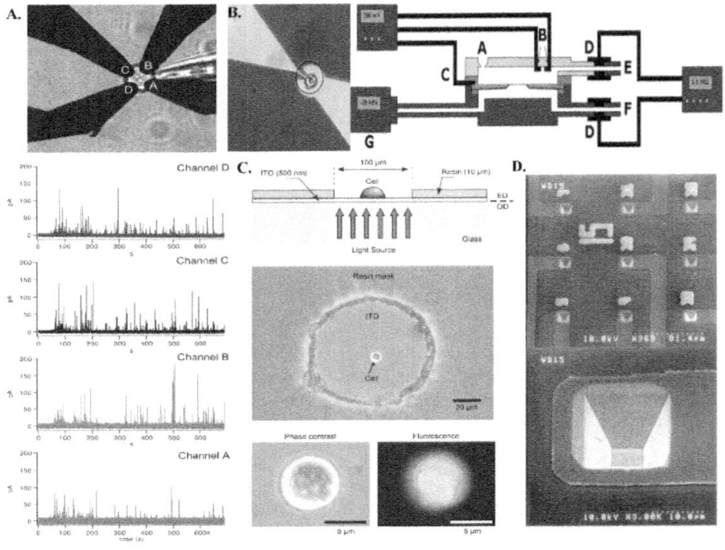

Figure 17. A) A microscopic view of a chromaffin cell in the middle of an array of four electrodes (upper panel) (denoted by A-D) that are used as an *electrochemical camera* to localize fusing vesicles during exocytotic events. Four simultaneous current-time traces recorded on the electrodes A-D (lower panel). Reprinted with permission from Ref. 198. Copyright (2005) National Academy of Sciences, U.S.A. B) A microscopic view of a PC12 cell on a ring-electrode with an orifice in the middle (right panel) and a schematic illustration of the structure of a microfluidic system (right panel) for cell handling in conjunction with monitoring of exocytosis. The system can be used to capture and retain single cells in the middle of a ring-electrode by an applied negative pressure of 2 kPA through the orifice. Reprinted with permission from Ref. 31, Copyright (2008), Reproduced by permission of The Royal Society of Chemistry. C) A schematic view of a system with ITO as electrode material used to monitor exocytosis under simultaneous optical and fluorescence microscopic observation (upper panel), a microscopic view of a single chromaffin cell on an ITO electrode (midpanel), a phasecontrast microscopic magnified view of a chromaffin cell on an ITO electrode (left lower panel) and fluorescence microscopic view of an exocytotic event visualized by the coexocytosed acriding orange that had initially been loaded in the cesicle (right lower panel). From Ref. 200, Copyright (2006) WILEY-VCH Verlag GmbH&Co KGaA. Reprinted with permission. D) A SEM image of an array of well-electrodes used to accommodate sigle cells during monitoring of exoytosis (upper panel) and a magnified view of a single well (lower panel). Reprinted with permission from Ref. 197. Copyright (2003) American Chemical Society.

camera, allowing monitoring of dopamine exocytosis from single chromaffin cells accommodated on top of the opening in the middle of the microelectrode array. With simultaneous measurements (Fig. 17A, lower panel) using each of the four microelectrodes and subsequent random walk simulations, the system was demonstrated to achieve a high spatial resolution in localization of an opening fusion pore. Figure 17B (left panel) shows a microelectrode chip having a single PC12 cell in the middle of a ring-shaped microelectrode.[31,198] The microelectrode chip formed the central part of a microfluidic cell handling system used for monitoring of dopamine exocytosis from PC12 cells. The system is schematically illustrated in Fig. 17B (right panel).[31] It facilitated single-cell capturing and positioning by application of a negative pressure of 2 kPa through an orifice with the diameter of 3 μm in the middle of the ring-electrode. After completed exocytosis measurements on a cell, this could be removed from the orifice by applying a pressure pulse of 15–25 kPa. After removal of a cell, another cell could be immediately captured and subjected to exocytosis measurements. The system provided a high spatial resolution for the localization of the active zone with the dopamine-releasing vesicle. Figure 17C shows transparent indium tin oxide (ITO) electrodes that allowed simultaneous phase contrast (left lower panel) and fluorescence (right lower panel) microscopic imaging of a cell while amperometric exocytosis monitoring was being conducted.[200] The fluorescence microscopic imaging that facilitated the visualization of catecholamine release from chromaffin cells was based on the application of acridine orange loaded into the vesicles and subsequently co-exocytosed with catecholamines. CFMEs as well as planar microelectrodes allow monitoring of exocytosis only on one side of a cell depending on the placement of the utilized electrode. Figure 17D shows a microelectrode chip with well-electrodes that have been reported to facilitate exocytosis monitoring from single cells to detect release from active zones covering a large fraction of a cell.[197]

The applications of microfabricated electrode chips are not only suitable for single-cell experiments. They have been demonstrated to facilitate exocytosis monitoring from adherently growing cell populations *in vitro*[19,21,203,204,212,214] and *in vivo.*[167] Monitoring of exocytosis from a cell population increases the length of the recording to several seconds, which is the consequence of that a

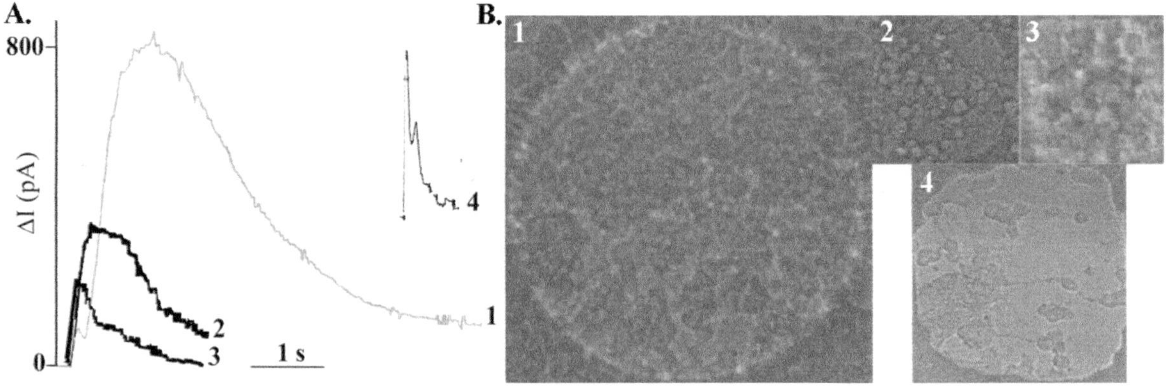

Figure 18. A) Recorded current-time traces during exocytotic events from populations of growing (traces 1 – 3) and differentiating (trace 4) PC12 cells. B) A microscopic view of the PC12 cell populations used to record traces 1 – 4. Reprinted with permission from Ref. 21, Copyright (2008) The Chemical and Biological Microsystems Society.

recorded event is an averaged sum of multiple single-vesicle events from the whole cell population. This measurement mode offers the possibility to widen the spectrum of detectable compounds, which primarily comprises electroactive catecholamines, serotonin and histamine,[171] by using enzyme-based biosensors instead of only non-modified or chemically modified electrodes.[20] facilitating, for instance, the detection of glutamate.[19,167] Figure 18A shows exocytosis recordings from populations of growing (traces 1-3) and differentiating (trace 4) PC12 cells.[13] Figure 18B shows the corresponding cell populations (marked with numbers 1-4). The current-time traces of recordings 1-3, featuring the behavior of non-differentiating PC12 cell, show t_{rise} in the regime of hundreds of milliseconds, $t_{½}$ of more than a second even for the smallest population (trace 3) and the decreasing phases have a long duration. This can be attributed to a non-simultaneous onset of the exocytotic events in the cells comprising the depicted populations. On the other hand, in the population of differentiating PC12 cells (trace 4), especially t_{rise} and the decreasing phase indicate faster exocytotic process from the population. This can be attributed to

a) a greater synchronization in the onset of the individual exocytotic events,
b) a closer proximity of the differentiated cells to the electrode surface and
c) the presence of active zones only in distinct parts of the differentiated cells.

These factors make the diffusion of dopamine to the electrode surface faster and the duration of the recorded process shorter. In contrast, the more roundish non-differentiated PC12 cells are not strongly adherent, and the catecholamine-loaded vesicles are distributed in the entire cell.

Although planar microelectrode chips in amperometric monitoring of exocytosis bring about a decrease in the spatial resolution, they, nevertheless, provide possibilities, which their counterparts, CFMEs, do not posses. Lithographic techniques give freedom to design and fabricate electrode arrays to allow the realization of diverse experimental procedures involved in monitoring of exocytosis. When the potentials of lithographic techniques in microelectrode fabrication are combined with techniques to fabricate microfluidic platforms, even a greater advancement is achieved in constructing monitoring systems that facilitate effective realization of cell handling, culturing and differentiation as prerequisites for exocytosis monitoring. Such systems are needed for, for instance, characterization of differentiating neuronal stem cells and their integration into brain tissue as a means to further develop the realization of stem cell therapy as a cure to Parkinson's disease (discussed more detailed below). Systems allowing cell culturing on the measurement electrodes facilitate experiments without prior trypsinization and handling of the cells, which makes them directly adjusted to the environment where the measurements are to be conducted. Other advantages of using microfabricated systems with cultured cells are automation and parallelization. Automation of measurements in conjunction with microfluidics can facilitate construction of systems suitable for HTS in, for instance, drug discovery related to studies on neurotransmitter release. Parallelization with the application of simultaneous measurements of cell populations on multiple electrodes allows monitoring

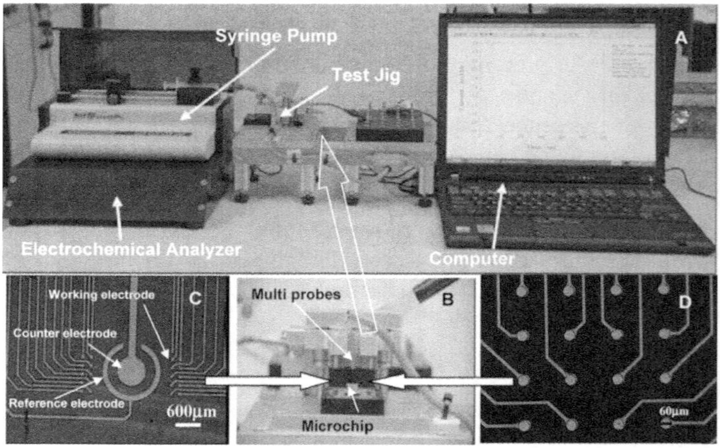

Figure 19. An overview of a microfluidic system with integrated microelectrode arrays used to monitor exocytosis from adherently growing cells (upper panel), microscopic view of the microelectrode arrays (lower left and right panel) and a close view of the microchip holder (midpanel). Reprinted with permission from Ref. 215, Copyright (2008) American Chemical Society.

with statistical control. Figure 19 illustrates a microfluidic system having an integrated microelectrode ship with a microelectrode array. The system is capable of the necessary liquid handling to introduce a buffer solution with an elevated K^+ ion concentration as well as automatic monitoring of exocytosis from cell populations adhering on the electrodes.[215]

Special challenge in fabricating cell handling and electrode systems for exocytosis studies. Aside from fundamental studies concerning the mechanism of exocytosis, a new application for monitoring exocytosis has emerged due to the development of new treatments for Parkinson's disease (PD), the second most common neurodegenerative disease,[216] which causes tremor, rigidity, *bradykinesia* (slow movement), *akinesia* (inability to initiate movement)[217] and *dyskinesia* (involuntary movement)[218] as well as *fatigue, depression* and *sleep disorder.*[30] Both oxidative stress[62] and genetic disorders[219,220] have been implicated as causative factors for PD. Neurologically, the impairment of motor function is caused by gradual deterioration and in the end total death of dopaminergic neurons in the pars compacta of *substantia nigra,*[216] which sends neural projections to the striatum. Alleviation of the symptoms involved in the impairment of motor function has been achieved by administering L-DOPA, an intermediate of DA synthesis in functional cells. This treatment has been used since the early 1960's when the first clinical tests were conducted.[221] Despite its positive effects, this treatment also suffers from severe side effects; long-term medication sensitizes DA receptors and results in chronic overstimulation. The consequence is that L-DOPA treatment is only useful in the early stage of PD but cannot provide a long-term cure. Attempts to reduce the need of medication include for instance deep-brain high frequency stimulation of ventral thalamic and basal ganglia nuclei,[222] gene transfer of the glial cell line derived neurotrophic factor[223] and cell replacement.[224]

Cell replacement treatment originates from the 1970's when transplantation of embryonic *substantia nigra* to reinnervate the regions with diminished or lacking DA supply was shown to be helpful[224,225] and since then hundreds of clinical tests have been conducted.[218] Despite its functionality, the major drawbacks of this approach are the lack of aborted fetuses as source of embryonic *substantia nigra* as well as the ethical issues involved in using them. A more recent approach in cell replacement is the transplantation of neuronal stem cells (NSCs) that can integrate into the damaged region of the brain, acquiring properties of the native dopaminergic neurons and form connections to the regions these are to innervate.[25,226] However, research in this area is still in its infancy, and fundamental studies are needed to learn how to control the differentiation of NSCs to dopaminergic neurons[218] as well as evaluation of the function of implanted already differentiated and integrated NSCs in brain tissue.[227]

These tasks impose new requirements, which are beyond microscopic observations and immunostaining to visualize the presence of cellular markers, such as tyrosine hydroxylase (TH), the rate limiting enzyme in DA synthesis, and β-tubulin, which is the characteristic cytoskeletal structure of neuronal axons.[228] Experimental evidence showing that NSCs have acquired TH and β-tubulin is not a sufficient proof for proper dopaminergic functions, i.e., active Ca^{2+} dependent exocytosis. Only by conducting studies on differentiated NSCs, monitoring exocytosis *in vitro,* it will be possible to evaluate how successful the procedures to promote differentiation toward dopaminergic neurons have been. Additionally, only by monitoring exocytosis from differentiated and integrated NSCs upon stimulation of brain tissue, it will be possible to determine whether integration really has brought about the desired *in vivo* functions that have been disrupted in the *substantia nigra* of PD patients. To achieve these goals, mere employment of the different monitoring techniques described above is not, however, sufficient. Fabrication of systems for handling NSC differentiation and integration into brain tissue is a prerequisite, indicating the significance of interplay between monitoring techniques and fabrication techniques. Ultimately, the realization of such systems providing capability for cell handling and monitoring can provide improved procedures for drug discovery.

IV. MONITORING OF CELLULAR ADHESION

1. Biological Functions of Cellular Adhesion

(*i*) *In-Vivo Functions*

Tissues in living organisms are composed of cells with different functions and morphologies. Cells are not isolated individual entities but form strictly controlled contacts with their environment. These contacts contribute to the overall organization and structural strength of the tissues as well as biological functions of the cells forming them. The cellular contacts with the surrounding environment, i.e., the matrix surrounding the cells and other cells constituting the tissue, can be divided into four main categories:[35] anchoring junctions, occluding junctions, channel-forming junctions and signal-relaying junctions. Anchoring junctions comprise both cell-to-cell contacts between adjacent cells and contacts between cells and the surrounding extracellular matrix. These junctions are connected to the intracellular cytoskeletal structures through transmembrane proteins, cadherins (cell-to-cell contacts) and integrins (cell-to-matrix contacts). Anchoring junctions have significance in transmitting both physical stress throughout a tissue and external signals to cells. Occluding junctions form a special sealing between adjacent cells, such as epithelial cells, contributing to formation of impermeable barriers. Channel-forming junctions and signal-relaying junctions have significance in direct communication between adjacent cells. Channel-forming junctions allow transfer of inorganic ions from the cytoplasm of one cell into the cytoplasm of another cell, coordinating the spreading of signals from one part of a tissue into the whole tissue. Signal-relaying junctions function in transmission of chemical signals between cells through a gap between the plasma membranes, for instance transmission of neuronal signals between two neurons through the action of neurotransmitters over the synaptic cleft.

Cells in different tissues are surrounded by varying combinations of biological macromolecules, which are divided into four classes, collagens, proteoglycans, structural glycoproteins and elastin, collectively referred to as the extracellular matrix (ECM).[229] Many cell types, such as epithelial, endothelial, and muscle cells, are dependent on adhesion to the ECM; if it is lost the consequence is apoptosis.[230] This is known as anchorage dependence.[35] Collagen refers to a family of proteins with related biological functions and histologically characteristic structures, having the main function to provide integrity to tissues.[229] Proteoglycans are composed of a protein core with one or more glycosaminoglycan side chains usually covalently bound to the core protein.[35] Proteoglycans surround cells as gels with different charge density and pore size. Structural glycoproteins form a diverse group of ECM components with such members as fibronectin, laminin and vitronectin. A characteristic feature of structural glycoproteins is that they possess binding sites where cells and other ECM components can bind independently. The interaction of structural glycoproteins with other ECM components serves in organizing the structure of ECM. The amino acid sequence recognized by the cellular transmembrane proteins, integrins, is arginin-glycin-aspartic acid (Arg-Gly-Asp or RGD).[231] Elastin is the main constituent of elastic fibers that form a major part the ECM of skin, blood vessels and lungs.[35]

The biological function of the ECM and its different components is far from only providing structural integrity and a static site for attachment for cells. Many biological effects, which are characteristic of growth factors and cytokines, such as regulation of growth, shape, differentiation, wound healing and metabolic responses, are also exerted by the ECM and mediated through integrins to the intracellular targets, ultimately resulting in changes in gene expression.[229] These effects are especially characteristic of the structural glycoproteins, fibronectin and laminin, the latter of which is significant in formation of neuronal outgrowths[232] and generally in the development of the central nervous system.[233] The fact that different tissues have varying isotypes of laminin controls the direction of differentiation and serves in maintenance of tissue phenotypes.[234] The gels formed by proteoglycans function as binding sites and selective sieves that control the residence time and diffusion of cellular signal molecules, such as growth factors, and hence directly affect cellular growth, differentiation and other responses.[35] The adhesion of cells to the ECM components also functions as a prerequisite for the more complex process of cell migration, which is an active interaction between different ECM components and integrins as well as other cell surface receptors.[235]

It is significant in developing tissues, where especially the structural glycoproteins serve as guides for cell movements. It is a highly controlled function and mutations that destroy anchorage dependence are the basis for the metastatic, invasive behavior of cancer cells [35]. Sufficiently strong but reversible binding to the ECM is necessary for successful cellular migration.[236,237] This is facilitated through allosteric control, i.e., when integrins bind to or detach from external ligands, they undergo conformational changes, which are translated to the cytoskeleton inside the cell. This is known as outside-in activation. Migration and cellular movements can also arise as the consequence of inside-out activation, which is mediated through the response of cell surface receptors, such as GPCRs to the binding of their ligands. The response of the receptors reaches the cytoskeleton, from which it is transmitted to the integrins, translating it into cellular movement.[35]

(*ii*) *In-Vitro Functions*

Conditions on a flat cell culture substrate do not correspond to those prevailing *in vivo* in a tissue. The three-dimensional structure of a tissue is exchanged for a two-dimensional one, disrupting cell-to-cell interactions that are characteristic of a tissue. Cells acquire a more spread morphology, become more mobile, and a greater fraction of a cell population undergoes growth. Despite the limitations imposed by the differences between an *in vivo* tissue and an *in vitro* cell culture, specialized cellular functions are expressed, making a cell culture a useful tool.[238] One of the crucial factors required for expression of specialized cellular functions is proper adhesion to the growth substrate. Cellular adhesion to planar growth substrate can be divided into two categories: close contacts (CCs) and focal adhesions (FAs).[239] CCs represent wide areas with weak adhesion, having a cell-substrate separation of 30–100 nm. FAs are formed by integrins that on the extracellular side anchor cells to the growth substrate and intracellularly mediate this contact to the cytoskeleton.[240] Figure 20 shows an illustration of the cellular contacts with a growth substrate.

Slightly negative net charge of the culture substrate has been found to enhance cellular adhesion.[238] Due to this, commercial polystyrene cell culture plastic ware is treated with ionizing electric discharge or high-energy radiation. Although sera, which are

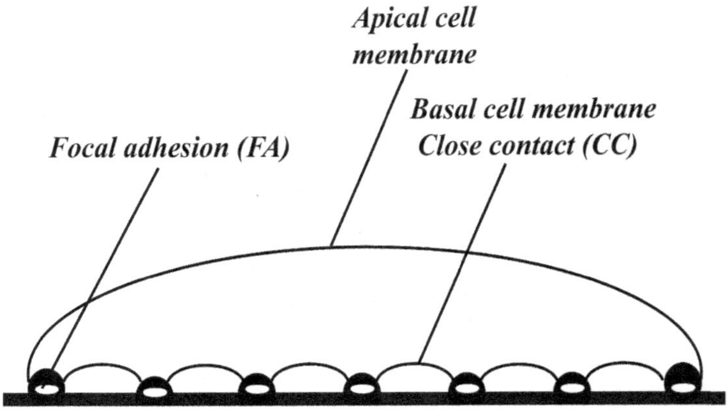

Figure 20. A schematic view of a cell on a growth substrate forming close contacts (CC) and focal adhesions (FA) with molecular contacts to the substrate.

used in preparation of cell culture media, contain the ECM components, fibronectin and vitronectin,[229] as well as that some cell lines are capable of producing their own ECM components required for adhesion, exogenously added ECM or different ECM components, such as fibronectin, laminin or collagen,[241] are most often needed. Different exogenously added polyelectrolytes with amine functionalities, such as poly-D-lysine (PDL)[242] and polyethyleneimine (PEI)[243] have also been employed as an alternative for ECM components. Commonly, the attachment of exogenously added adhesion factors on the growth substrate takes place through physisorption, where the forces of interaction range from ionic to hydrophobic, due to the negative net charge of cell culture plastic ware, and the often positive net charge of ECM components and other adhesion factors, such as polyelectrolytes. Research on the amino acid sequences recognized by cells on laminin and other structural glycoproteins have given the possibility to replace surface modifications using whole proteins and instead introducing, e.g., synthetic peptides containing the RGD recognition sequence. Massia and Hubbel found that for proper cellular adhesion such synthetic peptides needed to have an additional amino acid spacer to be used for covalent binding to the cell culture substrate.[244]

Research on synthetic biomaterials used in implants as well as the emergence of micro- and nanofabrication techniques in preparation of cell culture substrates have given rise to new approaches to produce growth substrates that promote cellular adhesion. Despite highly developed fabrication techniques, the purpose is not to avoid the usage of cellularily recognizable adhesion molecules, but instead improve their attachment and positioning on the surfaces through precise patterning. This approach makes it possible to avoid unspecific adhesion of cells and direct cellular growth and differentiation to defined areas. The two main directions are to fabricate either cytomimetic or ECM-mimetic materials.[245] Cytomimetic materials have a synthetic environment that mimics the three-dimensional environment of a tissue with contacts between the membranes of adjacent cells. In ECM-mimetic materials, the attachment of ECM components has been controlled through formation of, for instance, patterned silanized regions on quartz for attachment of vitronectin[246] and self-assembled monolayers (SAMs) of thiols on electrode surfaces for adsorption of fibronectin.[247, 248] The latter approach applied microcontact printing, where different thiols could be patterned on the electrode surface to facilitate adsorption of fibronectin on only those regions having thiols with an optimal functionalization for protein adsorption; the rest of the surface was made resistant to protein adsorption. Liu et al. applied electrochemistry to pattern a mixed SAM of carboxylic acid and hydroxyl group terminated thiols in order to obtain a suitable surface for creating a surface-density gradient of fibronectin.[249] The attachment of fibronectin was enhanced by covalent binding through carbodiimide activation of the carboxylic acid terminated thiols. This approach was utilized for the study of cell migration driven by a surface-density gradient of the adhesion factor. Cui et al. applied electrochemistry for modifying electrode surfaces through electropolymerization of polypyrrole (PPy) to be used for enhancing the attachment of fibronectin or synthetic peptide adhesion factors [250]. Lakard et al. also used electropolymerization to generate cell adhesion promoting electrode surfaces.[251] They polymerized PPy, PEI, polypropyleneimine (PPI) and poly(*p*-phenylenediamine) (PPPD) to be utilized directly without any exogenously added adhesion factors. Their results indicated that the adhesion promoting properties decreased in the order PEI>PPI>PPPD>PPy, PEI and PPI being far more superior for long term culturing than the two other materials. All of the materials had, however, better properties for cell adhesion than unmodified glass.

2. Techniques to Monitor Cellular Adhesion

Different approaches have been applied for studying cellular adhesion depending on what aspects of the phenomenon have been of interest. These range from quantification of cells adhering and migrating on different substrates, qualitative determination of morphological changes during adhesion, and cellular processes dependent on adhesion to the determination of the close contacts (CCs) and focal adhesions (FAs) with emphasis on the mechanism of cell migration as well as the force of adhesion.

(*i*) *Study of Close Contacts and Focal Adhesions*

Since the emergence of scanning electron microscopy (SEM), it has been an important tool in cell biological investigations. In the study of cell adhesion, it was the first method capable of giving a view of fibroblast adhesion and different stages of migration.[252] However, the images were obtained using fixed specimen, and considering the FAs with molecular contacts to the substrate, the information based on SEM was circumstantial. Interference reflection microscopy (IRM) was the first method able to yield conclusive evidence of the molecular contacts, locations and dimensions of FAs as well as the broader areas of CCs during migration of living fibroblasts.[239] IRM provides images that are created as interference between the reflections from the medium-substrate interface and medium-plasma membrane interface.[253] Originally, the method relied on light in the visible range of wavelengths. However, the high intensity of light that was needed had harmful effects on, especially migration of cells, only allowing short-term monitoring. With the introduction of near-infrared illumination, the method has been improved to eliminate the harmful effects high intensity visible light, now facilitating long-term monitoring of cellular adhesion and migration.[254] Development in fluorescence microscopy led to the emergence of total internal reflection fluorescence (TIRF), which with greater resolution could map membrane topography and cell-substrate contacts of living cells[255] by relying on the evanescent wave of a laser beam with an angle of incidence giving rise to total internal reflectance. This approach makes it possible to selectively excite fluorescent labels on the surface of the basal plasma membrane of an adhering cell without increased background fluorescence due to excitation of the non-adsorbed fluorophores in the bulk solution around the cell. Another technique that has become an important tool in cell biology with applications in the study of cellular adhesion and migration is confocal laser scanning microscopy (CLSM), which allows selective detection of excited fluorophores in one focal plane at a time.[256] Through the application of convolution techniques, images from individual focal planes can be combined to yield a three-dimensional view of the cellular adhesion zone. Further instrumental developments have merged CLSM and TIRF,[257] facilitating even more accurate studies of cellular adhesion.[258]

(*ii*) *Study of Cellular Morphology*

Aside from effects on the CCs and FAs, cellular adhesion changes the overall cell morphology. This is mediated through the integrins to the cytoskeleton. Phase contrast[259-261] and bright-field[262] microscopic observations have been applied as a method for evaluating the effect of different surfaces or biochemical responses on cell morphology. Changes in morphology can be registered as an increase or decrease in the footprint area of individual cells, i.e., cells become either more or less spread, as well as change the shape from fusiform to more roundish shape. Since morphological changes reflect the status of the cytoskeleton, fluorescence staining using suitable antibodies has been employed in evaluating cellular responses that are reflected in adhesion. Both conventional fluorescence (epifluorescence) microscopy[263,264] and CLSM[265,266] have been used for visualizing the stained cytoskeletons. Combination of microscopic techniques with time-lapse video monitoring[260, 263, 267, 268] has facilitated the development of four-dimensional monitoring of cellular adhesion and migration[269], i.e., the three-dimensional microscopic images obtained using CLSM can be generated with a temporal dimension. Aside from light and fluorescence microscopy, SEM has been utilized for observing the morphology of fixed cells.[265] A new technique that facilitates observation of morphological changes using live cells is atomic force microscopy (AFM), which allows probing of cellular morphology in either contact mode, using a constant vertical deflection of the AFM cantilever tip, or by performing indentation experiments, during which the cantilever tip presses the soft plasma membrane until it reaches the harder cytoskeleton.[270] The contact mode yields height images that correspond to the true cellular morphology, whereas indentation experiments yield force distance curves, primarily reflecting the structure of the cytoskeleton instead of the true morphology of the studied cell.[271] The AFM technique has been employed in studying the dynamics of the cytoskeletal organization of adhering cells,[272-276] the mechanical pulses in single cardiomyocytes,[275] the migration of fibroblasts[277] and the mitotic cycle of adhering cells.[278,279]

(*iii*) *Study of Cellular Adhesion Forces*

The study of forces involved in cellular adhesion elucidates the strength of the bonds formed between cell surface receptors, such as integrins, and the components of the ECM. Another area of interest is the interaction between cells and artificial surfaces. The developed methods can be divided into applications for cell populations and single cells. These approaches are used for either obtaining general knowledge regarding adhesion forces on different surfaces or studying the effects of different chemicals on the cytoskeletal organization and cell surface receptors. In the study of adhesion forces on population level, the general procedure is to apply different detaching forces and determine the fraction of cells remaining adherent. This yields a curve of percent adhesion as a function of the applied force. The detaching forces have been applied using, e.g., shearing in a laminar flow,[280] centrifugation in a specially constructed centrifuge[281,282] or spinning on a disc.[283] On single cell level, the utilized methods initially involves pulling or suction of cells from the growth substrate using, for instance, microneedles[284] or micropipettes,[285-287] allowing accurate determination of the adhesion force down to the order of magnitude of piconewton. The more recent instrumental development has made especially AFM significant in determining cellular adhesion forces by using cantilever tips modified with ligands, such as concanavalin A[288] and beads coated with vitronectin[289] or collagen,[290] that can bind to the cell surface more strongly than the cellular adhesion force to be measured.

(*iv*) *Study of Cell-Substrate Interactions*

The above presented techniques to study cellular adhesion have found wide usage in cell biology and biophysics. They have provided valuable information regarding the fundamental behavior of the phenomenon and the different factors affecting it. These techniques suffer, however, from labor-intensive procedures, low throughput, inability to be automated, and in many cases need of expensive instrumentation. Additionally, they primarily provide a possibility for endpoint detection instead of real-time detection of cellular dynamics involved in adhesion and migration. As a special example can be mentioned AFM-based monitoring of the dynamics in cellular migration.[277] Generally, the method facilitates observations on only one cell at a time, and despite attempts to improve the throughput by only scanning a small portion of the overall cell area, it is time consuming to monitor the organization of the cytoskeleton as a function of time. The temporal resolution is not by far sufficient for studying fast biochemical responses reflected in cellular adhesion. Hence, these methods cannot provide the necessary tools for activities involved in toxicological testing and HTS in drug discovery.[291]

Due to the involvement of cellular adhesion and reorganization of the cytoskeleton as response to different surfaces as well as cellular effectors that trigger, or modulate cellular signaling through the activity of, e.g., GPCRs, monitoring of cell-substrate interactions has gained significance in drug screening. Three methods, quartz crystal microbalance (QCM), optical waveguide lightmode spectroscopy (OWLS) and electrical cell-substrate impedance sensing (ECIS), have emerged since the 1980's, providing means for quantitative real-time monitoring of cell-substrate interactions without using fluorescence labeling.

(*a*) *Quartz crystal microbalance (QCM)*

The central functional part of a QCM system is a sensor made of an oscillating quartz crystal disc coated with metal, such as gold, on both sides. The oscillations of the disc at the fundamental frequency on megahertz level are driven by a sinusoidal potential poised between the metal surfaces. The change in the resonance frequency (Δf) of the disc is directly proportional to the added mass load (Δm) on the surface according to Sauerbrey equation,[292] which is strictly valid for thin films, such as chemical surface modifications[293] and adsorbed proteins.[294] In the case of layers with viscoelastic properties, such as cells, the oscillations undergo damping caused by dissipation of energy due to frictional forces[295] and described by a damping factor ΔR, measured as resistance. When QCM is used to monitor cellular adhesion, the observable effects are a decrease in Δf and an increase in ΔR. Aside from studies of cellular adhesion and spreading on different surfaces,[296-302] QCM has been employed in monitoring responses to cellular effectors, such as nocodazole,[303] which causes disintegration of the cytoskeleton, and paclitaxel, a chemotherapeutic substance initiating apoptosis.[304] Cellular responses that result in alteration of the cytoskeletal organization are reflected in Δf and ΔR in a concentration-dependent manner, yielding sigmoidal concentration-response curves. Instrumental development has made it possible to more accurately determine the behavior of cellular adhesion by monitoring Δf and the energy dissipation due to the viscoelasticity (denoted by ΔD) of the cell layer simultaneously at the fundamental frequency and several overtones (QCM-D).[294]

(*b*) *Optical waveguide lightmode spectroscopy (OWLS)*

In an OWLS system, a layer of an optically dense material (waveguide), such as TiO_2, with a refractive index higher than that of the medium above it, is coated on a support. At a certain critical angle of incidence, a plane-polarized laser beam is totally reflected at the interface and guided to propagate inside the waveguide.[305] At the surface of the waveguide, an optical grating diffracts and couples the light that propagates in the waveguide, generating a multitude of internal reflections. The light propagating inside the waveguide generates an electromagnetic field, referred to as an evanescent wave (the generated evanescent wave is based on the same physical principle as the one utilized in TIRF) that causes a perturbation in the medium-waveguide interface,[306] interacting with materials deposited at the interface. The evanescent wave can penetrate into the medium beyond the interface to a depth proportional to the wavelength of the utilized laser beam,[307] being about 100 nm for a laser beam with the wavelength of 633 nm.[291] Deposition of new material, such as proteins,[308] at the waveguide-medium interface causes a change in the refractive index of the interface, consequently generating a phase shift in the evanescent wave.[306] The change in the angle of incidence needed for maintaining the evanescent wave in phase serves as a measure of the change in the refractive index, giving a direct correlation with the amount of material deposited at the interface. The same principle applies to monitoring of cellular adhesion on the surface of an OWLS sensor. Cellular adhesion and spreading increase the refractive index due to an increase in the occupied area, whereas the response to compounds, such as colchicines that disrupts the cytoskeletal structures, shows the opposite effect, i.e., due to the disruption of the cytoskeleton, the cells become more roundish, which decreases the occupied area, and hence the measured refractive index.[309] Applications of OWLS in studying cellular adhesion range from adhesion and spreading of different cell lines[309-312] to monitoring of cellular growth[311,312] and the effect of toxic chemicals.[309,311,313]

(c) *Electrical cell-substrate impedance sensing (ECIS)*

Impedance-based monitoring of cellular adhesion, spreading and motion originates from the first published experiments by Giaever and Keese in the early 1980's.[314,315] The initial motivation was the lack of temporal and spatial resolution in the conventionally employed microscopic techniques to detect the subtle and transient morphological changes caused by biological phenomena. Impedance measurements are conducted by applying sufficiently low frequencies (f) of sinusoidal ac potential, where the modulation of the interface impedance is the primary contributor to the measured values. The basis for impedance measurements to monitor cellular adhesion, spreading and motion is related to the insulating properties of cell membranes. Generally, the behavior of a cell layer on electrodes is modeled as a resistor and capacitor in series. Impedance (Z), which is the transfer function relating the applied sinusoidal potential to the recorded current response, is

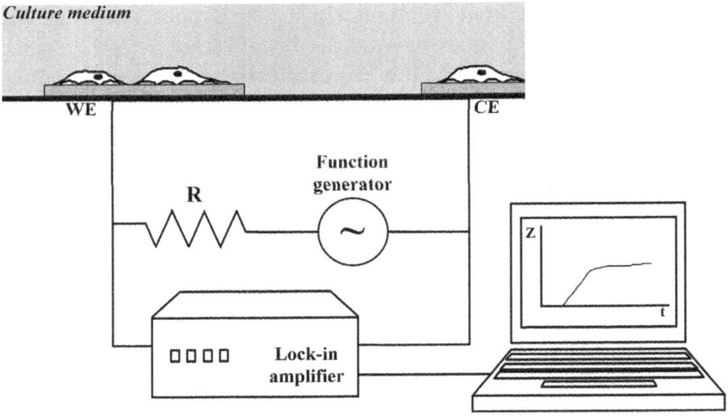

Figure 21. A schematic illustration of the usual measurement circuit for impedance monitoring of cellular adhesion, showing a cell culture platform with cells on a small WE and a large CE. The measurements are made using a lock-in amplifier with a function generator for applying the sinusoidal potential and a phase sensitive detector for detecting the in-phase and out-of-phase potentials that are proportional to the real and imaginary component of impedance, respectively. A computer is interfaced to the lock-in amplifier for collecting data. In order to limit the current cells encounter in the system, the sinusoidal potential is applied through a shunt resistor in series with the electrodes.

directly proportional to resistance (R) ($Z \propto R$) and inversely proportional to capacitance (C) ($Z \propto 1/C$) (Eq. 14):

$$Z = R - \frac{j}{\omega C}, \qquad (\omega = 2\pi f) \tag{14}$$

Resistance and capacitance, in turn, are inversely ($R \propto 1/A$) and directly proportional ($C \propto A$) to the unoccupied electrode area, respectively. Hence, an increase in the cellular coverage on an electrode causes an increase in the measured resistance (the real component of impedance) and decrease in the measured capacitance (the imaginary component of impedance), causing an overall increase in the interface impedance. The most commonly used configuration for measurements is depicted in Fig. 21, showing a large counter electrode (CE) and a small WE, making only the contribution of the WE significant for the overall impedance. In the initial studies of Giaever et al.,[314] the CEs had an area of 2 cm^2, while the area of the WEs was in the size regime of 10^{-4} cm^2. In a usual measurement protocol, the impedance changes are monitored at a constant frequency, such as 4 kHz, which primarily shows a manifestation of increasing resistance.

Experiments conducted and published at the early stage of impedance-based monitoring of cellular adhesion, such as characterization of adhesion on protein coated surfaces,[316,317] monitoring of morphological changes due to biochemical processes[317] and monitoring of cellular motion,[318] paved the way for future developments of the technique. The modeling of cellular motion was based on the following assumptions or simplifications:[318]

(*a*) cells are described as disks,
(*b*) only the current flowing radially between the basal cell membrane and the electrode and through the junctions between adjacent cells is considered significant,
(*c*) the current density between the basal cell membrane and the electrode is considered to remain constant in the vertical direction, and
(*d*) capacitive current can flow through the cells.

Figure 22 shows a schematic view of the cellular model and Eq. 15a-d are the individual differential equations that have been devised to describe the cellular behavior:

$$-dV = \frac{\rho dr}{h2\pi r} I \tag{15a}$$

$$V_n - V = \frac{Z_n}{2\pi r dr} dI_n \tag{15b}$$

$$V - V_m = \frac{Z_m}{2\pi r dr} dI_m \tag{15c}$$

$$dI = dI_n - dI_m \tag{15d}$$

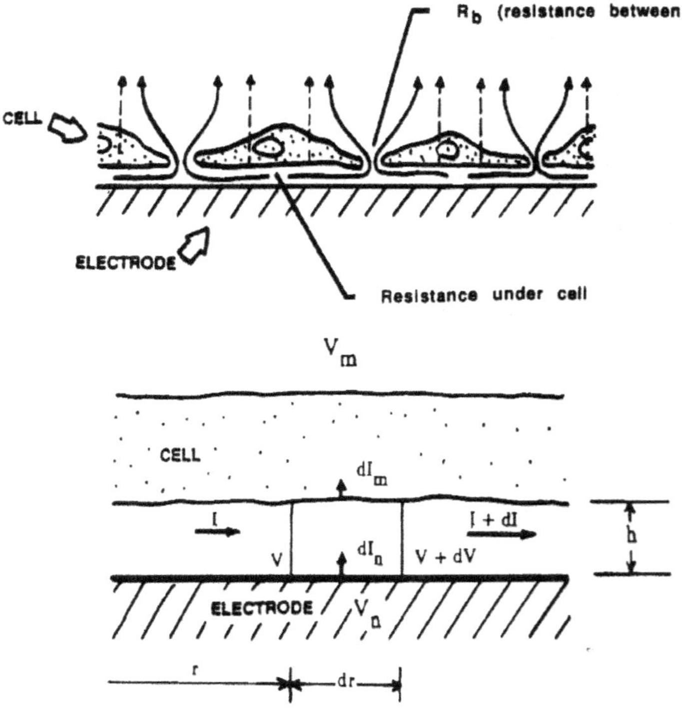

Figure 22. A schematic view of the model describing the behavior of a confluent monolayer of adherent cells. Shown are the different potentials, currents and resistances that have been utilized to derive the differential equation that describes the cellular interaction with an electrode surface. Reprinted with permission from Ref. 318, Copyright (1991) National Academy of Sciences, U.S.A.

The parameters in the differential equations, indicated in Fig. 22, have the following significance: V_n and V_m are the electrode potential and the potential in the medium outside the confluent monolayer of cells, respectively, V is the potential in the microenvironment under the basal cell membrane, I_n and I_m are the current at the electrode-medium interface and the capacitive current through the cell layer, respectively, I is the current flowing radially under the basal cell membrane and the electrode and further through the tight junctions between the cells, ρ is the resistivity of the culture medium and h is the height difference between the basal cell membrane and the electrode. The individual differential equations indicate the radial potential drop in the microenvironment between the basal cell membrane and the electrode (Eq. 15a), the potential differences between the electrode, the microenvironment under the basal cell membrane and the culture medium outside the confluent monolayer of cells (Eq. 15b-c) as well as the difference in current between that of the electrode-medium interface and the one through the cell layer (Eq. 15d). The overall differential equation (Eq. 16) is obtained by combining the ones presented as Eqs. (15a)-(15d):

$$\frac{d^2V}{dr^2}+\frac{1}{r}\frac{dV}{dr}-\gamma^2 V+\beta=0 \tag{16a}$$

$$\gamma^2=\frac{\rho}{h}\left(\frac{1}{Z_n}+\frac{1}{Z_m}\right) \tag{16b}$$

$$\beta=\frac{\rho}{h}\left(\frac{V_n}{Z_n}+\frac{V_m}{Z_m}\right) \tag{16c}$$

The mathematical solution of the differential equation (Eq. 16) yields the impedance (Z_c) that describes the cellular behavior in a confluent monolayer in terms of the devised model (Eq. 17):

$$\frac{1}{Z_c}=\frac{1}{Z_n}\left(\frac{Z_n}{Z_n+Z_m}+\frac{\dfrac{Z_m}{Z_n+Z_m}}{\dfrac{\gamma r_c}{2}\dfrac{I_0(\gamma r_c)}{I_1(\gamma r_c)}+R_b\left(\dfrac{1}{Z_n}+\dfrac{1}{Z_m}\right)}\right) \tag{17}$$

Based on the solution, the factors that contribute to cellular adhesion and motion are the resistance between adjacent cells (R_b), the average radius of cells (r_c) and the height difference between the basal cell membrane and the electrode (h), which is a part of the parameter γ. Other parameters that appear in the mathematical solution are the modified Bessel functions of the first kind of order 0 and 1, I_0 and I_1, respectively. This indicates that although the primary effect of cells on electrodes is related to the occupied electrode area, impedance measurements can provide detailed information of cellular adhesion far beyond merely monitoring the changes in the occupied electrode area. In comparison with QCM and OWLS that also have been applied to monitor cellular adhesion in terms of cell-substrate interaction, impedance measurements facilitate more diverse biological testing of cellular functions that influence adhesion. For instance, when cells undergo morphological changes, the only effect is not just a change in the occupied area, i.e., the footprint of the adhering cells. Instead, cells may retract the FAs transiently or irreversibly from the electrode surface, increasing the height difference between the basal cell membrane and the electrode, i.e., the microenvironment defining the CCs. Furthermore, in the case of cell types that form tight junctions between themselves, such as endothelial and epithelial cells, impedance measurements can contribute to evaluating biochemical processes that affect the barrier function of the tight junctions.

When utilizing the method for determining the influence of different cellular effectors on h and R_b, Z_n is initially measured at different frequencies (can be done by acquiring impedance spectra), Z_m is obtained by approximating the cell membrane capacitance as 1 $\mu F/cm^2$ and r_c as well as ρ are obtained from independent experiments. These initially defined values can then be utilized for fitting the impedance measurements on electrodes with a confluent monolayer of cells (can be done by acquiring impedance spectra) to the model. Figure 23 shows two impedance spectra, one for an electrode without cells and one with a confluent monolayer of fibroblasts (WI-38 VA13). As the result of curve fitting based on Eq. 17 one can obtain R_b and γr_c, which yields h. Values of h, calculated based on impedance measurements, have shown acceptable agreement with those performed using TIRF.[318]

Since the technique utilizes the interaction of the studied cells with the electrode surface and the measured impedance represents the electrode interface impedance, the technique was originally termed as the electric cell-substrate impedance sensing (ECIS®)[6] and commercialized by Applied BioPhysics, Inc. As the model based on the differential equations implies, the technique can provide information of several biologically significant factors related

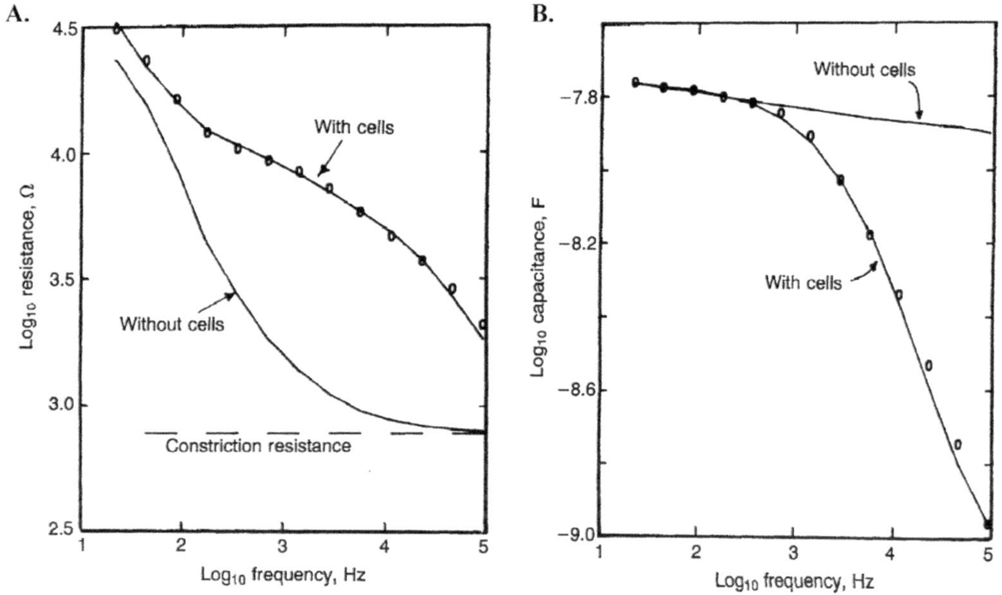

Figure 23. Impedance spectra of the electrode response without cells and in the presence of fibroblasts (WI-38 VA13 cells) that have been used for fitting to the model derived based on the illustration in figure 4.3. The spectra are shown as the logarithm of the A) real component and B) imaginary component of impedance as a function of the applied frequency on a logarithmic scale. Reprinted with permission from Ref. 318, Copyright (1991) National Academy of Sciences, U.S.A.

to cell adhesion. Applications have covered areas, such as monitoring of the endothelial barrier function upon widening of cell-cell junctions using α-thrombin,[6] effect of α-thrombin and histamine on focal adhesions of endothelial cells,[319] motion and morphological changes as a consequence of metabolic activity[320] and changes in external pH,[321] DNA transfection,[322] electroporation,[323] wound healing[7] and toxicological testing.[324-326] All of these biological processes comprise a combination of functional modes that affect the area each cell occupies as well as the distance from the growth substrate and the tightness of cell-cell contacts.

In the diverse application areas mentioned above, the common feature has been the utilization of cells, such as fibroblasts, endothelial and epithelial cells, that strongly adhere on the electrode surface and form tight junctions between themselves. In the case of cells that are poorly adherent and do not form tight junctions, such as PC12 cells, impedance measurements result in a low response. Furthermore, the formed cell layers are not necessarily confluent unless a high cell density is seeded on the electrodes. Different approaches have been tested to improve the sensitivity of impedance measurements using PC12 cells. Slaughter et al. have presented an approach to bind covalently ECM components, laminin, collagen and fibronectin, on electrodes modified with thiols having an amino endgroup.[327] The covalent binding was based on carbodiimide activation of carboxylic acid groups on the proteins and their subsequent reaction with the amino functionalities on the electrode surface. The best adherence was obtained using laminin covalently bound on cysteamine modified electrodes. During measurements using an initial cell density of 10^6 cells/well in an ECIS® platform from Applied BioPhysics, Inc., having WEs with a diameter of 250 μm, the real component of impedance measured at 4 kHz increased 1.5 fold with respect to the initial level during a 32.4-hour culture period. However, these measurements primarily mapped the initial adhesion due the high cell density, hence not facilitating long-term culturing. Heiskanen et al. utilized an ECIS® platform, where the electrodes were modified with physisorbed branched polyethyleneimine (M_w = 25 kDa), and the initially seeded cell density was 2×10^5 cells/well.[21] This approach facilitated long-term culturing for 168 hours, during which measurements were conducted every hour at the frequency of 4 kHz. In the end of the culture period, the real component of impedance closely reached a 1.4 fold increase with respect to the initial level.

When conducting impedance measurements to obtain biologically significant information regarding cellular behaviour, data presentation is a crucial part of a successful protocol. In toxicological testing conducted on fibroblasts utilizing ECIS® platforms, Xiao et al. noticed that toxic effects of different chemicals were concentration and time dependent,[326] affecting both the cellular growth rate and the interaction between the cells and the electrodes. This was reflected by the real component of the measured impedance. These studies led to development of the mathematical procedure for presenting the data obtained in impedance measurements, which is described by Eq. 18,

$$f(C,t)=k(C,t)[1+\eta(C,t)]^{t-\tau} \quad (18)$$

where (f(C,t)), (η(C,t)) and (k(C,t)) are the concentration and time dependent real component of the measured impedance, cellular growth rate and the change in interfacial resistance caused by an individual cell on an electrode, respectively. τ is the length of the lag phase between the cell seeding and the onset of the normal growth. This approach facilitated the evaluation of toxic effects that are either chronic or acute. In the case of chronic effects, $\eta(C,t)$ approaches zero, whereas for acute effects $k(C,t)$ approaches zero. $f(C,t)$ at different exposure times could then be plotted as a function of the applied concentration to obtain a sigmoidal concentration response curve that could be used for determining the IC_{50} of the introduced toxicant (Fig. 24).

Further development of ECIS® and other, mainly commercial systems, has shown that the main contribution of the technique is found in applications involved in toxicological, pharmacological and cell biological applications. A detailed review of systems and classification of applications for monitoring of cellular adhesion through impedance measurements can be found in a review by Spégel et al.[32] Other commercial systems are the real-time cell electronic sensing (RT-CES®) from ACEA Biosciences, Inc., CellKey™ from MDS Analytical Technologies and *adcon plate* from Bionas® GmbH. The common feature of these systems is that interdigitated electrodes (Fig. 25A), covering the majority of the

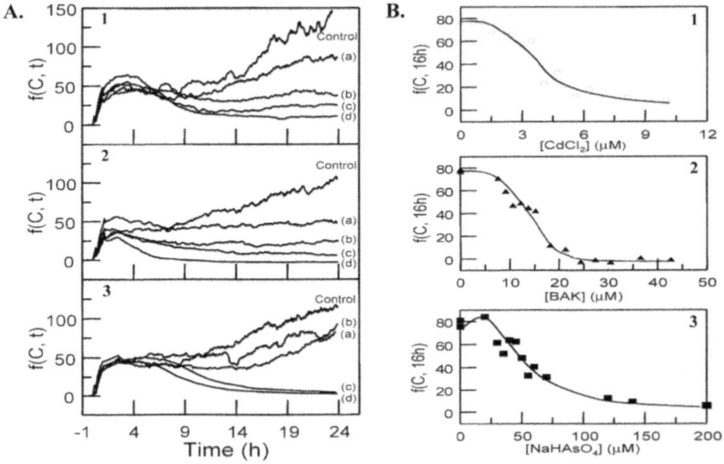

Figure 24. A) Impedance responses of fibroblast cells to $CdCl_2$ (panel 1), benzalkonium chloride (panel 2) and $NaHAsO_4$ (panel 3) presented as f(C,t) as a function of the exposure time (the details of the data presentation are discussed in the text). The applied concentrations (μM) were: $CdCl_2$ a) 2.9, b) 4.6, c) 6.2 and d) 8.1; benzalkonium chloride a) 15.2, b) 18.3, c) 21.3 and d) 30.4; $NaHAsO_4$ a) 45, b) 60, c) 140 and d) 200. B) Concentration-response curves at t = 16 h for the responses shown in B) featuring inhibition (IC_{50} in μM) by $CdCl_2$ (4.0) (panel 1), benzalkonium chloride (14.0) (panel 2) and $NaHAsO_4$ (50.1) (panel 3). Reprinted with permission from Ref. 326, Copyright (2002) American Chemical Society.

area of a cell culture well, are used for impedance measurements instead of a large CE and small WE. In this case, the behavior of a cell layer on electrodes is modeled as a resistor and capacitor in parallel. The consequence of using interdigitated electrodes is that both the CE and WE have an equal contribution to the measured interface impedance. This decreases the influence of the impedance associated with the bulk solution[328] and increases the reproducibility of measurements since even a small number of seeded cells (shown to be detectable down to 125 cells) causes a proportional contribution to the interface impedance[329] (Fig. 25B). Additionally, this approach provides better possibilities for miniaturization and parallelization necessary for effective utilization in, e.g., HTS in drug discovery.[291]

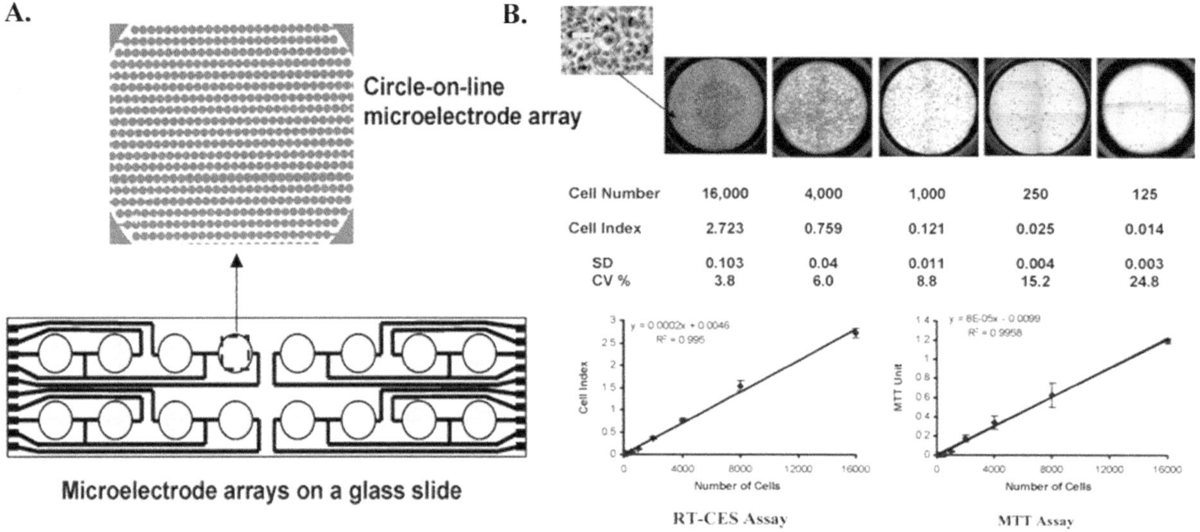

Figure 25. A) An enlarged microscopic view of a circle-on-line interdigitated microelectrode and a schematic view of a RT-CES 16X E-Plate® housing wells where each electrode is located. B) Microscopic images of Giemsa stained fibroblasts after 6 h of incubation on circle-on-line microelectrode arrays. The number of cells and the corresponding cell index (CI) is shown below each image (the details of the significance of CI are discussed in the text). CI vs. cell number for each of the microscopic images and a corresponding plot for a MTT assay are shown at the bottom. Reprinted with permission from Ref. 329, Copyright (2005) American Chemical Society.

Applications of the systems comprise, e.g., pharmacological (morphological changes associated with GPCR activation induced by ligand binding[9,330] and tyrosine kinase receptor activity[331]), drug discovery[332] and toxicological[329,333] studies. Figures 26A and B show results of impedance based monitoring of GPCR activation using the RT-CES® platform. In applications with RT-CES® platforms, the cellular responses are described with the Cell Index (CI), which is dependent on the parallel resistance of electrodes having adhering cells ($R_{cell}(f_i)$) and electrodes without cells ($R_0(f_i)$). The mathematical relation is shown in Eq. (19),

$$CI = \max_{i=1...N}\left\{\left[\frac{R_{cell}(f_i)}{R_0(f_i)}\right]-1\right\} \tag{19}$$

Each of the quantities, $R_{cell}(f_i)$ and $R_0(f_i)$, is evaluated at different frequencies and the frequency yielding the maximal response is chosen for presenting CI. Equation (19) indicates that in the absence of cells or in the case of no adherence, CI is zero. When cells are adhering, spreading and growing on the utilized electrode, CI increases until a confluent monolayer on the electrode is formed, resulting in a leveling-off of the calculated CI. When different cellular effectors are introduced, CI either decreases or increases from the plateau initially reached upon formation of a confluent monolayer. In the case of toxicants, cells become more rounded and /or retract from the electrode surface, which results in an observed decrease in the calculated values of CI. On the other

Figure 26. A) Impedance responses of CHO cells to GPCR ligands histamine (H1 receptor) (upper panel), SKF38393 (D1 receptor) (middle panel) and 8-OH-DPAT (5-HT1A receptor) (lower panel) in the form of normalized CI with respect to concentration and time,. The introduced concentrations were: histamine 30 pM, 30 nM and 30 μM; SKF38393 0.1 nM, 1 nM and 100 μM; 8-OH-DPAT 0.1 nM, 10 nM and 100 μM. B) Concentration response curves (based on normalized CI) and corresponding EC_{50} value for histamine (upper panel), SKF38393 (middle panel) and 8-OH-DPAT (lower panel). C) Concentration response curves (based on standard GPCR assays) and corresponding EC_{50} value for histamine (upper panel), SKF38393 (middle panel) and 8-OH-DPAT (lower panel). D) Impedance responses HeLa cells to histamine (upper panel), CHO cells to calcitonin (midpanel), N1E-115 cells to DPDPE (lower panel). Reprinted with permission from Ref. 9, Copyright (2006) American Chemical Society.

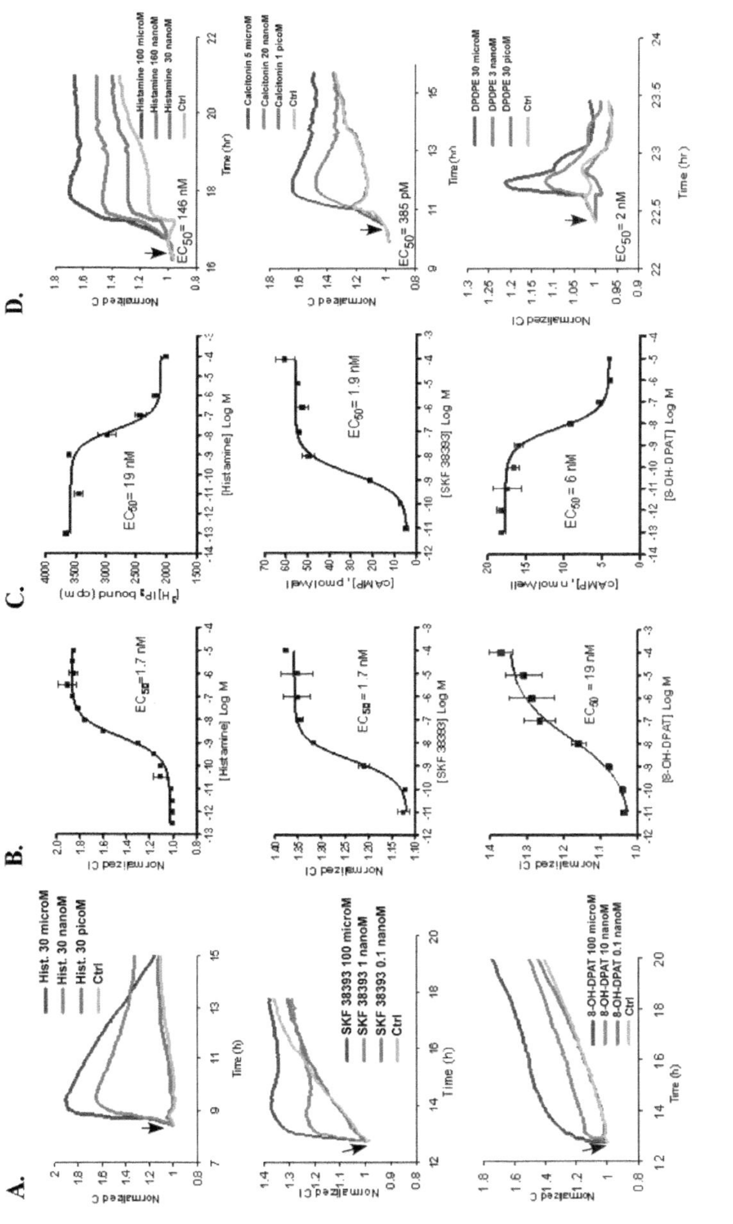
A.
B.
C.
D.
Hist. 30 microM
Hist. 30 nanoM
Hist. 30 picoM
Ctrl
SKF 38393 100 microM
SKF 38393 1 nanoM
SKF 38393 0.1 nanoM
Ctrl
8-OH-DPAT 100 microM
8-OH-DPAT 10 nanoM
8-OH-DPAT 0.1 nanoM
Ctrl
EC50=1.7 nM
EC50 = 1.7 nM
EC50 = 19 nM
EC50= 19 nM
EC50= 1.9 nM
EC50 = 6 nM
EC50= 146 nM
EC50= 385 pM
EC50= 2 nM
Histamine 100 microM
Histamine 160 nanoM
Histamine 30 nanoM
Ctrl
Calcitonin 5 microM
Calcitonin 20 nanoM
Calcitonin 1 picoM
Ctrl
DPDPE 30 microM
DPDPE 3 nanoM
DPDPE 30 picoM
Ctrl
[Histamine] Log M
[SKF 38393] Log M
[8-OH-DPAT] Log M
[cAMP], pmol/well
[cAMP], nmol/well
Normalized CI
Time (h)
Time (hr)

hand, if, for instance, a ligand is bound to a GPCR, the subsequent receptor activation results in a conformational change transiently increasing the membrane area in contact with the electrode surface. This causes an increase in the calculated values of CI. By evaluating CI at different concentrations of the introduced cellular effector, the values can be plotted as a function of the effector concentration on a logarithmic scale to obtain a sigmoidal concentration response curve and further the IC_{50} or EC_{50} describing the cellular response as shown in Fig. 26B (corresponding concentration response curves based on standard GPCR assays are shown in Fig. 26C).

Although electrode interfaces with adhering cells have generally been described as a resistor and capacitor in series or in parallel, more specialized equivalent circuits can be utilized to describe the behavior of cells. Wegener et al. have demonstrated the effect of isoproterenol (a β-adrenoceptor agonist) on aortic endothelial cells [334]. They found that the cellular behavior could be described with the equivalent circuit shown in Fig. 27A, where the parallel resistance and capacitance, R_{cell} and C_{cell}, respectively, represent the presence of a cell layer with resistive and capacitive properties (more detailed description of the assignment of the resistive and capacitive properties can be found in [334] and references therein), CPE is the constant phase element that comprises the contribution of the double layer capacitance and surface roughness of metal electrodes and R_{med} is the resistance of the bulk cell culture medium. Figure 27B shows two impedance spectra illustrating the difference between an electrode without cells and an electrode covered by a confluent monolayer of bovine aortic endothelial cells. Values of R_{cell} obtained by acquiring impedance spectra at different introduced concentrations of isoproterenol could be fitted to a four-parameter logistic equation of the type shown in Eq. (9) above to yield a concentration response curve for determination of EC_{50}. As a generalization, this approach facilitates the possibility to characterize the behavior of different cell types to obtain an equivalent circuit describing their interaction with an electrode surface. Upon introduction of a cellular effector and acquiring impedance spectra, values of a characteristic equivalent circuit component can be retrieved and related to the concentration of the effector. Alternatively, the same approach can be utilized when, for instance, monitoring cellular adhesion, spreading and growth as a function of culture time. For continuous evaluation of the value of the characteristic equivalent circuit component, such as R_{cell}, a tailor-made software interface can be utilized to fit automatically each acquired spectrum to an equation that describes the equivalent circuit as shown in Eq. (20a) for the equivalent circuit of Fig. 27A The contribution of the CPE has been comprised in the form of the empirical equation shown below (Eq. 20b), where the exponent n generally can acquire values between 0 and 1, the latter corresponding to pure capacitive behavior. In systems with electrodes modified with biological entities, such as cells, n ranges from 0.8 to 1:

$$Z = R_{med} + \frac{R_{cell}}{1+\omega^2 C_{cell}^2 R_{cell}^2} + \frac{\cos\left(\frac{n\pi}{2}\right)}{A\omega^n} - j\left[\frac{\omega C_{cell} R_{cell}^2}{1+\omega^2 C_{cell}^2 R_{cell}^2} + \frac{\sin\left(\frac{n\pi}{2}\right)}{A\omega^n}\right] \tag{20a}$$

$$Z_{CPE} = A(j\omega)^{-n} \tag{20b}$$

At present, all the commercial systems described above are available in a 96-well format, making the technique approach the needs of HTS. Realization of measurement platforms based on impedance monitoring provides a greater degree of freedom to design and fabricate systems that more easily can be combined with other electrochemical techniques as well as optical and fluorescence monitoring. Furthermore, the thoroughly modeled and understood impedimetric cellular behavior on electrodes inevitably fuels the development of multiparameter detection systems based on impedance monitoring of adhesion-related phenomena, representing one of the different relevant parameters of cellular dynamics. Figure 28A shows a multiparameter detection system, facilitating impedance monitoring of cellular adhesion together with capability for monitoring of changes in extracellular pH and oxygen level.[335] Figure 28B shows results correlating the three parameters

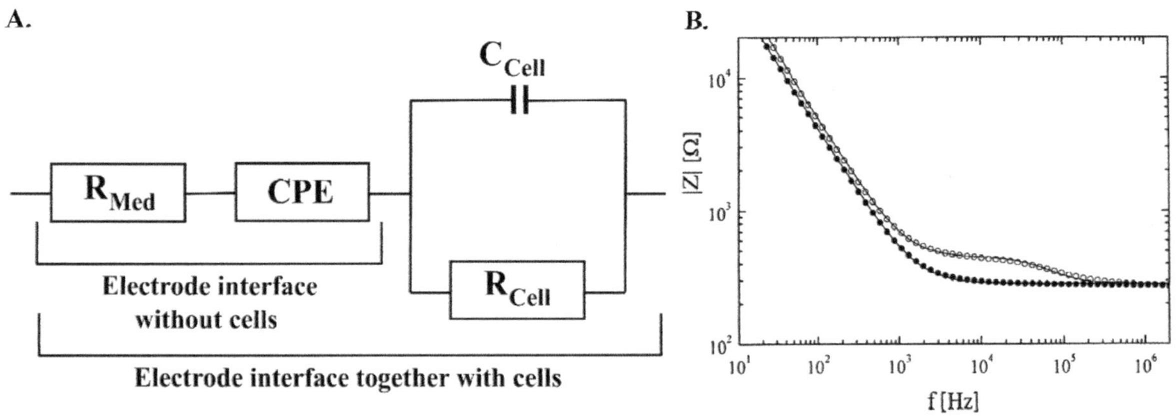

Figure 27. A) An equivalent circuit describing the behavior of bovine aortic endothelial cells at an electrode interface (details of the assignment of components is discussed in the text). B) Impedance spectra an electrode without cells (solid circles) and with a confluent monolayer of bovine aortic endothelial cells (open circles). Reprinted with permission from Ref. 334, Copyright (1999) Springer-Verlag.

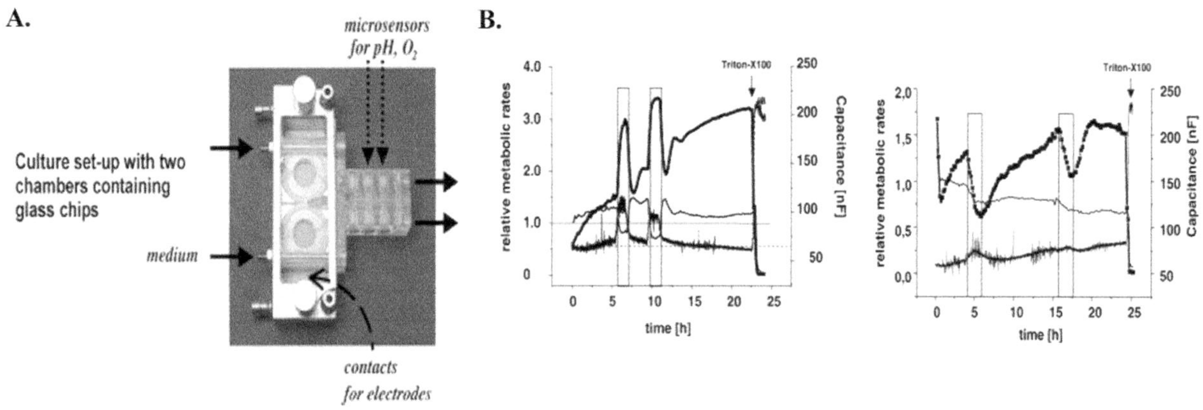

Figure 28 A) An image of the Cell-Monitoring-System® that can be used to monitor cellular adhesion as well as extracellular pH change and O_2 consumption. Responses to showing the effect of B) cytochalasin B (1 µg/ml) (left panel) and C) chloroacetaldehyde (50 µM) (right panel) on metabolism (O_2 consumption: thick line; pH change: thin line; and morphology: jagged line at the bottom). The exposure to the cellular effectors lasted during the periods indicated by the open bars. Reprinted with permission from Ref. 335, Copyright (2003) International Society for Preventive Oncology, published by Elsevier Ltd.

of cellular behavior to the effect of cytochalasin B (left panel) and chloroacetaldehyde (right panel).

V. CONCLUSIONS AND FUTURE PERSPECTIVES

This chapter has provided an overview of electrochemical applications in real-time monitoring of the dynamics of different cellular processes, which generally represent the cellular interaction with the growth environment either in the form of release of substances or interaction with growth surfaces or other cells. These areas cover diverse applications that are relevant in cell biological and medical research as well as in drug discovery activities. The function of amperometry and impedance spectroscopy in such applications is well characterized and as detection techniques they have reached a maturation clearly paving the way for future developments that not only can lift up the potentials of the techniques but moreover provide the development of advanced tools for biological and medical research.

In order for electrochemical techniques to find even more valuable biological applications in, for instance, drug discovery, they have to be utilized more strongly in parallelized systems that facilitate high throughput screening of cellular functions with respect to different cellular effectors. Considering the development of lithographic fabrication techniques, parallelization based on an increased number of electrodes is easily realizable. At present, however, the limitation is caused by the necessary instrumentation needed for the measurements. Aside from unavailability of instruments that facilitate simultaneous measurements on a large number of electrodes, an even greater obstacle is caused by interfacing of miniaturized multielectrode systems to existing external benchtop instrumentation. In order to realize an effective parallelization of electrochemical measurement systems, miniaturization of the instrumentation needed for electrochemical monitoring is significant and more research is needed to realize, for instance, construction of single-chip potentiostats using advanced CMOS technology.[336] Furthermore, considering the ongoing development of parallelized microfluidic systems for cell handling and culturing,[28] the only possible way to find an effective adaptation of electrochemical transducers to such systems is through fabrication of integrated

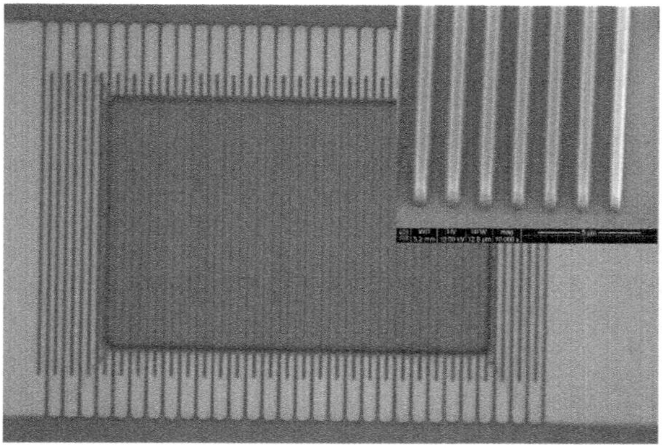

Figure 29. A SEM image of a carbon interdigitated electrode (IDE) with width/gap in the nanometer size regime. Reprinted with permission from Ref. 339, Copyright (2010) Lars Henrik Skjolding.

measurement systems to eliminate the increasing difficulty in interfacing of external instrumentation.

Development in fabrication techniques to construct more advanced electrode systems has, however, significance for future applications in biology and medicine. One significant area of development is miniaturization of planar interdigitated electrodes (IDEs) to achieve the gap/width size regime of ~100 nm. When approaching such dimensions, a considerable increase in signal amplification is achieved.[337] However, the already constructed nanometer sized IDEs are based on metal structures, which due to the formation of a galvanic system between the gold or platinum based electrode material, and the chromium or titanium adhesion layer, have a very short life time, making them incapable of sustaining the conditions during biological experiments. Development of carbon based electrode materials by pyrolization is a new area of development which in combination with electron beam lithographic definition of electrode areas[338] (Fig. 29) has shown promising future perspectives that could facilitate more advanced biological experiments. Aside from better stability in comparison with corresponding metal electrodes, carbon as electrode material is also superior to metals in, for instance, DA exocytosis monitoring

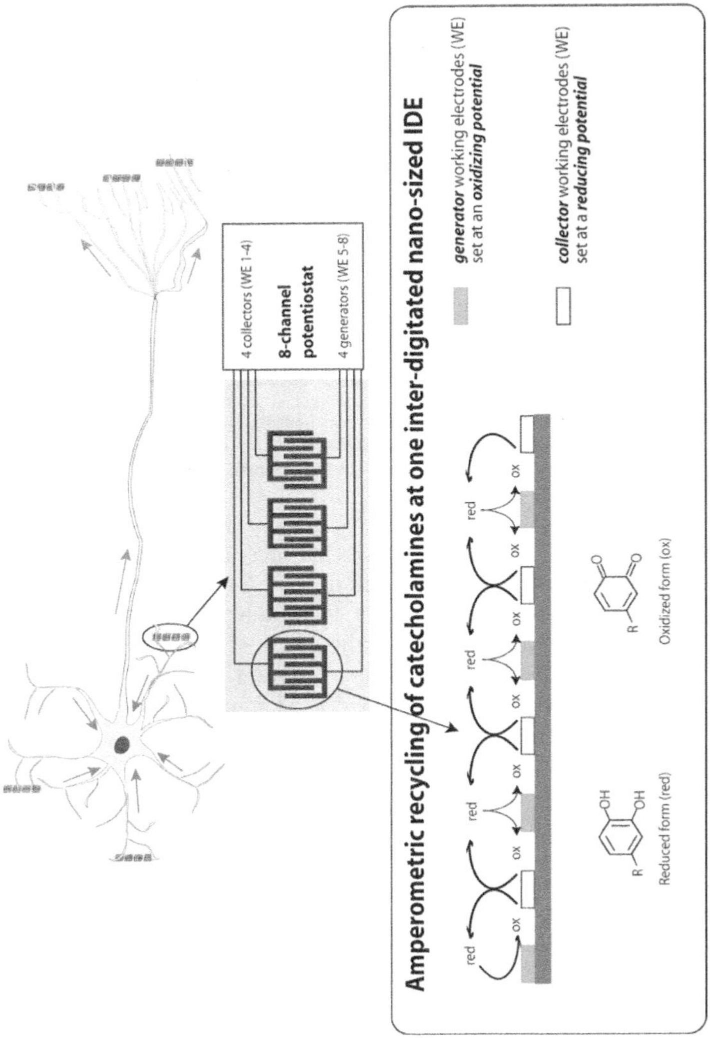
4 collectors (WE 1-4)
8-channel potentiostat
4 generators (WE 5-8)
Amperometric recycling of catecholamines at one inter-digitated nano-sized IDE
generator working electrodes (WE) set at an oxidizing potential
collector working electrodes (WE) set at a reducing potential
red
ox
Reduced form (red)
Oxidized form (ox)

since no additional electrode modification is needed to facilitate a nearly reversible DA electrochemistry, which in the case of metal electrodes is not easily achievable despite chemical modifications.[20] Upon fabrication of carbon based IDE arrays with dimensions in the nanometer size regime combined with surface patterning, exocytosis studies on neurons or differentiated NSCs could potentially be conducted by individually addressing different dendritic structures to achieve enhanced spatial control over the exocytotically active zones as schematically visualized in Figure 30.

Development of electrochemical techniques and electrode systems has predominantly aimed at utilizing planar electrodes to detect, e.g., extracellularly secretion of biologically significant molecules. Although such systems and experiments can provide specialized information regarding cellular dynamics, they are, nevertheless, limited in the scope of experiments that can be conducted, and hence in the depth of biological queries. A future development that ultimately can improve the depth of biological information gained by electrochemical measurements will involve direct electrochemical experiments in the intracellular environment using 3D electrodes in the nanometer size regime suitable for inquiries on the single cell level. Figure 31 illustrates a vision to conduct intracellular electrochemical measurements using 3D electrodes. If these are utilized in amperometric and impedance spectroscopic measurements, they could facilitate experiments that can provide direct information regarding synthesis of biomolecules, such as mRNA and proteins, in real-time to reveal the turnover on single-cell level. Furthermore, if such electrodes are combined with an actuation system to guide their insertion into cells, they could facilitate intracellular experiments on single cells in a population of adherently growing cells.

Figure 30. (Upper panel) A schematic view of a neuron being addressed with individual nanometer sized IDEs, the dimensions of which are in the size regime of the neuronal dendrites. (Lower panel) A schematic view of the functional principle of dopamine (DA) detection using IDEs: The digits on one side of an IDE are poised at a potential, at which DA is oxidized, while those on the other side are poised at a potential, at which the oxidized form of DA is reduced. When DA is oxidized, the oxidized form diffuses to the digits beside to undergo reduction to DA. This recycling between oxidation and reduction continues, amplifying the generated amperometric current, hence increasing the sensitivity when very small quantities are to be detected.

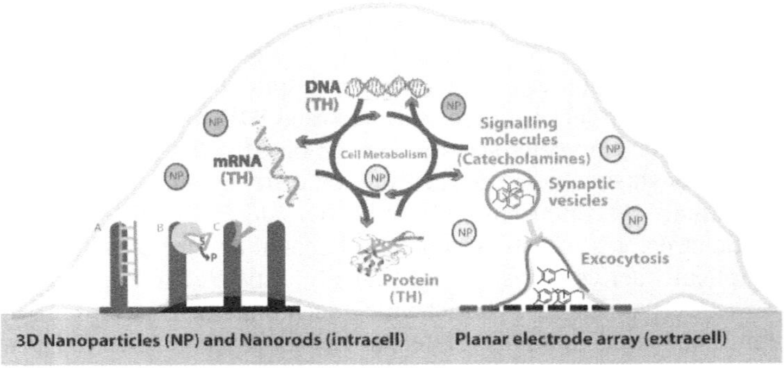

Figure 31. A schematic view of a future approach, where the dynamics of cellular processes on single-cell level can be studied using both extra- and intracellularly placed electrodes. The former approach applies nanometer sized electrodes capable of addressing a part of a cell, while the latter utilizes nanometer sized 3D electrodes capable of penetrating the cell membrane to address directly the intracellular chemical species, facilitating monitoring of processes, such as mRNA and protein synthesis. When the utilized 3D electrodes are combined with an actuation technique to move them vertically, the technique provides possibility to address single cells in a confluent monolayer of adherent cells.

ACKNOWLEDGEMENT

The EU FP7 NMP project EXCELL and the Ørsted Postdoc program, Denmark, are kindly acknowledged for financial support as well as financial support from the Ørsted Postdoc program, Denmark. Roberto Raiteri and co-workers (University of Genova, Dept. Biophysical and Electronic Engineering, Genova, Italy) are kindly acknowledged for derivation of Eq. 20a based on the equivalent circuit in Fig. 27A.

REFERENCES

[1]C. Gota, K. Okabe, T. Funatsu, Y. Harada, and S. Uchiyama, *J. Am. Chem. Soc.*, **131** (2009) 2766.

[2]B. Tang, F. Yu, P. Li, L. Tong, X. Duan, T. Xie, and X. Wang, *J. Am. Chem. Soc.*, **131** (2009) 3016.

[3]A. V. Zhdanov, M. W. Ward, J. H. M. Prehn, and D. B. Papkovsky, *J. Biol. Chem.*, **283** (2008) 5650.

[4]M. Brischwein, E. R. Motrescu, E. Cabala, A. M. Otto, H. Grothe, and B. Wolf, *Lab Chip*, **3** (2003) 234.
[5]G. T. Baxter, L. J. Bousse, T. D. Dawes, J. M. Libby, D. N. Modlin, J. C. Owicki, and J. W. Parce, *Clin. Chem.*, **40** (1994) 1800.
[6]C. Tiruppathi, A. B. Malik, P. J. Delvecchio, C. R. Keese, and I. Giaever, *Proc. Nat. Acad. Sci. U.S.A.*, **89** (1992) 7919.
[7]C. R. Keese, J. Wegener, S. R. Walker, and L. Giaever, *Proc. Nat. Acad. Sci. U.S.A.*, **101** (2004) 1554.
[8]K. Kotarsky, C. Owman, and B. Olde, *Anal. Biochem.*, 288 (2001) 209-215.
[9]N. C. Yu, J. M. Atienza, J. Bernard, S. Blanc, J. Zhu, X. B. Wang, X. Xu, and Y. A. Abassi, *Anal. Chem.*, **78** (2006) 35.
[10]J. H. Kelly, and N. L. Sussman, *J. Biomolec. Screen.*, **5** (2000) 249.
[11]Y. Y. Lee, A. H. Westphal, L. H. J. de Haan, J. Aarts, I. Rietjens, and W. J. H. van Berkel, *Free Radical Biol. Med.*, **39** (2005) 257.
[12]N. V. Kostesha, J. R. M. Almeida, A. R. Heiskanen, M. F. Gorwa-Grauslund, B. Hahn-Hagerdal, and J. Emneus, *Anal. Chem.*, **81** (2009) 9896.
[13]A. Heiskanen, C. Spegel, N. Kostesha, S. Lindahl, T. Ruzgas, and J. Emneus, *Anal. Biochem.*, **384** (2009) 11.
[14]J. D. Rabinowitz, J. F. Vacchino, C. Beeson, and H. M. McConnell, *J. Am. Chem. Soc.*, **120** (1998) 2464.
[15]J. S. Zhao, Z. Y. Yang, Q. T. Gong, L. Yao, and W. Min, *Anal. Lett.*, **38** (2005) 89.
[16]C. F. Spegel, A. R. Heiskanen, N. Kostesha, T. H. Johanson, M. F. Gorwa-Grauslund, M. Koudelka-Hep, J. Emneus, and T. Ruzgas, *Anal. Chem.*, **79** (2007) 8919.
[17]N. Kostesha, A. Heiskanen, C. Spegel, B. Hahn-Hagerdal, M. F. Gorwa-Grauslund, and J. Emneus, *Bioelectrochemistry*, 76 (2009) 180-188.
[18]C. Amatore, S. Arbault, M. Guille, and F. Lemaitre, *Chem. Rev.*, **108** (2008) 2585.
[19]J. Castillo, A. Blochl, S. Dennison, W. Schuhmann, and E. Csoregi, *Biosens. Bioelectron.*, **20** (2005) 2116.
[20]C. Spégel, A. Heiskanen, J. Acklid, A. Wolff, R. Taboryski, J. Emnéus, and T. Ruzgas, *Electroanalysis*, **19** (2007) 263.
[21]A. Heiskanen, C. Spégel, J. Tønnesen, Z. Fohlerova, L. Goulart, J. Hansen, M. Kokaia, T. Ruzgas, M. Dufva, and J. Emnéus, *Proceeding of the Twelfth International Conference on Miniaturized Systems for Chemistry and Life Sciences, October 12-16, 2008, San Diego, CA.*
[22]D. P. Bratu, in *Cell imaging techniques: Methods and protocols, Vol. 319, Methods in molecular biology*, Ed. by D. J. Taatjes, and B. T. Mossman, Humana Press, Inc., Totowa, 2006, p. 1.
[23]K. Jakubs, A. Nanobashvili, S. Bonde, C. T. Ekdahl, Z. Kokaia, M. Kokaia, and O. Lindvall, *Neuron*, **52** (2006) 1047.
[24]P. M. Jones, M. L. Courtney, C. J. Burns, and S. J. Persaud, *Drug Discov. Today*, **13** (2008) 888.
[25]Y. Wang, S. Chen, D. Yang, and W. D. Le, *J. Neuroimm. Pharmacol.*, **2** (2007) 243.
[26]P. S. Dittrich, and A. Manz, *Nature Reviews Drug Discov.*, **5** (2006) 210.
[27]J. El-Ali, P. K. Sorger, and K. F. Jensen, *Nature*, **442** (2006) 403.
[28]R. Gomez-Sjoberg, A. A. Leyrat, D. M. Pirone, C. S. Chen, and S. R. Quake, *Anal. Chem.*, **79** (2007) 8557.

[29]J. M. Finnegan, K. Pihel, P. S. Cahill, L. Huang, S. E. Zerby, A. G. Ewing, R. T. Kennedy, and R. M. Wightman, *J. Neurochem.*, **66** (1996) 1914.
[30]C. H. Adler, *Mov. Disord.*, **20** (2005) S23.
[31]C. Spegel, A. Heiskanen, S. Pedersen, J. Emneus, T. Ruzgas, and R. Taboryski, *Lab Chip*, **8** (2008) 323.
[32]C. Spegel, A. Heiskanen, L. H. D. Skjolding, and J. Emneus, *Electroanalysis*, **20** (2008) 680.
[33]F. Q. Schafer, and G. R. Buettner, *Free Radical Biol. Med.*, **30** (2001) 1191-1212.
[34]D. Voet, J. G. Voet, and C. W. Pratt, *Fundamentals of biochemistry*, John Wilkey & Sons, Inc., New York, 1999.
[35]B. Alberts, A. Johnson, J. Lewis, M. Raff, K. Roberts, and P. Walter, *Molecular biology of the cell*, Garland Science, Taylor & Francis Group, New York, 2008.
[36]T. S. Wong, and U. Schwaneberg, *Curr. Opin. Biotechnol.*, **14** (2003) 590.
[37]F. W. Scheller, U. Wollenberger, C. Lei, W. Jin, B. Ge, C. Lehmann, F. Lisdat, and V. Fridman, *Rev. Molec. Biotechnol.*, **82** (2002) 411-424.
[38]D. G. Nichols, and S. J. Ferguson, *Bioenergetics 3*, Academic Press, London, 2002.
[39]A. K. Groen, R. van der Meer, H. V. Westerhoff, R. J. A. Wanders, T. P. M. Akerboom, and J. M. Tager, in *Metabolic compartmentation*, Ed. by H. Sies, Academic Press, London, 1982, p. 9.
[40]H. Sies, in *Metabolic compartmentation*, Ed. by H. Sies, Academic Press, London, 1982, p. 1.
[41]J. G. Reich and E. E. Sel'kov, *Energy metabolism of the cell - a theoretical treatise*, Academic Press, London, 1981.
[42]B. Bloom, and D. Stetten, *J. Am. Chem. Soc.*, **75** (1953) 5446.
[43]R. L. Veech, L. V. Eggleston, and H. A. Krebs, *Biochem. J.*, **115** (1969) 609.
[44]Y. J. Suzuki, H. J. Forman, and A. Sevanian, *Free Radical Biol. Med.*, **22** (1997) 269.
[45]G. Powis, J. R. Gasdaska, and A. Baker, in *Antioxidants in disease mechanisms and therapy*, Ed. by H. Sies, Academic Press, San Diego, 1997, p. 329-359.
[46]C. K. Sen, and L. Packer, *FASB J.*, **10** (1996) 709.
[47]R. E. Shackelford, W. K. Kaufmann, and R. S. Paules, *Free Radical Biol. Med.*, **28** (2000) 1387.
[48]Y. Makino, N. Yoshikawa, K. Okamoto, K. Hirota, J. Yodoi, I. Makino, and H. Tanaka, *J. Biol. Chem.*, **274** (1999) 3182.
[49]C. Abate, L. Patel, F. J. Rauscher, and T. Curran, *Science*, **249** (1990) 1157.
[50]S. Raha, and B. H. Robinson, *Trends Biochem. Sci.*, **25** (2000) 502.
[51]B. Chance, and G. R. Williams, in *Advances in Enzymology and Related Subjects of Biochemistry*, Vol. 17, Ed. by E. F. Nordi, John Wiley & Sons, Ltd., Hoboken, 1956, p. 65.
[52]J. F. Turrens, and A. Boveris, *Biochem. J.*, **191** (1980) 421.
[53]A. Boveris, E. Cadenas, and A. O. M. Stoppani, *Biochem. J.*, **156** (1976) 435.
[54]E. Cadenas, A. Boveris, C. I. Ragan, and A. O. M. Stoppani, *Arch. Biochem. Biophys.*, **180** (1977) 248.
[55]J. F. Turrens, A. Alexandre, and A. L. Lehninger, *Arch. Biochem. Biophys.*, **237** (1985) 408.
[56]L. Kussmaul, and J. Hirst, *Proc. Nat. Acad. Sci. U.S.A.*, **103** (2006) 7607.
[57]S. Droese, and U. Brandt, *J. Biol. Chem.*, **283** (2008) 21649.
[58]J. F. Turrens, *Biosci. Rep.*, **17** (1997) 3.

[59]A. C. Mello, and R. Meneghini, *Biochim. Biophys. Acta*, **781** (1984) 56.
[60]J. L. Cadet, and C. Brannock, *Neurochem. Int.*, **32** (1998) 117.
[61]W. R. Markesbery, *Free Radical Biol. Med.*, **23** (1997) 134.
[62]T. G. Hastings, and M. J. Zigmond, in *Neurodegenerative diseases: Molecular and cellular mechanisms and therapeutic advances*, Ed. by G. Fiskum, Plenum Press, New York, 1996, p. 37.
[63]A. P. Arrigo, *Free Radical Biol. Med.*, **27** (1999) 936.
[64]R. S. Balaban, S. Nemoto, and T. Finkel, *Cell*, **120** (2005) 483.
[65]E. Beutler, W. Kuhl, J. L. Vivescorrons, and J. T. Prchal, *Blood*, **74** (1989) 2550.
[66]J. M. Dypbukt, M. Ankarcrona, M. Burkitt, A. Sjoholm, K. Strom, S. Orrenius, and P. Nicotera, *J. Biol. Chem.*, **269** (1994) 30553.
[67]A. Okado, Y. Kawasaki, Y. Hasuike, M. Takahashi, T. Teshima, J. Fujii, and N. Taniguchi, *Biochem. Biophys. Res. Commun.*, **225** (1996) 219.
[68]C. S. I. Nobel, M. Kimland, B. Lind, S. Orrenius, and A. F. G. Slater, *J. Biol. Chem.*, **270** (1995) 26202.
[69]H. Sies, in *Glutathione conjugation*, Ed. by H. Sies, and B. Ketterer, Academic Perss, London, 1988, p. 175.
[70]J. A. Du, D. H. Daniels, C. Asbury, S. Venkataraman, J. R. Liu, D. R. Spitz, L. W. Oberley, and J. J. Cullen, *J. Biol. Chem.*, **281** (2006) 37416.
[71]C. Lind, P. Hochstein, and L. Ernster, *Arch. Biochem. Biophys.*, **216** (1982) 178.
[72]E. Cadenas, *Biochem. Pharmacol.*, **49** (1995) 127..
[73]M. Belinsky, and A. K. Jaiswal, *Cancer Metast. Rev.*, **12** (1993) 103..
[74]D. L. Dehn, D. Siegel, K. S. Zafar, P. Reigan, E. Swann, C. J. Moody, and D. Ross, *Molec. Cancer Therap.*, **5** (2006) 1702..
[75]S. Y. Sharp, L. R. Kelland, M. R. Valenti, L. A. Brunton, S. Hobbs, and P. Workman, *Molec. Pharmacol.*, **58** (2000) 1146..
[76]J. Fourie, C. J. Oleschuk, F. Guziec, L. Guziec, D. J. Fiterman, C. Monterrosa, and A. Begleiter, *Cancer Chemother. Pharmacol.*, **49** (2002) 101.
[77]W. F. Hodnick, and A. C. Sartorelli, *Anal. Biochem.*, **252** (1997) 165..
[78]H. Pelicano, D. S. Martin, R. H. Xu, and P. Huang, *Oncogene*, **25** (2006) 4633..
[79]A. Aguilera, *Curr. Gen.*, **11** (1987) 429..
[80]E. Boles, W. Lehnert, and F. K. Zimmermann, *Eur. J. Biochem.*, **217** (1993) 469..
[81]A. Petersson, J. R. M. Almeida, T. Modig, K. Karhumaa, B. Hahn-Hagerdal, M. F. Gorwa-Grauslund, and G. Liden, *Yeast*, **23** (2006) 455..
[82]C. Larroy, M. R. Fernandez, E. Gonzalez, X. Pares, and J. A. Biosca, *Biochem. J.*, **361** (2002) 163.
[83]H. A. Krebs, and T. Gascoyne, *B. J.*, **108** (1968) 513.
[84]D. W. Piston, and S. M. Knobel, *Trends Endocrinol. Metab.*, **10** (1999) 413.
[85]K. C. Reinert, W. C. Gao, G. Chen, and T. J. Ebner, *J. Neurosci. Res.*, **85** (2007) 3221.
[86]Z. Liu, R. Cai, and J. Wang, in *Reviews in fluorescence 2006*, Ed. by C. D. Geddes, and J. R. Lacowicz, Springer, New York, 2006, p. 107.
[87]F. Wolbers, H. Andersson, A. van den Berg, and I. Vermes, *Apoptosis*, **9** (2004) 749.
[88]B. D. Bennett, T. L. Jetton, G. T. Ying, M. A. Magnuson, and D. W. Piston, *J. Biol. Chem.*, **271** (1996) 3647.
[89]M. Dellinger, M. Geze, R. Santus, E. Kohen, C. Kohen, J. G. Hirschberg, and M. Monti, *Biotechnol. Appl. Biochem.*, **28** (1998) 25.

[90]L. A. Reinke, R. G. Thurman, and F. C. Kauffman, *Biochem. Pharmacol.*, **28** (1979) 2381.
[91]A. L. Nieminen, A. M. Byrne, B. Herman, and J. J. Lemasters, *Am. J. Physiol.-Cell Physiol.*, **272** (1997) C1286.
[92]K. C. Reinert, R. L. Dunbar, W. C. Gao, G. Chen, and T. J. Ebner, *J. Neurophysiol.*, **92** (2004) 199.
[93]A. M. Brennan, J. A. Connor, and C. W. Shuttleworth, *J. Cereb. Blood Flow Metab.*, **26** (2006) 1389.
[94]B. R. Masters, P. T. C. So, and E. Gratton, *Ann. New York Acad. Sci.*, **838** (1998) 58.
[95]B. Rajwa, T. Bernas, H. Acker, J. Dobrucki, and J. P. Robinson, *Microsc. Res. Techn.*, **70** (2007) 869.
[96]D. W. Piston, *Trends Cell Biol.*, **9** (1999) 66.
[97]H. A. O. Hill, and I. J. Higgins, *Phil- Trans. Royal Soc. London A*, **302** (1981) 267.
[98]L. Gorton, A. Lindgren, T. Larsson, F. D. Munteanu, T. Ruzgas, and I. Gazaryan, *Anal. Chim. Acta*, **400** (1999) 91.
[99]E. Ferapontova, and L. Gorton, *Electroanalysis*, **15** (2002) 484.
[100]L. Stoica, R. Ludwig, D. Haltrich, and L. Gorton, *Anal. Chem.*, **78** (2006) 393.
[101]R. Matsumoto, M. Mochizuki, K. Kano, and T. Ikeda, *Anal. Chem.*, **74** (2002) 3297.
[102]Y. Ogino, K. Takagi, K. Kano, and T. Ikeda, *J. Electroanal. Chem.*, **396** (1995) 517.
[103]D. F. Xing, Y. Zuo, S. A. Cheng, J. M. Regan, and B. E. Logan, *Environ. Sci. Technol.*, **42** (2008) 4146.
[104]D. Prasad, S. Arun, A. Murugesan, S. Padmanaban, R. S. Satyanarayanan, S. Berchmans, and V. Yegnaraman, *Biosens. Bioelectron.*, **22** (2007) 2604.
[105]L. P. Hadjipetrou, T. Gray-Young, and M. D. Lilly, *J. Gen. Microbiol.*, **45** (1966) 479.
[106]J. Racek, and J. Musil, *Clin. Chim. Acta*, **162** (1987) 129.
[107]T. Ikeda, T. Kurosaki, K. Takayama, K. Kano, and K. Miki, *Anal. Chem.*, **68** (1996) 192-198.
[108]K. Takayama, T. Kurosaki, and T. Ikeda, *J. Electroanal. Chem.*, **356** (1993) 295.
[109]T. Ikeda, K. Kato, M. Maeda, H. Tatsumi, K. Kano, and K. Matsushita, *J. Electroanal. Chem.*, **430** (1997) 197.
[110]S. Timur, U. Anik, D. Odaci, and L. Lo Gorton, *Electrochem. Commun.*, **9** (2007) 1810.
[111]V. Coman, T. Gustavsson, A. Finkelsteinas, C. von Wachenfeldt, C. Hagerhall, and L. Gorton, *J. Am. Chem. Soc.*, **131** (2009) 16171.
[112]D. M. Rawson, A. J. Willmer, and A. P. P. Turner, *Biosensors*, **4** (1989) 299.
[113]N. Pasco, K. Baronian, C. Jeffries, and J. Hay, *Appl. Microbiol. Biotechnol.*, **53** (2000) 613.
[114]M. E. Polak, D. M. Rawson, and B. G. D. Haggett, *Biosens. Bioelectron.*, **11** (1996) 1253.
[115]B. Liu, S. A. Rotenberg, and M. V. Mirkin, *Proc. Nat. Acad. Sci. U.S.A.*, **97** (2000) 9855.
[116]S. A. Rotenberg, and M. V. Mirkin, *J. Mammary Gland Biol. Neopl.*, **9** (2004) 375.
[117]A. Heiskanen, J. Yakovleva, C. Spegel, R. Taboryski, M. Koudelka-Hep, J. Emneus, and T. Ruzgas, *Electrochem. Commun.*, **6** (2004) 219.

[118]E. Boles, and F. K. Zimmermann, *Molec. Gen. Gen.*, **243** (1994) 363.
[119]J. Mauzeroll, and A. J. Bard, *Proc. Nat. Acad. Sci. U.S.A.*, **101** (2004) 7862.
[120]T. L. Schwarz, in *Fundamental neuroscience*, Ed. by L. Squire, D. Berg, F. Bloom, S. Du Lac, A. Ghosh, and N. Spitzer, Academic Press, San Diego, 2008, p. 157.
[121]S. Sugita, *Acta Physiol.*, **192** (2008) 185..
[122]D. Purves, G. J. Augustine, D. Fitzpatric, W. C. Hall, A.-S. LaMantia, J. O. McNamara, and L. E. White, *Neuroscience*, Sinauer Associates, Inc., Sunderland, 2008, p. 119.
[123]M. K. Bennett, and R. H. Scheller, *P. Nat. Acad. Sci. U.S.America*, **90** (1993) 2559.
[124]M. N. Waxham, in *Fundamental neuroscience*, Ed. by L. Squire, D. Berg, F. Bloom, S. Du Lac, A. Ghosh, and N. Spitzer, Academic Press, San Diego, 2008, p. 181.
[125]B. Falkowska-Hansen, M. Falkowski, P. Metharom, D. Krunic, and S. Goerdt, *Experim. Cell Res.*, **313** (2007) 1745.
[126]M. R. Rad, H. L. Phan, L. Kirchrath, P. K. Tan, T. Kirchhausen, C. P. Hollenberg, and G. S. Payne, *J. Cell Sci.*, **108** (1995) 1605.
[127]A. A. Kolokoltsov, D. Deniger, E. H. Fleming, N. J. Roberts, J. M. Karpilow, and R. A. Davey, *J. Virol.*, **81** (2007) 7786.
[128]R. M. Wightman, and C. L. Haynes, *Nature Neurosci.*, **7** (2004) 321.
[129]E. Neher, and A. Marty, *Proc. Nat. Acad. Sci. U.S.A.*, **79** (1982) 6712.
[130]J. A. Steyer, and W. Almers, *Biophys. J.*, **76** (1999) 2262.
[131]A. Albillos, G. Dernick, H. Horstmann, W. Almers, G. A. deToledo, and M. Lindau, *Nature*, **389** (1997) 509.
[132]S. P. Gandhi, and C. F. Stevens, *Nature*, **423** (2003) 607.
[133]B. Granseth, B. Odermatt, S. J. Royle, and L. Lagnado, *Neuron*, **51** (2006) 773.
[134]V. A. Klyachko, and M. B. Jackson, *Nature*, **418** (2002) 89.
[135]P. Verstreken, O. Kjaerulff, T. E. Lloyd, R. Atkinson, Y. Zhou, I. A. Meinertzhagen, and H. J. Bellen, *Cell*, **109** (2002) 101.
[136]A. M. Aravanis, J. L. Pyle, N. C. Harata, and R. W. Tsien, *Neuropharmacol.*, **45** (2003) 797.
[137]G. A. Detoledo, R. Fernandezchacon, and J. M. Fernandez, *Nature*, **363** (1993) 554.
[138]R. G. W. Staal, E. V. Mosharov, and D. Sulzer, *Nature Neurosci.*, **7** (2004) 341.
[139]K. Lollike, N. Borregaard, and M. Lindau, *J. Cell Biol.*, **129** (1995) 99.
[140]L. Siksou, P. Rostaing, J. P. Lechaire, T. Boudier, T. Ohtsuka, A. Fejtova, H. T. Kao, P. Greengard, E. D. Gundelfinger, A. Triller, and S. Marty, *J. Neurosci.*, **27** (2007) 6868.
[141]R. M. Wightman, J. A. Jankowski, R. T. Kennedy, K. T. Kawagoe, T. J. Schroeder, D. J. Leszczyszyn, J. A. Near, E. J. Diliberto, and O. H. Viveros, *Proc. Nat. Acad. Sci. U.S.A.*, **88** (1991) 10754.
[142]K. Pihel, E. R. Travis, R. Borges, and R. M. Wightman, *Biophys. J.*, **71** (1996) 1633.
[143]E. R. Travis, and R. M. Wightman, *Ann. Rev. Biophys. Biomolec. Structure*, **27** (1998) 77.
[144]T. L. Colliver, E. J. Hess, and A. G. Ewing, *J. Neurosci. Meth.*, **105** (2001) 95.
[145]C. Amatore, S. Arbault, I. Bonifas, Y. Bouret, M. Erard, A. G. Ewing, and L. A. Sombers, *Biophys. J.*, **88** (2005) 4411.
[146]T. K. Chen, G. O. Luo, and A. G. Ewing, *Anal. Chem.*, **66** (1994) 3031.

[147]S. E. Zerby, and A. G. Ewing, *Brain Res.*, **712** (1996) 1.
[148]B. B. Anderson, S. E. Zerby, and A. G. Ewing, *J. Neurosci. Meth.*, **88** (1999) 163.
[149]L. A. Sombers, M. M. Maxson, and A. G. Ewing, *J. Neurochem.*, **93** (2005) 1122.
[150]R. T. Kennedy, L. Huang, M. A. Atkinson, and P. Dush, *Anal. Chem.*, **65** (1993) 1882.
[151]L. Huang, H. Shen, M. A. Atkinson, and R. T. Kennedy, *Proc. Nat. Acad. Sci. U.S.A.*, **92** (1995) 9608.
[152]C. A. Aspinwall, J. R. T. Lakey, and R. T. Kennedy, *J. Biol. Chem.*, **274** (1999) 6360.
[153]C. D. Paras, W. J. Qian, J. R. Lakey, W. H. Tan, and R. T. Kennedy, *Cell Biochem. Biophys.*, **33** (2000) 227.
[154]D. Purves, G. J. Augustine, D. Fitzpatric, W. C. Hall, A.-S. LaMantia, J. O. McNamara, and L. E. White, Neuroscience, Sinauer Associates, Inc., Sunderland, 2008, p.85.
[155]S. C. Taylor, and C. Peers, *Biochem. Biophys. Res. Commun.*, **248** (1998) 13.
[156]J. A. Umbach, Y. Zhao, and C. B. Gundersen, *J. Neurochem.*, **94** (2005) 1306.
[157]K. D. Kozminski, D. A. Gutman, V. Davila, D. Sulzer, and A. G. Ewing, *Anal. Chem.*, **70** (1998) 3123.
[158]J. P. Bressler, L. BelloniOlivi, S. Forman, and G. W. Goldstein, *J. Neurosci. Res.*, **46** (1996) 678.
[159]R. H. S. Westerink, A. A. Klompmakers, H. G. M. Westenberg, and H. P. M. Vijverberg, *Brain Res.*, **957** (2002) 25.
[160]R. H. S. Westerink, and H. P. M. Vijverberg, *Neurosci. Lett.*, **326** (2002) 81.
[161]R. H. S. Westerink, and H. P. M. Vijverberg, *Toxicol. Appl. Pharmacol.*, **183** (2002) 153.
[162]B. B. Anderson, G. Y. Chen, D. A. Gutman, and A. G. Ewing, *Brain Res.*, **788** (1998) 294.
[163]D. Sulzer, T. K. Chen, Y. Y. Lau, H. Kristensen, S. Rayport, and A. Ewing, *J. Neurosci.*, **15** (1995) 4102.
[164]A. G. Ewing, K. D. Alloway, S. D. Curtis, M. A. Dayton, R. M. Wightman, and G. V. Rebec, *Brain Res.*, **261** (1983) 101.
[165]M. A. Hebert, and G. A. Gerhardt, *J. Pharmacol. Experim. Therap.*, **288** (1999) 879.
[166]M. Lundblad, F. Pomerleau, G. A. Gerhardt, M. A. Cenci, and I. Stromberg, *13th Annual Conference of the American-Society-for-Neural-Therapy-and-Repair, Apr 11-May 14, 2006, Clearwater Beach, FL.*
[167]J. J. Burmeister, T. D. Coates, and G. A. Gerhardt, *Proceedings of the 26th Annual International Conference of the IEEE EMBS, Sep 01-05, 2004, San Francisco, CA.*
[168]M. F. Salvatore, O. Hudspeth, L. E. Arnold, P. E. Wilson, J. A. Stanford, C. F. Mactutus, R. M. Booze, and G. A. Gerhardt, *Neurosci.*, **123** (2004) 481.
[169]O. P. Hamill, A. Marty, E. Neher, B. Sakmann, and F. J. Sigworth, *Pfl. Arch.-Eur. J. Physiol.*, **391** (1981) 85.
[170]F. T. Horrigan, and R. J. Bookman, *Neuron*, **13** (1994) 1119.
[171]R. Borges, M. Camacho, and K. D. Gillis, *Acta Physiol.*, **192** (2008) 173.
[172]S. Hallermann, C. Pawlu, P. Jonas, and M. Heckmann, *Proc. Nat. Acad. Sci. U.S.A.*, **100** (2003) 8975-8980.
[173]P. Chen, T. C. Hwang, and K. D. Gillis, *J. Gen. Physiol.*, **118** (2001) 135.

[174]T. A. Ryan, H. Reuter, and S. J. Smith, *Nature*, **388** (1997) 478.
[175]T. Lang, I. Wacker, J. Steyer, C. Kaether, I. Wunderlich, T. Soldati, H. H. Gerdes, and W. Almers, *Neuron*, **18** (1997) 857.
[176]M. W. Allersma, L. Wang, D. Axelrod, and R. W. Holz, *Molec. Biol. Cell*, **15** (2004) 4658.
[177]D. N. Bowser, and B. S. Khakh, *Proc. Nat. Acad. Sci. U.S.A.*, **104** (2007) 4212.
[178]G. Miesenbock, D. A. De Angelis, and J. E. Rothman, *Nature*, **394** (1998) 192.
[179]A. Hengstenberg, A. Blochl, I. D. Dietzel, and W. Schuhmann, *Angew. Chem.-Int. Ed.*, **40** (2001) 905.
[180]L. P. Bauermann, W. Schuhmann, and A. Schulte, *Phys. Chem. Chem. Phys.*, 6 (2004) 4003.
[181]D. Bruns, and R. Jahn, *Nature*, **377** (1995) 62.
[182]W. Z. Wu, W. H. Huang, W. Wang, Z. L. Wang, J. K. Cheng, T. Xu, R. Y. Zhang, Y. Chen, and J. Liut, *J. Am. Chem. Soc.*, **127** (2005) 8914.
[183]E. V. Mosharov, and D. Sulzer, *Nature Meth.*, **2** (2005) 651.
[184]Z. Zhang, and M. B. Jackson, *J. Gen. Physiol.*, **131** (2008) 117.
[185]R. H. Chow, L. Vonruden, and E. Neher, *Nature*, **356** (1992) 60.
[186]L. A. Sombers, H. J. Hanchar, T. L. Colliver, N. Wittenberg, A. Cans, S. Arbault, C. Amatore, and A. G. Ewing, *J. Neurosci.*, **24** (2004) 303.
[187]J. W. Barclay, M. Aldea, T. J. Craig, A. Morgan, and R. D. Burgoyne, *J. Biol. Chem.*, **279** (2004) 41495.
[188]R. D. Burgoyne, and J. W. Barclay, *Trends Neurosci.*, **25** (2002) 176.
[189]C. P. Grabner, S. D. Price, A. Lysakowski, and A. P. Fox, *J. Neurophysiol.*, **94** (2005) 2093.
[190]E. L. Ciolkowski, B. R. Cooper, J. A. Jankowski, J. W. Jorgenson, and R. M. Wightman, *J. Am. Chem. Soc.*, **114** (1992) 2815.
[191]K. Pihel, S. Hsieh, J. W. Jorgenson, and R. M. Wightman, *Anal. Chem.*, **67** (1995) 4514.
[192]J. E. Baur, E. W. Kristensen, L. J. May, D. J. Wiedemann, and R. M. Wightman, *Anal. Chem.*, **60** (1988) 1268.
[193]M. D. Hawley, Tatawawa.Sv, Piekarsk.S, and R. N. Adams, *Journal of the American Chemical Society*, **89** (1967) 447-450.
[194]G. Dernick, L. W. Gong, L. Tabares, G. A. de Toledo, and M. Lindau, *Nature Meth.*, **2** (2005) 699.
[195]L. Tabares, M. Lindau, and G. A. de Toeldo, *Biochem. Soc. Trans.*, **31** (2003) 837.
[196]A. F. Dias, G. Dernick, V. Valero, M. G. Yong, C. D. James, H. G. Craighead, and M. Lindau, *Nanotechnol.*, **13** (2002) 285.
[197]P. Chen, B. Xu, N. Tokranova, X. J. Feng, J. Castracane, and K. D. Gillis, *Anal. Chem.*, **75** (2003) 518.
[198]I. Hafez, K. Kisler, K. Berberian, G. Dernick, V. Valero, M. G. Yong, H. G. Craighead, and M. Lindau, *Proc. Nat. Acad. Sci. U.S.A.*, **102** (2005) 13879.
[199]X. H. Sun, and K. D. Gillis, *Anal. Chem.*, **78** (2006) 2521.
[200]C. Amatore, S. Arbault, Y. Chen, C. Crozatier, F. Lemaitre, and Y. Verchier, *Angew. Chem.-Int. Ed.*, **45** (2006) 4000.
[201]C. Amatore, S. Arbault, F. Lemaitre, and Y. Verchier, *Biophys. Chem.*, **127** (2007) 165.
[202]W. H. Huang, W. Cheng, Z. Zhang, D. W. Pang, Z. L. Wang, J. K. Cheng, and D. F. Cui, *Anal. Chem.*, **76** (2004) 483.

[203]Y. Chen, H. F. Cui, J. S. Ye, S. C. Chong, T. M. Lim, F. S. Sheu, and W. C. Hui, *J. Phys. Conference Series*, **34** (2006) 198.
[204]H. F. Cui, J. S. Ye, Y. Chen, S. C. Chong, X. Liu, T. M. Lim, and F. S. Sheu, *Sens. Act. B*, **115** (2006) 634.
[205]O. Niwa, T. Horiuchi, R. Kurita, H. Tabei, and K. Torimitsu, *Anal. Sci.*, **14** (1998) 947.
[206]O. Niwa, T. Horiuchi, and K. Torimitsu, *Biosens. Bioelectron.*, **12** (1997) 311.
[207]O. Niwa, R. Kurita, T. Horiuchi, and K. Torimitsu, *Electroanalysis*, **11** (1999) 356.
[208]O. Niwa, K. Torimitsu, M. Morita, P. Osborne, and K. Yamamoto, *Anal. Chem.*, **68** (1996) 1865.
[209]R. Kurita, H. Tabei, K. Hayashi, T. Horiuchi, K. Torimitsu, and O. Niwa, *Anal. Chim. Acta*, **441** (2001) 165.
[210]R. Kurita, K. Hayashi, K. Torimitsu, and O. Niwa, *Anal. Sci.*, **19** (2003) 1581.
[211]O. Niwa, R. Kurita, K. Hayashi, T. Horiuchi, K. Torimitsu, K. Maeyama, and K. Tanizawa, *Sens. Act. B*, **67** (2000) 43.
[212]S. Iwaki, M. Ogasawara, R. Kurita, O. Niwa, K. Tanizawa, Y. Ohashi, and K. Maeyama, *Anal. Biochem.*, **304** (2002) 236.
[213]R. Kurita, K. Hayashi, T. Horiuchi, O. Niwa, K. Maeyama, and K. Tanizawa, *Lab Chip*, **2** (2002) 34.
[214]H. F. Cui, J. S. Ye, Y. Chen, S. C. Chong, and F. S. Sheu, *Anal. Chem.*, **78** (2006) 6347.
[215]Y. Chen, C. Guo, L. Lim, S. Cheong, Q. Zhang, K. Tang, and J. Reboud, *Anal. Chem.*, **80** (2008) 1133.
[216]W. Dauer, and S. Przedborski, *Neuron*, **39** (2003) 889.
[217]J. Jankovic, *J. Neurol. Neurosurg. Psych.*, **79** (2008) 368.
[218]O. Lindvall, in *Neural transplantation in neurodegenerative disease: Current status and new directions*, Ed. by D. J. Chadwick, and J. A. Goode, John Wiley & Sons, Ltd., Chichester, 2000, p. 110.
[219]T. M. Dawson, and V. L. Dawson, *J. Clin. Invest.*, **111** (2003) 145.
[220]V. M. Y. Lee, and J. Q. Trojanowski, *Neuron*, **52** (2006) 33.
[221]T. Nagatsu, and M. Sawada, *Parkinsonism Relat. Disord.*, **15** (2009) S3.
[222]L. Garcia, G. D'Alessandro, B. Bioulac, and C. Hammond, *Trends Neurosci.*, **28** (2005) 209.
[223]M. C. Bohn, B. Connor, D. A. Kozlowski, and M. H. Mohajeri, in *Neural transplantation in neurodegenerative disease: Current status and new directions*, Ed. by D. J. Chadwick, and J. A. Goode, John Wiley & Sons, Ltd., Chichester, 2000, p. 70.
[224]A. Björklund, in *Neural transplantation in neurodegenerative disease: Current status and new directions*, Ed. by D. J. Chadwick, and J. A. Goode, John Wiley & Sons, Ltd., Chichester, 2000, p. 7.
[225]A. Bjorklund, and U. Stenevi, *Brain Res.*, **177** (1979) 555.
[226]L. M. Bjorklund, R. Sanchez-Pernaute, S. M. Chung, T. Andersson, I. Y. C. Chen, K. S. McNaught, A. L. Brownell, B. G. Jenkins, C. Wahlestedt, K. S. Kim, and O. Isacson, *Proc. Nat. Acad. Sci. U.S.A.*, **99** (2002) 2344.
[227]U. Englund, A. Bjorklund, K. Wictorin, O. Lindvall, and M. Kokaia, *Proc. Nat. Acad. Sci. U.S.A.*, **99** (2002) 17089.
[228]I. Liste, E. Garcia-Garcia, and A. Martinez-Serrano, *J. Neurosci.*, **24** (2004) 10786.

[229]M. A. Haralson, and J. R. Hassel, in *Extracellular matrix - a practical approach*, Ed by M. A. Haralson, and J. R. Hassel, Oxford University Press, Oxford, 1995, p. 1.
[230]S. M. Frisch, and H. Francis, *J. Cell Biol.*, **124** (1994) 619.
[231]R. L. Juliano, and S. Haskill, *J. Cell Biol.*, **120** (1993) 577.
[232]K. Beck, I. Hunter, and J. Engel, *FFASEB J.*, **4** (1990) 148.
[233]L. F. Reichardt, *Ann. Rev. Neurosci.*, **14** (1991) 531.
[234]H. Colognato, and P. D. Yurchenco, *Develop. Dynam.*, **218** (2000) 213.
[235]P. Spessotto, E. Giacomello, and R. Perris, in *Extracellular matrix protocols (methods in molecular biology,* Vol. 139), Ed. by C. H. Streuli, and M. Grant, Humana Press, Inc., Totowa, 2000, p. 321-343.
[236]P. A. Dimilla, J. A. Stone, J. A. Quinn, S. M. Albelda, and D. A. Lauffenburger, *J. Cell Biol.*, **122** (1993) 729.
[237]S. P. Palecek, J. C. Loftus, M. H. Ginsberg, D. A. Lauffenburger, and A. F. Horwitz, *Nature*, **385** (1997) 537.
[238]R. I. Freshney, *Culture of animal cells: A manual of basic technique, 5th ed.*, John Wiley & Sons, Inc., Hoboken, 2005.
[239]C. S. Izzard, and L. R. Lochner, *J. Cell Sci.*, **21** (1976) 129.
[240]S. Yamaji, A. Suzuki, Y. Sugiyama, Y. Koide, M. Yoshida, H. Kanamori, H. Mohri, S. Ohno, and Y. Ishigatsubo, *J. Cell Biol.*, **153** (2001) 1251.
[241]C. de Aguiar, B. Lobao-Soares, M. Alvarez-Silva, and A. G. Trentin, *BMC. Cell Biol.*, **6** (2005) doi:10.1186/1471-2121-6-31.
[242]R. Milner, and I. L. Campbell, *J. Neurosci.*, **22** (2002) 1562.
[243]A. R. Vancha, S. Govindaraju, K. V. L. Parsa, M. Jasti, M. Gonzalez-Garcia, and R. P. Ballestero, *BMC Biotechnol.*, **4** (2004) doi:10.1186/1472-6750-4-23.
[244]S. P. Massia, and J. A. Hubbell, *Anal. Biochem.*, **187** (1990) 292.
[245]A. K. Dillow, and M. Tirrell, *Curr. Opin. Solid State Mater. Sci.*, **3** (1998) 252.
[246]C. H. Thomas, C. D. McFarland, M. L. Jenkins, A. Rezania, J. G. Steele, and K. E. Healy, *J. Biomed. Mater. Res.*, **37** (1997) 81.
[247]M. Mrksich, L. E. Dike, J. Tien, D. E. Ingber, and G. M. Whitesides, *Exp. Cell Res.*, **235** (1997) 305.
[248]M. Mrksich, C. S. Chen, Y. N. Xia, L. E. Dike, D. E. Ingber, and G. M. Whitesides, *Proc. Nat. Acad. Sci. U.S.A.*, **93** (1996) 10775.
[249]L. Y. Liu, B. D. Ratner, E. H. Sage, and S. Y. Jiang, *Langmuir*, **23** (2007) 11168.
[250]X. Y. Cui, V. A. Lee, Y. Raphael, J. A. Wiler, J. F. Hetke, D. J. Anderson, and D. C. Martin, *J. Biomed. Mater. Res.*, **56** (2001) 261.
[251]S. Lakard, G. Herlem, A. Propper, A. Kastner, G. Michel, N. Valles-Villarreal, T. Gharbi, and B. Fahys, *Bioelectrochemistry*, **62** (2004) 19.
[252]M. Abercrom, J. E. Heaysman, and S. M. Pegrum, *Exp. Cell Res.*, **67** (1971) 359.
[253]H. Verschueren, *J. Cell Sci.*, **75** (1985) 279.
[254]M. S. Zand, and G. Albrechtbuehler, *Cell Mot. Cytoskelet.*, **13** (1989) 94.
[255]D. Axelrod, *J. Cell Biol.*, **89** (1981) 141.
[256]S. Inoué, in *Handbook of biological confocal microscopy*, Ed. by J. B. Pawley, Springer, New York, 2006, p. 1.
[257]*Materials Today*, **9** (2006) 62.
[258]J. Drazba, P. Liljelund, C. Smith, R. Payne, and V. Lemmon, *Develop. Brain Res.*, **100** (1997) 183.
[259]M. Kishi, Y. Emori, Y. Tsukamoto, and K. Abe, *Biosci. Biotechnol. Biochem.*, **66** (2002) 484.

[260]C. R. Wittmer, J. A. Phelps, W. M. Saltzman, and P. R. Van Tassel, *Biomaterials*, **28** (2007) 851.
[261]J. Watanabe, T. Eriguchi, and K. Ishihara, *Biomacromolec.*, **3** (2002) 1375.
[262]C. Picart, R. Elkaim, L. Richert, T. Audoin, Y. Arntz, M. D. Cardoso, P. Schaaf, J. C. Voegel, and B. Frisch, *Adv. Funct. Mater.*, **15** (2005) 83.
[263]T. Korb, K. Schluter, A. Enns, H. U. Spiegel, N. Senninger, G. L. Nicolson, and J. Haier, *Exp. Cell Res.*, **299** (2004) 236.
[264]M. Schuler, G. R. Owen, D. W. Hamilton, M. De Wilde, M. Textor, D. M. Brunette, and S. G. P. Tosatti, *Biomaterials*, **27** (2006) 4003.
[265]S. Teixeira, M. P. Ferraz, and F. J. Monteiro, *J. Mater. Sci.-Mater. Med.*, **19** (2008) 855.
[266]M. Navarro, E. M. Benetti, S. Zapotoczny, J. A. Planell, and G. J. Vancso, *Langmuir*, **24** (2008) 10996.
[267]J. Grunwald, *Biotechniques*, **5** (1987) 680.
[268]C. E. Krull, A. Collazo, S. E. Fraser, and M. Bronnerfraser, *Development*, **121** (1995) 3733.
[269]C. F. Thomas, and J. G. White, *Trends Biotechnol.*, **16** (1998) 175.
[270]J. Domke, S. Dannohl, W. J. Parak, O. Muller, W. K. Aicher, and M. Radmacher, *Colloid. Surface. B*, **19** (2000) 367.
[271]M. Radmacher, in *Atomic force microscopy in cell biology (methods in cell biology)*, Vol. 68, Ed. by B. P. Jena, and J. K. H. Hörber, Academic Press, New York, 2002, p. 67.
[272]F. Braet, C. Seynaeve, R. De Zanger, and E. Wisse, *J, Microscopy*, **190** (1998) 328.
[273]U. G. Hofmann, C. Rotsch, W. J. Parak, and M. Radmacher, *J. Structur. Biol.*, **119** (1997) 84.
[274]E. A-Hassan, W. F. Heinz, M. D. Antonik, N. P. D'Costa, S. Nageswaran, C. A. Schoenenberger, and J. H. Hoh, *Biophys. J.*, **74** (1998) 1564.
[275]J. Domke, W. J. Parak, M. George, H. E. Gaub, and M. Radmacher, *Eur. Biophys. J.*, **28** (1999) 179.
[276]A. Simon, T. Cohen-Bouhacina, J. P. Aime, M. C. Porte, J. Amedee, and C. Baquey, *Cell. Molec. Biol.*, **50** (2004) 255.
[277]C. Rotsch, K. Jacobson, and M. Radmacher, *Proc. Nat. Acad. Sci. U.S.A.*, **96** (1999) 921.
[278]J. A. Dvorak, and E. Nagao, *Exp. Cell Res.*, **242** (1998) 69.
[279]R. Matzke, K. Jacobson, and M. Radmacher, *Nature Cell Biol.*, **3** (2001) 607.
[280]J. Visser, in *Surface and colloid science,* Vol. 8, Ed by. E. Matijevic, John Wiley & Sons, Inc., New York, 1978, p. 121-141.
[281]J. N. George, R. I. Weed, and C. F. Reed, *J. Cell. Physiol.*, **77** (1971) 51.
[282]A. Trommler, D. Gingell, and H. Wolf, *Biophys. J.*, **48** (1985) 835.
[283]A. J. Garcia, P. Ducheyne, and D. Boettiger, *Biomaterials*, **18** (1997) 1091.
[284]D. R. Coman, *Cancer Res.*, **21** (1961) 1436.
[285]E. A. Evans, *Colloid. Surface.*, **10** (1984) 133.
[286]J. Y. Shao, and R. M. Hochmuth, *Biophys. J.*, **71** (1996) 2892.
[287]Y. S. Chu, W. A. Thomas, O. Eder, F. Pincet, E. Perez, J. P. Thiery, and S. Dufour, *J.Cell Biol.*, **167** (2004) 1183.
[288]M. Tulla, J. Helenius, J. Jokinen, A. Taubenberger, D. J. Muller, and J. Heino, *FEBS Lett.*, **582** (2008) 3520.
[289]H. Kim, H. Arakawa, T. Osada, and A. Ikai, *Ultramicroscopy*, **97** (2002) 359.

[290]H. Kim, H. Arakawa, T. Osada, and A. Ikai, *Applied Surface Science*, **188** (2001) 493-498.
[291]T. S. Hug, *Assay Drug Dev. Technol.*, **1** (2003) 479.
[292]G. Sauerbrey, *Z. Phys.*, **155** (1959) 206.
[293]K. A. Marx, T. Zhou, and R. Sarma, *Biotechnol. Progr.*, **15** (1999) 522.
[294]M. Rodahl, F. Hook, C. Fredriksson, C. A. Keller, A. Krozer, P. Brzezinski, M. Voinova, and B. Kasemo, *Farad. Disc.*, **107** (1997) 229.
[295]M. Rodahl, F. Hook, A. Krozer, P. Brzezinski, and B. Kasemo, *Rev. Scient. Instrum.*, **66** (1995) 3924.
[296]X. H. Zhang, A. Chen, D. De Leon, H. Li, E. Noiri, V. T. Moy, and M. S. Goligorsky, *Am. J. Physiol.-Heart Circulat. Physiol.*, **286** (2004) H359.
[297]J. Wegener, A. Janshoff, and H. J. Galla, *Eur. Biophys. J.*, **28** (1998) 26.
[298]J. Wegener, J. Seebach, A. Janshoff, and H. J. Galla, *Biophys. J.*, **78** (2000) 2821.
[299]T. Zhou, K. A. Marx, M. Warren, H. Schulze, and S. J. Braunhut, *Biotechnol. Progr.*, **16** (2000) 268.
[300]K. A. Marx, T. Zhou, M. Warren, and S. J. Braunhut, *Biotechnol. Progr.*, **19** (2003) 987.
[301]D. Le Guillou-Buffello, G. Helary, M. Gindre, G. Pavon-Djavid, P. Laugier, and V. Migonney, *Biomaterials*, **26** (2005) 4197.
[302]Z. Fohlerova, P. Skladal, and J. Turanek, *Biosens. Bioelectron.*, **22** (2007) 1896.
[303]K. A. Marx, T. A. Zhou, A. Montrone, H. Schulze, and S. J. Braunhut, *Biosens. Bioelectron.*, **16** (2001) 773.
[304]X. L. Wei, Z. H. Mo, B. Li, and J. M. Wei, *Colloid. Surface. B*, **59** (2007) 100.
[305]K. Tiefenthaler, *Adv. Biosens.*, **2** (1992) 261.
[306]L. Marcotte, and A. Tabrizian, *IRBM*, **29** (2008) 77.
[307]N. J. Harric, *Internal reflection spectroscopy*, Harric Scientific Corporation, New York, 1979.
[308]S. Pasche, S. M. De Paul, J. Voros, N. D. Spencer, and M. Textor, *Langmuir*, **19** (2003) 9216.
[309]T. S. Hug, J. E. Prenosil, and M. Morbidelli, *Biosens. Bioelectron.*, **16** (2001) 865.
[310]J. J. Ramsden, S. Y. Li, E. Heinzle, and J. E. Prenosil, *Cytometry*, **19** (1995) 97.
[311]T. S. Hug, J. E. Prenosil, P. Maier, and M. Morbidelli, *Biotechnol. Bioeng.*, **80** (2002) 213.
[312]T. S. Hug, J. E. Prenosil, P. Maier, and M. Morbidelli, *Biotechnol. Progr.*, **18** (2002) 1408.
[313]J. Voros, R. Graf, G. L. Kenausis, A. Bruinink, J. Mayer, M. Textor, E. Wintermantel, and N. D. Spencer, *Biosens. Bioelectron.*, **15** (2000) 423.
[314]I. Giaever, and C. R. Keese, *Proc. Nat. Acad. Sci. U.S.A.*, **81** (1984) 3761.
[315]I. Giaever, and C. R. Keese, *IEEE Trans. Biomed. Eng.*, **33** (1986) 242.
[316]M. Kowolenko, C. R. Keese, D. A. Lawrence, and I. Giaever, *J. Immunol. Meth.*, **127** (1990) 71.
[317]P. Mitra, C. R. Keese, and I. Giaever, *Biotechniques*, **11** (1991) 504.
[318]I. Giaever, and C. R. Keese, *Proc. Nat. Acad. Sci. U.S.A.*, **88** (1991) 7896.
[319]A. B. Moy, J. VanEngelenhoven, J. Bodmer, J. Kamath, C. Keese, I. Giaever, S. Shasby, and D. M. Shasby, *J. Clin. Invest.*, **97** (1996) 1020.
[320]C. M. Lo, C. R. Keese, and I. Giaever, *Exp. Cell Res.*, **204** (1993) 102.
[321]C. M. Lo, C. R. Keese, and I. Giaever, *Exp. Cell Res.*, **213** (1994) 391.
[322]Z. Pei, C. R. Keese, I. Giaever, H. Kurzawa, and D. E. Wilson, *Exp. Cell Res.*, **212** (1994) 225.

[323]P. M. Ghosh, C. R. Keese, and I. Giaever, *Biophys. J.*, **64** (1993) 1602.
[324]C. Xiao, and J. H. T. Luong, *Biotechnol. Progr.*, **19** (2003) 1000.
[325]C. Xiao, and J. H. T. Luong, *Toxicol. Appl. Pharmacol.*, **206** (2005) 102.
[326]C. D. Xiao, B. Lachance, G. Sunahara, and J. H. T. Luong, *Anal. Chem.*, **74** (2002) 5748.
[327]G. E. Slaughter, E. Bieberich, G. E. Wnek, K. J. Wynne, and A. Guiseppi-Elei, *Langmuir*, **20** (2004) 7189.
[328]R. Ehret, W. Baumann, M. Brischwein, A. Schwinde, K. Stegbauer, and B. Wolf, *Biosens. Bioelectron.*, **12** (1997) 29.
[329]J. Z. Xing, L. Zhu, J. A. Jackson, S. Gabos, X. J. Sun, X. b. Wang, and X. Xu, *Chem. Res. Toxicol.*, **18** (2005) 154.
[330]J. Ciambrone Gary, F. Liu Vivian, C. Lin Deborah, P. McGuinness Ryan, K. Leung Gordon, and S. Pitchford, *J. Biomolec. Screen.*, **9** (2004) 467.
[331]E. Verdonk, K. Johnson, R. McGuinness, G. Leung, Y.-W. Chen, H. R. Tang, J. M. Michelotti, and V. F. Liu, *Assay Drug Dev. Technol.*, **4** (2006) 609.
[332]K. Solly, X. Wang, X. Xu, B. Strulovici, and W. Zheng, *Assay Drug Dev. Technol.*, **2** (2004) 363.
[333]J. Z. Xing, L. Zhu, S. Gabos, and L. Xie, *Toxicol. in Vitro*, **20** (2006) 995.
[334]J. Wegener, S. Zink, P. Rosen, and H. J. Galla, *Pfl. Arch.-Eur. J. Physiol.*, **437** (1999) 925.
[335]A. M. Otto, M. Brischwein, A. Niendorf, T. Henning, E. Motrescu, and B. Wolf, *Cancer Det. Prev.*, **27** (2003) 291.
[336]M. Carminati, Thesis, Politecnico di Milano, 2010.
[337]L. H. D. Skjolding, C. Spegel, A. Ribayrol, J. Emneus, and L. Montelius, *J. Phys*, **100** (2008) 052045.
[338]G. T. Teixidor, R. A. Gorkin, P. P. Tripathi, G. S. Bisht, M. Kulkarni, T. K. Maiti, T. K. Battacharyya, J. R. Subramaniam, A. Sharma, B. Y. Park, and M. Madou, *Biomed. Mater.*, **3** (2008) 034116.
[339]L. H. D. Skjolding, Thesis, Lund University, 2010.

Arto Heiskanen

Arto Heiskanen received his PhD in April 2009 at the Department of Analytical Chemistry at Lund University, Sweden, in research focused on monitoring of cellular dynamics of living cells using electrochemical techniques. Since May 2009, he has held an Ørsted Postdoctoral Fellow to conduct research in the Biomedical Microsystems (BIOMICS) section at the Department of Micro- and Nanotechnology at the Technical University of Denmark (DTU). His present research is focused on the development of microfluidic systems and electrode microchips for the study of stem cell differentiation to dopaminergic neurons and subsequent integration into brain tissue to evaluate the properties of the achieved innervation using electrochemical and microscopic detection techniques. He has 10 peer reviewed publications on electrode fabrication and modification with applications in electrochemical monitoring of cellular functions. At present he is a co-supervisor of 3 PhD students and 3 master students as well as a work package leader in the EU FP 7 project "Exploring Cellular Dynamics at Nanoscale" (EXCELL).

Jenny Emnéus

Jenny Emnéus was recruited as new Professor and head of the Biomedical Microsystems (BIOMICS) section at the Department of Micro- and Nanotechnology at Technical University of Denmark (DTU) in March 2007. She previously pursued her academic career at the Department of Analytical Chemistry at Lund University in Sweden, where she obtained her PhD in 1992. She has been involved in numerous European projects of which she has coordinated six. She has served as an evaluator of the Swedish Research Council and EU FP 7 projects for the environment program. She has 101 publications in peer reviewed international journals and an H-index of 25. At present she is supervising 6 PhD students at DTU. She has many years of experience in development and application of electrochemical biosensors and microchip based biological systems. Her main research interest is now focused on development of microfluidic bioassay systems based on enzymes, antibodies and cells, using multi-parameter detection systems (optical and electrochemical) for dynamic monitoring of cellular events in real-time to gain insight into biological processes at cell level. This research is also the main focus of the EU FP 7 project "Exploring Cellular Dynamics at Nanoscale" (EXCELL) presently coordinated by Professor Emnéus.

2

Fundamental Studies of Long- and Short-Range Electron Exchange Mechanisms between Electrodes and Proteins

David H. Waldeck[1] and Dimitri E. Khoshtariya[1,2]

[1]*Department of Chemistry, University of Pittsburgh, Pittsburgh, Pennsylvania 15260, USA*
[2]*Institute for Biophysics and Bionanosciences, Department of Physics, I. Javakhishvili Tbilisi State University, 0128 Tbilisi, and Research Institute of Molecular Biology and Biophysics, 0160 Tbilisi, Georgian Republic*

1. INTRODUCTION

Electron transfer[1-7] by thermally activated hopping through localized centers is a common motif that occurs over a broad spectrum of matter such as non-conducting crystalline solids, amorphous glasses, fluidic/viscous liquids, and biomolecules. This basic kinetic motif is an essential element for a broad variety of vital biological and technological processes.[8-10] New directions in electrochemistry, a discipline which has contributed greatly to developing the theoretical underpinnings of charge-transfer phenomena,[1-7,11-13] promise to reveal important features of electron transfer (ET) in biomolecular redox chains. This new frontier in electrochemistry, which has been developing over the past two decades,[14-53] uses

N. Eliaz (ed.), *Applications of Electrochemistry and Nanotechnology in Biology and Medicine I*, Modern Aspects of Electrochemistry 52,
DOI 10.1007/978-1-4614-0347-0_2, © Springer Science+Business Media, LLC 2011

chemically tunable, nanoscopic electrode structures for addressing fundamental questions in charge exchange between electrodes and biomolecules.[14-21,54-88]

Electrochemical methods offer several well-recognized (traditional) advantages over *homogeneous* approaches for elucidating charge transfer processes. Two of the more important advantages are:

- the ability to study a single redox-active component by using an electrode to act as the other reactant;[1-13] and
- the ability to gradually vary the redox potential (electrode overvoltage); this feature allows for the smooth variation of the reaction free energy, ΔG_0.[1-13]

Classical heterogeneous approaches also have some clear disadvantages. Noteworthy are

- the strong impact of the electrode/solution interfacial potential drop on the ET process, especially in the free diffusion regime for which it complicates the over-all kinetic analysis;[2,11-13] and
- possible changes in the redox-active molecules upon adsorption to an electrode; in particular the adsorption of biomacromolecules, such as proteins, can lead to their denaturation/inactivation.[16-18]

The above mentioned disadvantages can be largely eliminated through self-assembly techniques that coat the electrode with a biocompatible surface, *vide infra*. One common method uses thiolate groups, which form weak covalent bonds with surface-exposed atoms of metal electrodes; e.g., of Au, Ag, Hg, etc.,[14-88] and lead to the formation of almost perfect (defect-free) quasi-crystalline self-assembled monolayers (films)—SAMs. The chemical and physical properties of these films can be modified to provide additional control over the electron transfer. These include the variation of

- the chemical identity and composition (mixing/dilution) of the SAM terminal groups that are presented to the solution (electrolyte); e.g., they can be manipulated to control the binding position and orientation of biomolecules;[54-88]

- the strength of the protein/SAM terminal group interaction, which can change the mode for the protein's electron exchange with the electrode (e.g., freely diffusing *versus* irreversible binding regimes);[54-88] and
- the SAM thickness, which can be varied by changing the length of the chemical modifiers through a homologous series (e.g., $-CH_2-$), implying variation of physical properties like electrical conductivity (insulating ability).[54-88] The SAM's conductivity can be varied through the chemical properties of the SAM chains (e.g. their degree of conjugation, their mixing, their cross-linking, etc.) also.[48,49]

The development of SAM-coated electrodes has enabled fundamental and applied studies of biomolecular ET mechanisms, and offers many promising opportunities for creating arrays of redox active biomolecules. Recent reviews[16-22] provide an excellent introduction and overview of how electrochemical methods bring an important new capability to the exploration of protein mechanism. As methodologies to immobilize proteins[14,84-88] improve so that one has better control over protein orientation and spatial placement on an electrode surface, the detail and precision of fundamental questions that can be addressed will improve and the applications of *electrically wired* protein arrays will increase. This review focuses on studies by the authors and others that use electrode/SAM assemblies to explore fundamental aspects of the electron exchange between electrodes and redox active molecules; most particularly, proteins. These new findings identify a need to generalize modern charge transfer theory and bear on the use of electrochemical methods in modern bioengineering and biomedical applications.

Even the very first applications of SAM-based electrochemical systems enabled fundamental testing of basic features of contemporary charge-transfer theory.[1,4-6] For example, studies of the ET rate as a function of SAM thickness revealed that the nonadiabatic (diabatic) mechanism, or electron tunneling, is the rate controlling step for long range ET. In this limiting case, theoretical considerations predict a distinct experimental signature, an exponential decay of the ET rate constant, k_{et}, with the ET distance:

$$k_{ET(NA)} \propto \exp\left[-\beta(R_e - R_o)\right] \tag{1}$$

where R_e is the ET distance, R_o is the ET distance at the closest approach of the redox-active couple to the electrode, and β is the decay parameter, normally of the order of ca. 1 $Å^{-1}$ (within a series of congruent systems). Indeed, numerous experiments involving small redox species and redox-active proteins, which are immobilized onto electrodes with SAMs of variable thickness, have confirmed this fundamental prediction.[1,5,18,23-46,51-53,57-63,66-79] We note that the constraints imposed by *minor* SAM defects, which can be very important for small freely diffusing redox species, is less severe for *bulky* biomolecules and/or irreversibly attached redox species.

Until recently, the mechanism(s) controlling short-range ET, either at bare conductive electrodes (mostly implying small redox species), or SAM-modified metal electrodes (both *simple* model systems and redox-active biomolecules), has remained more controversial. A generalized version of the classical ET theory[89-98] accounts for a change in the inherent reaction mechanism with a gradual increase of electrode-reactant electronic coupling. The model systems allow the coupling to be tuned by changing the SAM thickness from thick (weak coupling/long range ET) to thin (strong coupling/short range ET) and in essence explore the connection between different ET limits.[23-47,51-53,58-63,66-71,74-83] Provided that the ET process remains a single barrier-crossing event, the ET kinetics can display a dependence on nuclear dynamics and medium friction (so called dynamic control), which in the extreme becomes the rate limiting step. In this limit, one may write the rate constant as

$$k_{ET(DC)} \propto \nu_{eff} \tag{2}$$

where ν_{eff} is the effective nuclear vibrational frequency for the barrier crossing. In any case, the friction control in soft-matter (in contrast to non-conductive crystalline solids, e.g., Ref. 99) is intrinsically related to *viscosity control* in the sense that the same relaxation modes affect the phenomena of ET and viscous flow (see Section II). This feature is captured by Eq. (3):

$$k_{ET(DC)} \propto \eta^{-\delta} \tag{3}$$

where η is the zero-frequency, bulk shear viscosity and δ is an *empirical* parameter with values between 0 and 1 (see Sections II and V for further details). Thus, a correlation of the rate constant with the macroscopic (or microscopic) viscosity is a signature for a friction-controlled ET mechanism, given that the ET event is not complicated by other barrier crossing processes such as mass transport, conformational rearrangement,[59,100-104] or proton coupled ET.[61,62,105] For biological ET processes, the bulk (macroscopic) solution viscosity[106-111] or applied hydrostatic pressure[112,113] affect the proteins' internal (local) viscosity (certainly, temperature has a comparable impact,[106] however it can be masked by thermally activated barrier-crossing along other degrees of freedom) and ET control occurs through a *protein friction* mechanism[80,81,83,111] (see Sections II, V, VI and VIII below), which is the biological counterpart of the friction controlled mechanism.

A viscosity dependence (Eq. 3) can also be accounted for by a *conformational gating* mechanism,[100-104] which arises from an extra barrier-crossing event (e.g., macromolecular intra- or intermolecular rearrangement) that is not intrinsically coupled with ET but may be rate limiting. In this mechanism the effective rate constant, k_{eff}, is also proposed to become independent of R_e (Eq. 2) and dependent on η.[58,59,66,67,101-103,112-114] A number of examples of rigorously confirmed conformationally gated ET have been reported[101,103] and it appears to be of considerable biological importance; however, conformational gating should be considered as a special case compared to the more general adiabatic ET mechanism that has an essentially *universal* origin.[89-97] Regrettably, as far as biological ET is concerned, except for several cases limited to different cytochrome *c* assemblies[74,76,78,80-82] and azurin,[83] the friction controlled mechanism has not been discussed or demonstrated unequivocally. In fact, viscosity-dependent kinetics (also exhibiting a weak dependence of the rate on the SAM thickness) is often interpreted as arising from a conformational-gating mechanism without relevant experimental cross-testing or analysis to ascertain whether it could arise from a friction controlled ET mechanism. Given the numerous model electrochemical processes occurring at bare metal electrodes (i.e., for the case of maximally strong electronic coupling, for which the dynamically controlled (friction controlled) nature of the ET event has been reported), the friction controlled mechanism should be considered for biological ET.[114-136]

This chapter reviews details of the theoretical background for a mechanistic understanding of heterogeneous ET, some selected (most commonly exploited) methodological aspects, and the results from some systematic studies of nanoscopic electrode/SAM/protein film, from which a mechanism change from tunneling control to friction control has been identified. It should be mentioned that the ability to observe a mechanism change through a series of congruent systems is unusual, as compared to the case of homogeneous solution systems (see Refs.[108,137]) for which systematic variation of fundamental ET parameters is very restricted.

II. THEORETICAL BACKGROUND

1. General Mechanistic Survey

In a general form, the rate constant for nonbiological and biological ET may be cast in a thermally activated form that is valid within the range of commonly used temperatures:[1-7,138-145]

$$k_{ET} = A \ \exp(-\Delta G_{a(EXP)} / RT) \tag{4}$$

where $\Delta G_{a(EXP)}$ is an experimental activation free energy, and A is a pre-exponential factor whose form is specific for the ET mechanism (see e.g. Eqs. 1 and 2 above, *vide infra*). The first successful theoretical treatments of the electrochemical electron transfer used either a potential energy curve approach[3,140] or a density of states approach.[138,141-146] In both treatments, the Condon approximation was applied and the rate constant contained a pre-exponential factor that could be written as the product of a nuclear frequency factor (associated with nuclear motion through the transition state), and an electronic factor (associated with the strength of the electronic coupling); see Refs. 1-7, 138-140 Some later workers[89-92,94-97] (see Refs. 5, 6, 93 for reviews) extended the electrochemical ET theory, in the spirit of a Kramers stochastic approach,[147] to account for frictional coupling and searched the conditions for a changeover between the limiting mechanisms. Recently, Mishra and Waldeck[98] have compared various theoretical approaches to electrode ET, and proposed a unified model that accounts for the electrochemical ET rate in all four of the commonly discussed limiting cases, namely (*i*) TST adiabatic, (*ii*) non-adiabatic, (*iii*) friction controlled barrier, and (*iv*) friction controlled cusp case, and interpolates among them. For biochemical ET, however, the more general term dynamical control is preferred and is taken to account for peculiar features of the heterogeneous protein/solvent environment, *vide infra*).

To simplify, yet retain physical sharpness, we use an approach that interpolates between the nonadiabatic (weak coupling limit) and dynamically controlled limit to write the rate constant as:[76,80-83]

$$k_{ET}^{o} = \frac{(H_{if})^2}{\hbar}\,\frac{\rho_m}{1+g}\left(\frac{\pi^3 RT}{\lambda}\right)^{1/2}\exp\left(-\frac{\Delta G_a^*}{RT}\right) \quad (5)$$

where H_{if} is the electronic coupling matrix element, λ is the overall reorganization free energy (to be specified below), and ρ_m is the density of electronic states in the metal (electrode). The activation free energy ΔG_a^* is written as:[1-7,138,139]

$$\Delta G_a^* = \frac{(\lambda + \Delta G_0)^2}{4\lambda} - H_{if} \quad (6)$$

where ΔG_o is the reaction free energy. Note that we distinguish between ΔG_a^* and $\Delta G_{a(EXP)}$ because the latter term may contain contributions different from that shown by Eq. (6), *vide infra*. Unless otherwise noted, this review considers the special case of $\Delta G_o = 0$, which is also indicated by the superscript 'o' for k^o_{ET} in Eq. (5) and hereafter; see also Section III below.

The relative importance of the electron tunneling and the nuclear friction (solvent/medium relaxation) within this simplified model can be gauged by an adiabaticity parameter g, which is given by[76,80-83]

$$g = \frac{\pi^3 RT (H_{if})^2 \rho_m}{\hbar \nu_{eff} \lambda} \tag{7}$$

where the effective frequency ν_{eff} is related to a single or several relaxation process(es) in the vicinity of the reaction zone that is (are) intrinsically coupled to electron transfer (see also Refs. 5, 6, 89, 92-94). In the nonadiabatic (NA) limit, $g << 1$, one finds that:

$$k^o_{ET(NA)} = \frac{(H_{if})^2}{\hbar} \rho_m \left(\frac{\pi^3 RT}{\lambda}\right)^{1/2} \exp\left(-\frac{\Delta G_a^*}{RT}\right) \tag{8}$$

in which

$$H_{if} = H_{if}^o \exp\left[-\frac{\beta}{2}(R_e - R_o)\right] \tag{9}$$

where H_{if} is the electronic coupling magnitude at the electron transfer distance R_e, and H_{if}^0 is the electronic coupling at the electron transfer distance R_o. This result leads directly to Eq. (1). In the dynamically controlled (DC) limit, $g >> 1$, one arrives at the equation:

$$k^o_{ET(DC)} = \nu_{eff} \left(\frac{\lambda}{\pi^3 RT}\right)^{1/2} \exp\left(\frac{-\Delta G_a^*}{RT}\right) \tag{10}$$

While ν_{eff} is most often correlated with the solvent, or medium, viscosity, this relation may not always be simple. Mishra[98] has suggested using a Kramers form for ν_{eff} because it interpolates naturally to the transition state theory limit; however, it has been more common to use a simple Debye form,[148] namely:

$$\nu_{eff} = \left(\frac{\varepsilon_s}{\varepsilon_\infty}\right)\frac{RT}{3\eta V_m} \tag{11}$$

Here, η is the solvent viscosity, ε_s is the static dielectric constant, ε_∞ is the high-frequency dielectric constant, and V_m is the molar volume. For the case of Debye-like solvents ν_{eff} is associated with the longitudinal dielectric relaxation time, whereas for the case of slowly relaxing media with several or a distribution of relaxation times (e.g., viscous molecular liquids, ionic liquids, biomolecules, etc.), it can be associated with a combination of slower diffusional or quasi-diffusional modes that are linked to the local and/or bulk viscosity.[149] From this formalism Eq. (2) can be deduced with $\delta =$ 1. Given the likely coupling of structural and dynamical features in the polarization response of SAM coated electrode assemblies, it may not scale linearly with the viscosity; hence, it is common to use a power law dependence, $0 < \delta < 1$, Eq. (3), when correlating rates with the viscosity (see Sections V, VI and VIII below).

2. The Reorganization Free Energy Parameter

The reorganization free energy λ, which is defined in Fig. 1, is a basic physical parameter of contemporary ET theory.[1-7,138-140] It arises from the complete Franck–Condon weighted density of states, encompassing the inner-sphere $\lambda_{(IS)}$ (reactants' body) and outer-sphere $\lambda_{(OS)}$ (reactants' immediate and distant environment) components:[1-7,139]

$$\lambda = \lambda_{(IS)} + \lambda_{(OS)} \tag{12}$$

In the harmonic approximation, the inner-sphere component may be written as:

$$\lambda_{(IS)} = \sum_j \frac{f_j^r f_j^p}{f_j^r + f_j^p}\left(\Delta q_j\right)^2 \tag{13}$$

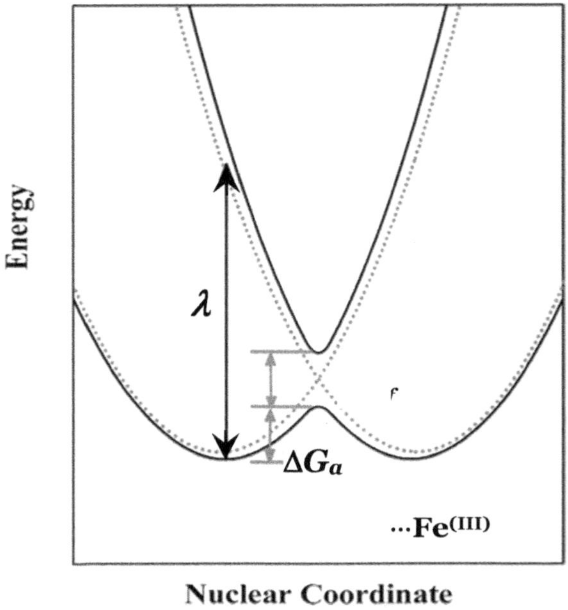

Figure 1. Schematic illustration showing key physical parameters that determine the probability and mechanistic character of thermally activated electron transfer under the condition of zero driving force $\Delta G_o = 0$ (implying electron exchange between the reactant's redox core and the electrode's Fermi level.)[1-7,138,140]

where f_j^r and f_j^p are, respectively, the jth normal mode force constant in the reactants and products, and Δq_j is the change in equilibrium value of the j-th normal coordinate. The outer-sphere component $\lambda_{(OS)}$ is connected with the medium's polarization response, and without specification of the medium's composition one may write it as:[1-7,139]

$$\lambda_{(OS)} \propto (\Delta e)^2\, \mathbf{F}[\varepsilon_{\text{op(i)}}, \varepsilon_{\text{s(i)}}] \int (D_R - D_P)^2 \, \mathrm{d}V \tag{14}$$

where Δe is the net charge transferred in the course of the elementary step, $\varepsilon_{\text{op(i)}}$ and $\varepsilon_{\text{s(i)}}$ are optical and static dielectric constants of the ith component of the medium (including SAM interior, protein, solvating water, etc.), D_R and D_P are dielectric displacement vectors for the reactant and product states, respectively, and dV is the reactant/product cavity differential volume. In a simple model, which represents the reactant species by spherical cavities immersed in a dielectric continuum,[1-7,139] the outer sphere reorganization energy is given by:

$$\lambda_{(OS)} = \frac{(\Delta e)^2 N_A}{4\pi \ \varepsilon_o} \left(\frac{1}{\varepsilon_{op}} - \frac{1}{\varepsilon_s} \right) \left(\frac{1}{2a} - \frac{1}{4R_e} \right) \tag{15}$$

where N_A is Avogadro's number, Δe is the net charge transferred, and a is the effective radius of the reactant sphere (e.g., $a = 4.3$ Å for $Fe(CN)_6^{3-/4-}$). A more sophisticated treatment of reorganization effects, developed by Liu and Newton,[150] models the redox couple by a sphere of radius a and the electrode/SAM/electrolyte junction by a three-zone pattern to specifically account for the role of alkanethiol spacers. This model gives rise to the following expression

$$\lambda_{(OS)} = \left(\frac{1}{\varepsilon_I^{op}} - \frac{1}{\varepsilon_I^{st}} \right) \frac{(\Delta e)^2}{2a} - \left(\frac{\eta_{II,I}^{op}}{\varepsilon_I^{op}} - \frac{\eta_{II,I}^{st}}{\varepsilon_I^{st}} \right) \frac{(\Delta e)^2}{4d}$$
$$+ \sum_{n=1}^{\infty} \left\{ \frac{\varepsilon_{II}^{op} \left(\eta_{II,I}^{op} \right)^{n-1} \left(\eta_{II,III}^{op} \right)^n}{\left(\varepsilon_{II}^{op} + \varepsilon_I^{op} \right)^2} - \frac{\varepsilon_{II}^{st} \left(\eta_{II,I}^{st} \right)^{n-1} \left(\eta_{II,III}^{st} \right)^n}{\left(\varepsilon_{II}^{st} + \varepsilon_I^{st} \right)^2} \right\} \frac{(\Delta e)^2}{d + nL} \tag{16}$$

where

$$\eta_{i,j} = \frac{\varepsilon_i - \varepsilon_j}{\varepsilon_i + \varepsilon_j} \tag{17}$$

d is the distance from the redox species to the SAM/solution surface (for adsorbed redox species, $a = d$), L is the film thickness, and ε is the dielectric constant, where the superscripts of op and st

correspond to optical and static, and the subscripts I, II, III correspond to the bulk solvent, the film, and the electrode, respectively. For a Au/alkanethiol SAM/aqueous electrolyte assembly, reasonable parameters are $\varepsilon_I^{st} = 50$ or 120, $\varepsilon_I^{op} = 1.8$, $\varepsilon_{II}^{st} = \varepsilon_{II}^{op} = 2.25$, $\varepsilon_{III}^{st} = \varepsilon_{III}^{op} = \infty$. Some estimates of the reorganization energy by this model with values of $\varepsilon_S^{I} = \infty$ (metal), $\varepsilon_S^{II} = 2.25$ (SAM film), and $\varepsilon_S^{III} = 78$ (substitution of the latter value by $\varepsilon_S^{III} = 20$ for the diffuse part of the double layer, which is taken to begin at the outer border of SAM films, does not change the numerical result significantly) are depicted in Fig.11A.

These two models for $\lambda_{(OS)}$ (through Eqs. 15, 16, and 17) give similar results for the case of *small* model redox-active species that exchange electrons with SAM-coated metal electrodes at comparable reactant-electrode separation distances, R_e (or d).[47] The reason for this outcome probably originates from the small contribution from the SAM and metal components (since, $\varepsilon_{II}^{st} = \varepsilon_{II}^{op} = 2.25$ and $\varepsilon_{III}^{st} = \varepsilon_{III}^{op} = \infty$, *vide supra*). Because of the heterogeneous and irregular dielectric properties of protein, protein/electrolyte, protein/SAM, or protein/electrolyte/SAM assemblies, it is more challenging to make rigorous theoretical estimates of $\lambda_{(OS)}$. Recent work shows that calculations of $\lambda_{(OS)}$, and the impact of temperature and pressure on it, are becoming available for globular proteins in an aqueous environment.[151-154] Taking into account results of that work and those discussed above[150] (yielding Eqs. 16 and 17), we postulate that the overall medium reorganization energy for triple junctions can be approximated by a summation over individual components:

$$\lambda_{(OS)} = \sum_k C_k \left(\frac{1}{\varepsilon_{o(k)}} - \frac{1}{\varepsilon_{s(k)}} \right) \tag{18}$$

where k indicates different contributing domains (here, solvent, protein and SAM interiors). Each component is described by its characteristic geometrical constant, C_k, and the effective dielectric constants, $\varepsilon_{o(k)}$ and $\varepsilon_{s(k)}$. Hence, the overall value can be given by $\lambda_{(OS)} = \lambda_{Protein} + \lambda_{Solvent} + \lambda_{SAM}$, where the three contributions represent the protein, solvent, and SAM interior, respectively. While this approximation is limited, an insightful segregation (e.g., inclusion of some water molecules into the protein contribution) of a multicomponent system can lead to this approximation being quite reasonable.

3. Temperature, Viscosity, and Pressure Effects

Given the considerations above, a viscosity dependence of the ET rate constant is an experimental signature that can be used to distinguish between the nonadiabatic and solvent controlled ET mechanisms. It is well-established that a solution's microscopic viscosity in the vicinity of dissolved redox species can be readily affected through the variation of temperature and pressure, or by adding viscogenic materials like glycerol, sugars, etc. Numerous systematic studies by a variety of authors[106-111,155-163] indicate that the internal dynamical properties of proteins, such as characteristic relaxation time distribution, can be altered by such factors also. Indeed, these three factors have a significant impact on the intrinsic protein viscosity, hence on $k_{ET(DC)}$ and $\Delta H_{a(EXP)}$, and provide the necessary information for distinguishing the two extreme ET mechanisms. However, a viscosity dependence does not constitute a proof that the solvent controlled ET mechanism is operative, because some other process, e.g., *conformational gating*, that is not directly coupled to ET may exhibit a similar kinetic signature regarding the effective rate constant of ET, $k_{ET(eff)}$. While it is difficult to distinguish the two mechanisms in *homogeneous* (solution) systems,[107-109,112-114] electrochemical methods provide a way to distinguish them, e.g., through an overpotential dependence, a scan rate dependence of the voltammetry, etc.[164-167] Such complementary information is needed and often can be provided by linking our modern understanding of ET theory with the electrochemical standard data analysis, including the basic concept of a *finite* reorganization free energy (related to the parabolic free energy surfaces, Fig. 1, that is the *calling card* of the Marcus theory).[1-7,138,139]

From Eq. (5) it follows that when the nonadiabatic mechanism operates, the dependence of the pre-exponential term on the temperature is weak (proportional to $T^{1/2}$), and in comparison to the exponential term can be neglected. Under this assumption and in the spirit of the Arrhenius equation, one can write:[168,169]

$$\Delta H_{a(EXP)} = -R \left[\frac{\partial(\ln k_{ET})}{\partial(1/T)}\right]_P \tag{19}$$

in which $\Delta H_{a(EXP)}$ is an experimentally determined (effective) value of the activation enthalpy. Under the specified conditions, it holds that $\Delta H_{a(EXP)} = \Delta H_{a(NA)}$ where the latter term is directly related to ΔG_a ($\Delta G_a = \Delta H_a - T\Delta S_a$) and λ in Eqs. (6) and (8). Both theoretical estimates and experimental studies[170-175] indicate that $\lambda_{(OS)}$ (along with its inner-sphere counterpart, $\lambda_{(IS)}$) has a weak temperature dependence. In the friction controlled limit the preexponential term contains the parameter ν_{eff}, which depends on the solution's microscopic fluidity and/or a protein's internal fluidity (more precisely, relaxation rates of the protein modes that are intrinsically coupled to the ET), hence in this case: $\Delta H_{a(EXP)} = \Delta H_{a(DC)} + \Delta H_{a(\eta)}$, where $\Delta H_{a(\eta)}$ denotes the contribution to the activation enthalpy from ν_{eff} (i.e., η^{-1} or $\eta^{-\delta}$).[76,106-109,135,168,169,185,187]

High-pressure kinetic studies provide unique information about the activation volumes and enable the construction of reaction volume profiles, adding a new dimension to the development of our fundamental mechanistic understanding.[176-187] It has been demonstrated that high pressure kinetic studies of heterogeneous bioelectrochemical processes are advantageous for identifying the intrinsic reaction mechanism for the cases of both model[176-187] and bioelectrochemical[78,79,81,111] ET processes. Let us consider pressure effects on the ET rate constant for both the nonadiabatic and dynamically controlled regimes. A phenomenological expression for the activation volume of any kind of microscopic barrier-crossing process, including ET can be defined as:[177-187]

$$\Delta V_{a(EXP)} = -RT \left[\frac{\partial(\ln k_{ET})}{\partial P}\right]_T \tag{20}$$

After the substitution of Eqs. (8) and (9) and neglecting minor terms, one obtains:[80,81,83,111,181-184]

$$\Delta V_{a(NA)} = \beta RT\left(\frac{\partial R_e}{\partial P}\right)_T + \frac{1}{4}\left(\frac{\partial \lambda}{\partial P}\right)_T \tag{21}$$

for the non-adiabatic case. Equation (21) indicates that the experimentally measurable volume of activation, Eq. (20), may originate from the effects of pressure on the ET distance (here - SAM/ protein) and/or a change in the medium's (SAM/protein/solvent) reorganization energy (Franck-Condon factor).

For the friction controlled ET limit, Eqs. (10) and (11), we can write that[80,81,83,111,181-184]

$$\Delta V_{a(DC)} = RT\left(\frac{\partial \ln \eta}{\partial P}\right)_T + \frac{1}{4}\left(\frac{\partial \lambda}{\partial P}\right)_T \tag{22}$$

where some minor terms are neglected. Eq. (22) reveals that in the case of full dynamic (viscosity) control, Eq. (10), viscosity changes arising from an increase in pressure cause a notable positive contribution to the activation volume, provided that viscosity is affected by pressure. All known liquids, with the notable exception of water (see Fig. 2)[182,184,188-190] display such a dependence. The peculiar behavior of water is probably connected with its unusual structure.[191-193] Indeed, moderately high pressure (up to 200–300 MPa) affects mostly those structural components that are not responsible for transport properties such as fluidity and diffusion.[194,195] For most solvents η increases exponentially with pressure,[182,184] and yields a maximum net viscosity-related contribution as high as +20 cm^3 mol^{-1} (as an upper limit; see also sections below). Previous work[111] pointed out how the insensitivity of water's viscosity to pressure change around room temperature, Fig. 2, provides an opportunity to use high pressure kinetic studies to vary a protein's intrinsic viscosity without altering the viscosity of the external (bulk) water (note that the solvating or *bound* water is considered to be part of the protein molecule.)[196] Numerous experimental and theoretical studies indicate that the protein, including its peripheral regions as part of a unified, albeit dynamic, system, can be viewed as a highly heterogeneous viscous liquid rather than a dissolved, solid-like macromolecular substance[152-163,197-201] and that the effective viscosity of such a complex *liquid* should increase with increasing pressure (density), *vide infra*. At the same time, it has been well-established that proteins may denature under pressure conditions above 200–300 MPa,[202,203] yet they normally retain native-like conformations for moderate pressures (below 150-200 MPa).[204-207] In particular, cytochrome *c* has been shown to be stabilized by moderate pressures (under 220 MPa), provided that the pH is controlled.[206] This fact is consistent with an increase in the internal protein friction with increasing pressure, *vide infra* (Sections V and VIII). Accordingly, the increased hydrostatic pressure and stabilizing additive (glycerol) both cause a decrease in the protein *free* volume.[204-207] We note that the protein-liquid analogy is a rough assumption. Certainly, because of rather dense packing of the protein interior,[202-207] in the case of ET controlled by the protein's intrinsic friction (local viscosity) one cannot expect to observe activation volumes as large as for low molecular weight (organic) liquids (see Sections V and VIII below).

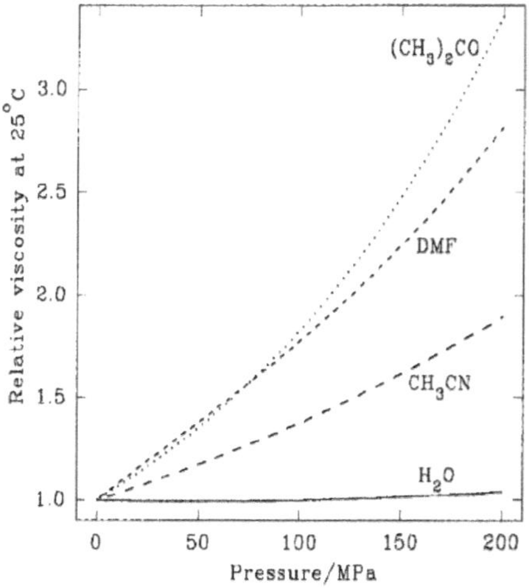

Figure 2. Pressure dependence of relative viscosity, η_r, for typical solvents at 25 °C.[188-190] Reprinted with permission from Refs. 182, Copyright (1999) American Chemical Society.

4. Electron Transfer for a Freely Diffusing Redox Couple

Although electrochemical rate constant measurements of SAM coated electrodes with a freely diffusing redox couple have important technical challenges, they remain an important subject of study because they provide valuable complementary information, especially where direct comparison with systems involving uncoated (bare) electrodes is desirable. Within the framework of the conventional encounter pre-equilibrium model, the experimentally determined standard heterogeneous rate constant (see Section III) for freely diffusing reactant species can be written as:[208-212]

$$k^o_{het} = k^o_{ET} K = k^o_{ET} \delta R_e \exp\left(\frac{-\Delta G_A}{RT}\right) = k^o_{ET} \delta R_e \exp\left(\frac{-Z_{eff} F \Phi_r}{RT}\right) \quad (23)$$

where k^o_{ET} can be viewed as the intrinsic unimolecular ET rate constant, identical to that defined by Eqs. (5), (8), or (10) (that is the ET rate constant for fixed electrode-reactant distance), and K_A is a statistically averaged equilibrium constant proportional to the probability of finding the reactant species at the reactive site near the electrode, usually considered as the outer Helmholtz plane (OHP), δR_e is the *effective thickness* of the planar reaction zone (which is expected to have a value of the order of 2×10^{-9} to 10^{-8} cm),[209-212] reflecting the major portion of the space integral over the intrinsic charge transfer constant (k^o_{ET}), and ΔG_A is the equilibrium free energy required to bring the reactant ion to the active site near the electrode (presumably to the OHP). Furthermore, Z_{eff} is the effective charge of a reactant ion and Φ_r is the effective potential at the average distance of the electron transfer (OHP). The formalism reflected by Eq. (23) allows one to consider heterogeneous ET occurring in the freely diffusing and adsorbed regimes almost on the same theoretical basis, albeit the methodology for determining k^o_{ET} in these two cases is different (see the next Section). In the approximation that a variation in SAM thickness does not change the structure and potential at the SAM solution boundary, we expect that the equilibrium parameter, K_A, is constant throughout the series with SAMs of variable thickness. In general, however, this is not the case and corrections for variable Φ_r and ΔG_A, hence K_A, are necessary, *vide infra*. Anyway, because the value of the scaling parameter, δR_e, commonly is not known, or is typically evaluated by rough estimates, either the raw, or partially (comparatively) corrected experimental values of k^o_{het} are correlated with well-defined variables (e.g. the SAM thickness, the temperature, or the pressure) to determine the inherent physical parameters (e.g., β, $\Delta H_{a(EXP)}$, $\Delta V_{a(EXP)}$) with an acceptable accuracy (*vide infra*).

III. Methodology

1. SAM-modified Electrodes

For most of the kinetic studies reviewed in this chapter the standard procedures for the SAM deposition onto Au electrodes were applied with minor modifications (see Refs. 47, 73-76, 80-83 for the detailed description). The working electrodes were either custom-made Au ball[47,73-76] or commercial Au disk[80-83] electrodes, sealed into glass or Teflon, respectively. The equipment was standard electrochemical instrumentation unless otherwise indicated. The details of the particular systems can be found in the original papers.[47,73-76, 80-83] In nearly all cases the auxiliary electrode was a platinum wire and the reference electrode was Ag/AgCl. While some studies were performed under an inert atmosphere, the presence of ambient oxygen did not significantly affect the experimental results. For temperatures studies, the electrochemical cell contained a water jacket and the temperature range was chosen to avoid freezing of the aqueous electrolyte solution at the lowest temperature and denaturation of the protein at the highest temperature. For the high pressure experiments, the high pressure vessel and electrochemical cell have been described.[80,81,83,114,185,187] In brief, the working electrode (a 1.6 mm diameter gold disk electrode sealed in a Teflon cylinder, BAS), the auxiliary electrode (platinum wire), and the reference electrode (Ag/AgCl/4M KCl) were sealed into the cell cap by two O-rings. The working volume of the high pressure electrochemical cell was 5 ml in most cases.

The assembled pressure vessel containing the cell was placed in a thermostated water jacket equilibrated at 25.0±0.1 °C. All the high pressure experiments were performed using the Tris-HCl buffer which is known to withstand pressure-induced pH changes.[213,214]

The typical voltammetric response for freely diffusing redox species at thinner SAMs (short-range ET) results in well-defined anodic and cathodic peaks, at low scan rates separated at least by 57 mV (related to the diffusion-controlled fast electron exchange; for representative cases, see Fig. 3).[215-218] ET in the freely diffusing regime can be justified by verifying that the background-corrected peak current is proportional to the square root of the potential scan rate[217,218] (see Eq. 25 and Fig. 5 in the next Section). The dramatic change in the measured current density and the shape of the voltammogram with the increase in the methylene number, n, indicates a slowing of the electron transfer rate, Figs. 3a and 3b. In some instances, the application of a steady-state approach[217,218] has been reported, either in combination with the rotating disk electrode (RDE) technique (applicable for the case of bare electrodes with fast electron exchange), or in the case of thicker SAMs in which the peak-shaped feature disappears because of the dominance of ET kinetics over that of diffusion (see e.g., Refs. 47 and 81). For voltammograms that do not have a well-defined peak (the thicker SAMs in Fig. 3), the rate constant typically is calculated using data from the low overpotential region (initial portions of i-V curves) where the mass transport effect and other effects resulting in non-linearity of the dependence of $\log(k_{het})$ on $\Delta E = E - E^o$ (Tafel plot) are negligible.[217,218]

In contrast to the freely diffusing case, redox species that are irreversibly immobilized on SAM-coated electrodes, at small reactant-electrode separations and low scan rates, display symmetrically situated well-defined anodic and cathodic peaks that arise from the Faradaic process of electron exchange.[217-220] The surface coverage by electrochemically active species immobilized at SAM-coated electrodes can be determined by considering the voltammetric peak areas (θ), through Eq. (24):[221]

$$\Gamma = \theta / (n F S v) \tag{24}$$

where n is the number of transferred electrons, F is the Faraday constant, S is the geometric area of the electrode, and v is the po-

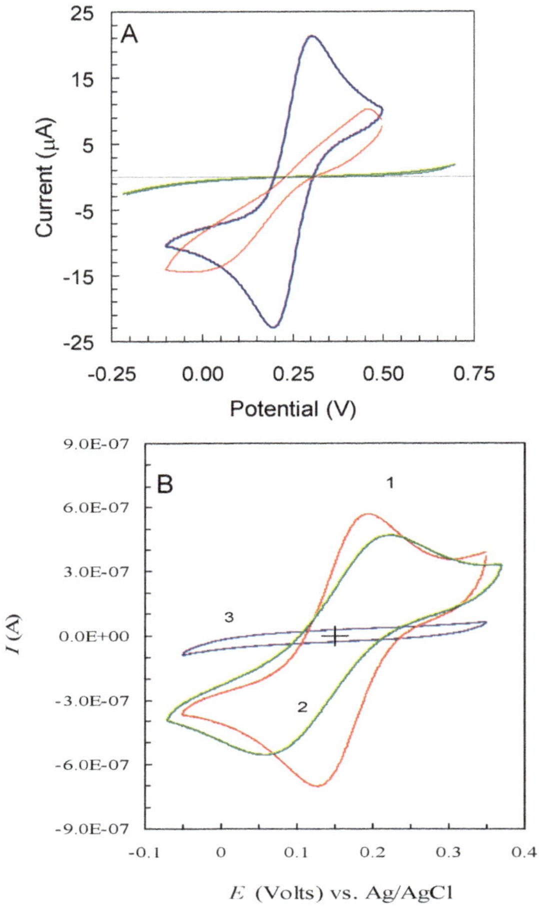

Figure 3. Cyclic voltammograms for the electron exchange of freely diffusing species: (A) the model redox couple, $Fe(CN)_6^{3-/4-}$ at a Au electrode coated by $-CH_3$ terminated *n*-alkanethiol SAM with $n = 2$ (curve 1), $n = 4$ (2) and $n = 8$ (3), scan rate: 0.1 V s^{-1};[47] and (B) cytochrome *c* at a Au electrode coated by –OH terminated *n*-alkanethiol SAM with $n = 3$ (curve 1), $n = 6$ (2) and $n = 11$ (3), scan rate: 0.05 V s^{-1}.[81] Reprinted with permission from Ref. 47, Copyright (2001) American Chemical Society and Ref. 81, Copyright (2006) Wiley-VCH Verlag Gmbtt & Co.

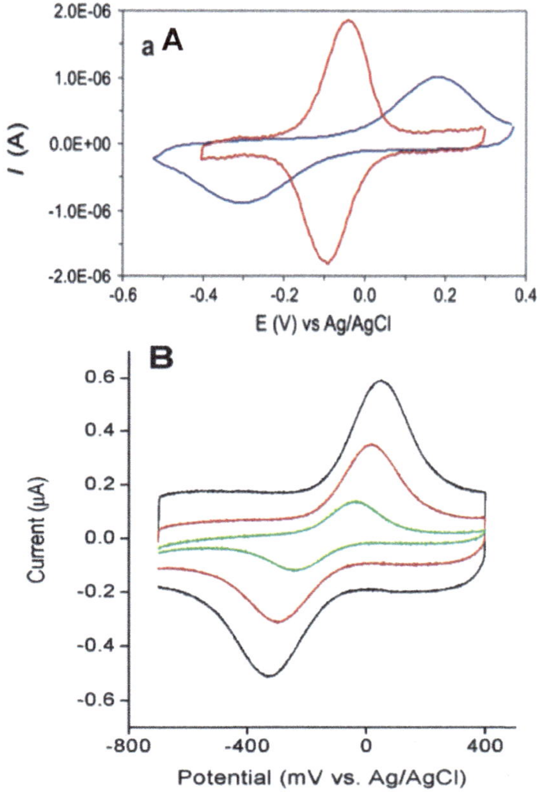

Figure 4. (A) Cyclic voltammograms for azurin electron exchange at Au electrodes modified by CH_3-terminated alkanethiol SAMs with variable *n*. (a) Red curve is 1-pentanethiol (*n* = 4); blue curve is *n*-hexadecanethiol (*n* = 15), scan rate: 5 V s^{-1}.[83] (B) Representative cyclic voltammograms of rat mutant K13A cytochrome *c* immobilized on C20Py/C19 mixed monolayer modified gold electrodes.[77] Scan rates are 0.2 V/s (green), 0.6 V/s (red), and 1.0 V/s (black). Reprinted with permission from Ref. 77, Copyright (2004) American Chemical Society and Ref. 83, Copyright (2010) Proc. Natl. Acad. of Sci.

tential scan rate. An increase of the ET distance (SAM thickness) and/or potential scan rate causes a separation of peak positions, Figs. 4a and 4b. In this case, plots of the background-corrected peak current *versus* the potential scan rate should display straight lines intercepting the origin (see Fig. 8b). This linear dependence can be used to verify that the redox species is immobilized, however, one should bear in mind that a decrease of the surface coverage, Γ, by desorption of molecules may occur over time. Such a process can obfuscate the linear dependence of the type given in Fig. 8b and may confuse the interpretation.

In the following subsections the most widely used electrochemical experimental technique of cyclic voltammetry (CV) and fast scan cyclic voltammetry (FSCV) will be considered in some detail.[16,18,21,22,164-166,215-220] Most of the kinetic data discussed here have been collected through this methodology. A detailed review of electrochemical techniques is beyond the scope of this chapter; instead we refer to other sources in which applicability of these and other electrochemical or hybrid (electrochemical/ spectroscopic) methods (e.g., AC voltammetry, electrochemical impedance spectroscopy, chronoamperometry,[11,12,15,22,217,218] indirect laser-induced temperature jump technique,[50-53] time-resolved surface-enhanced resonance Raman spectroelectrochemistry[19,59-61]) as regarding bioelectrochemical devices are thoroughly described.

2. Freely Diffusing Species

Conventional fast scan cyclic voltammetry and the Nicholson method[215-219] can be applied for the accurate and prompt calculation of rate constants. For simple (single step) electron transfer reactions a study of the *i*-V curve as a function of the potential scan rate over the range where it exhibits kinetic behavior, an increase in separation of the cathodic and anodic peak potentials (ΔE_p), can be used to calculate the standard rate constant of electron transfer (k^o_{het}). Typically, one uses the numerically evaluated relationship between (ΔE_p) and the Ψ function, given by Eq. (25),

$$\Psi = \frac{(D_O / D_R)^{\alpha/2} (RT)^{1/2} k^o_{het}}{(\pi n F D_O v)^{1/2}} \tag{25}$$

where α is the transfer coefficient, D_O and D_R are the diffusion coefficients of the oxidized and reduced forms, respectively, v is the scan rate and n is the number of transferred electrons. The diffusion coefficients for the reduced (initial) forms of reactant ions, D_R, can be calculated from the slope of the dependence of $I_{p(corr)}$ on $v^{1/2}$ using Eq. (26):[217,218]

$$I_{p(corr)} = -0.446nF \left(\frac{nF}{RT}\right)^{1/2} C_o D^{1/2} v^{1/2} \tag{26}$$

Because the ratio for $I_{p(Red)}$ and $I_{p(Ox)}$ is always close to unity, especially for the case where bulky proteins are the freely diffusing reactants, it is reasonable to simplify by assuming that $D_R \approx D_O \approx D$. This assumption introduces minor errors in the subsequent calculations using Eq. (25). Figure 5 depicts typical dependencies of the background corrected anodic peak current on the square root of the scan rate obtained for freely diffusing species, allowing for the determination of the diffusion coefficient, D, from slopes of linear plots and Eq. (26). Note that the plot's intercept is in the origin, indicating a single ET process and the reactants' free diffusion.

Recently, Nicholson's approach was extended to the case of any temperature for which the freely diffusing regime may operate.[185] The numerical data (Ψ *versus* ΔE_p) of Nicholson[215] originally obtained for 25°C were fitted by the following analytical function (Fig. 6a) reported by Swaddle:[184]

$$\ln \Psi = 3.69 - 1.16 \ln (\Delta E_p - 59) \tag{27}$$

and subsequently recalculated to generate the curves depicted in Fig. 6a.[185] This procedure allows for the determination of k^o_{het} at different temperatures provided that the values of D are determined beforehand (either by use of Eq. 25 and Fig. 5 or by another method; see Refs. 217 and 218).

Alternatively, the values of heterogeneous rate constants, and simultaneously of α, can be calculated from background-corrected voltammograms recorded at slow scan rates, through the application of the full computer fitting procedure (see, for example, Ref. 184). Figure 6b illustrates the application of the standard fitting

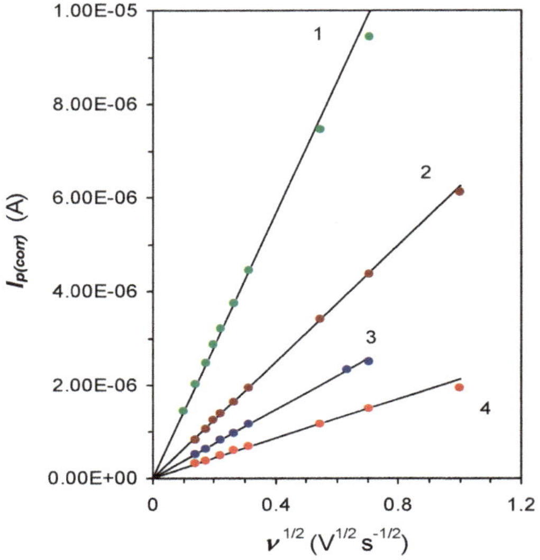

Figure 5. Typical dependencies of the background-corrected anodic peak current on the square-root of the scan rate obtained for freely diffusing $[Fe(cp)_2]^{+/0}$ (1) and $[Fe(bipy)_3]^{3+/2+}$ (2-4) with the Au electrode, in the [BMIM][BTA] ionic liquid as solvent, under different pressure and temperature conditions: (1) $T = 25°C$, $P = 0.1$ MPa, $c = 5 \times 10^{-3}$ M; (2) $T = 85°C$, $P = 0.1$ MPa, $c = 2 \times 10^{-3}$ M; (3) $T = 20°C$, $P = 50$ MPa, $c = 3 \times 10^{-3}$ M; (4) $T = 20°C$, $P = 100$ MPa, $c = 3 \times 10^{-3}$ M. Rerpinted with permission from Ref. 185, Copyright (2008) American Chemical Society.

program available through the Autolab software. Actually, the original method of Nicholson can be considered as a simplified fit procedure using two fixed key points (peak maximum positions.) However, it is important to realize that the standard Nicholson method[215,216] is based on the Butler-Volmer model[11-13,217] for the relation between the electron-transfer rate and the electrode potential, which is a limiting case of the more general Marcus theory.[1-5] In the case of relatively small peak-to-peak separations, i.e., when the reaction free energy is substantially less than the Marcus reorganization energy, λ , i.e., when $\Delta E_p/2 << \lambda$, the models give simi-

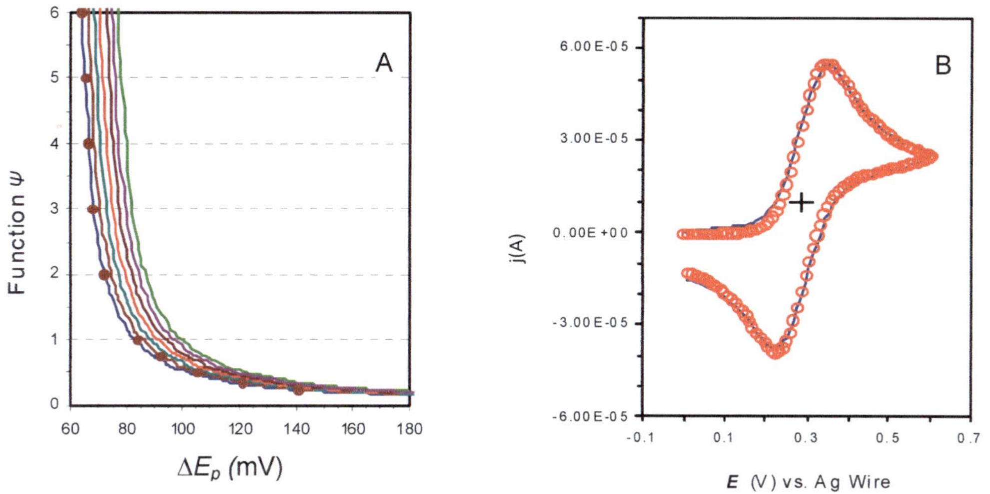

Figure 6. (A) The dependence of the Ψ function on the cyclic voltammogram's peak potential separation (ΔE_p) at different temperatures.[185] The data of Nicholson[215] obtained for 25°C (given as closed circles) are fitted by Eq. 27,[184] and for other temperatures were calculated as described in the text. From left to right: recorded at 25, 35, 45, 55, 65, 75 and 85 °C; (B) Experimental voltammogram of $[Fe(cp)_2]^{+/0}$ (5×10^{-3} M) recorded with the Au electrode, in a [BMIM][BTA] ionic liquid as solvent, at $P = 0.1$ MPa, $T = 65$°C and $v = 4$ V s^{-1} (calculated value of $k^0_{het} = 8.1 \times 10^{-3}$ cm s^{-1} using Eq. 25), along with fitted quasi-reversible one electron transfer model (red circles); results obtained from the fitting procedure: $k^0_{fitt} = 9 \times 10^{-3}$ cm s^{-1}; transfer coefficient $\alpha = 0.5$.[185] Rerpinted with permission from Ref. 185, Copyright (2008) American Chemical Society.

lar results. In a more sophisticated approach developed by Savéant[166,217] the values of k_{het} as a function of reaction driving force, ΔG_o ($\Delta G_o = -e\xi$, where $\xi = E - E_0$ is the overpotential) can be calculated through a convolution procedure. For this purpose, background-subtracted voltammograms are converted to sigmoidal-shaped curves by a semi-integral convolution technique and rate constants at different overvoltages are calculated using Eq. (28): [166,217]

$$k_{EXP}(E) = D^{1/2}\frac{i(E)}{I_{\lim} - I(E)\left[1+\exp\left(nF/RT\right)\left(E-E_0\right)\right]} \tag{28}$$

where $i(E)$ is the experimental current, $I(E)$ is the convoluted current and $I_{\lim}$ is the convoluted limiting current $I_{\lim} = nFAD^{-1/2}C_0$. Furthermore, estimation of the value of λ is possible through a Gaussian fitting of the dependence k_{het} on ΔG_o.[167,185] This data processing procedure is best applicable in cases where the diffusion impact is less pronounced; that is, in cases of intermediate or thicker SAMs rather than the thinner ones which may display effects of diffusional control (electrochemically reversible shape).[217,185]

3. Immobilized Species

For the case of redox species irreversibly attached to (immobilized at) bare or SAM-coated electrodes, Laviron[220] developed a formalism that calculates the rate constant of electron transfer from the relationship between the peak position and the scan rate. In this model, like in the previous case described above, the overpotential is assumed to be much smaller than the solvent reorganization energy (peak separation is smaller than 200 mV). In this limit, the rate constant for the reaction can be determined from the equation:

$$\log k_{ET}^{o} = \alpha\log(1-\alpha) + (1-\alpha)\log\alpha - \log\left(\frac{-RT}{n_a F\nu}\right) - \frac{\alpha(1-\alpha)n_a F\Delta E_p}{2.3RT} \tag{29}$$

where ΔE_p is the peak separation between the cathodic and anodic waves. This model is consistent with the more general Marcus theory of electron transfer under the following conditions: $e\xi << \lambda$ and $k_BT << \lambda$ (where ξ is the overpotential).[164,165,217] Weber and Creager[164] and Murray et al.[165] developed a numerical treatment, which is based on the Marcus model. In combination with a fitting algorithm, it can simultaneously determine the values of unimolecular standard rate constants and reorganization energies from the dependence of the cathodic and anodic peak potentials on the potential scan rate. This model also accounts for the population of electronic states in the metal electrode around the Fermi level:[3,7,138,140]

$$k_{ET} \propto \int_{-\infty}^{\infty} f(\epsilon) \exp\left[-\frac{(e\xi + \epsilon_F - \epsilon \pm \lambda)^2}{4\lambda RT}\right] d\epsilon \tag{30}$$

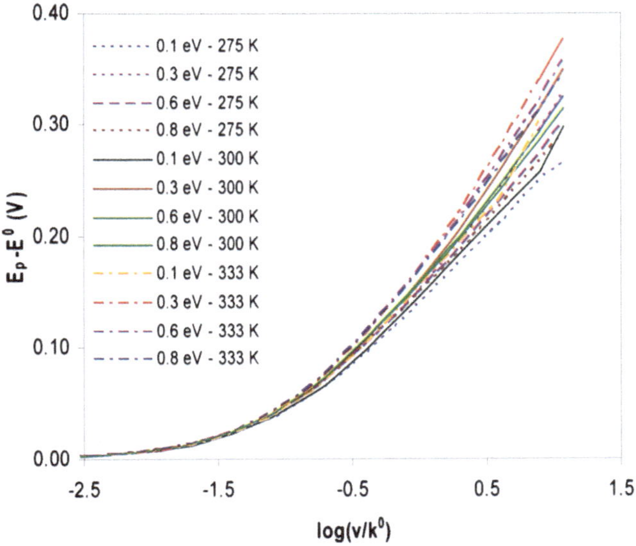

Figure 7. Numerically generated model fitting curves taking into the account Eq. 30, corresponding to different values of λ and T (K. Davis and D. H. Waldeck, unpublished work).

where $f(\epsilon)$ is the Fermi-Dirac distribution function and ϵ_F is the Fermi energy. Numerically generated model fitting curves corresponding to different values of λ and T are presented in Fig. 7 (K. Davis and D. H. Waldeck, unpublished work; for the application, see Ref. 83).

Figure 8 presents typical data and fits for the blue copper protein azurin electron exchange at Au electrodes coated by undecanethiol ($n = 10$) SAMs that display high surface coverage (ca. 10% error in ΔE, at higher scan rates).[83] This figure also demonstrates the impact of the post-measurement IR_S correction (according to Eq. 31, *vide infra*) on the data fitting procedure and sets a framework for appreciating how much such corrections are needed for a proper determination of true (unperturbed) values for parameters k^o_{ET} and λ (see also Refs. 217, 218, 111,185,187 for representative book chapters and original papers, respectively). Note that the uncorrected values of ΔE_{UC} (open symbols) cannot be fitted properly with any reasonable theoretical curve depicted in Fig. 6. In addition, it should be noted that for the case of redox species irreversibly immobilized at bare or SAM-coated electrodes, both the Laviron theory[220] and its Marcus theory-based extensions[164,165] predict a linear dependence of the peak current on the potential scan rate (see Fig. 8b), in contrast to the case of freely diffusing redox species (see Eq. 26 and Fig. 5). This distinct difference in experimental signatures can be used to distinguish the two classes of ET in equivocal cases with less well-studied redox species involved as reactants.[217,218]

The voltammetry for cases of low electrolyte concentration (below 0.05 M), high reactant concentration (above ca. 5×10^{-4} M; freely diffusing regime), and high surface coverage (above ca. 10 pmol cm^{-2}; immobilized regime), large separation of working and reference electrodes, can be significantly distorted, especially at high potential scan rates (effect of Ohmic polarization) causing notable enhancement of the peak-to-peak separation and, consequently, yielding lowered values of rate constants (k^o_{het} or k^o_{ET}).[217,218] When the conditions leading to large voltammogram distortions are unavoidable, the post measurement IR_S drop corrections (where R_S is the solution resistance between the surface of the working electrode and the tip of the reference electrode) can be successfully applied through the electrochemical standard software programs to eliminate an error caused by the voltammograms dis-

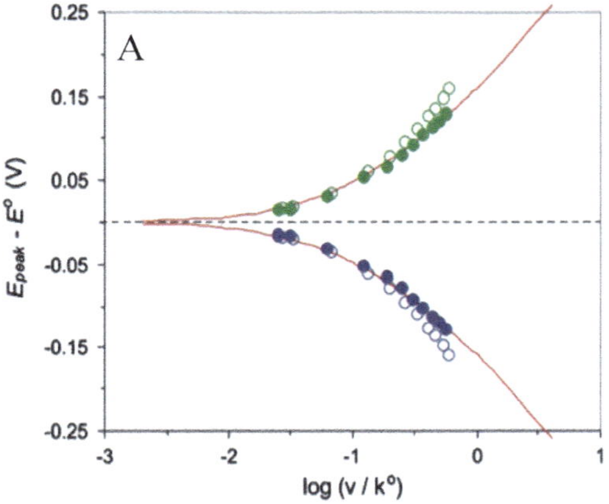

Figure 8. (A) The data fitting procedure for azurin electron exchange at Au electrodes coated by undecanethiol (n = 10) SAMs that display high surface coverage (ca. 10% error in ΔE, at higher scan rates). This figure demonstrates the impact of the post-measurement IR_S correction (according to Eq. 31) on the data fitting procedure. Standard errors for experimental points directly match the symbol size. Closed symbols: ΔE_{CO} after the proper IR_S correction, yielding the values k^o_{ET} = 170 ± 5 s^{-1} and λ_o = 0.3 ± 0.06 eV (at 25°C). Note that the uncorrected values of ΔE_{UC} (open symbols) cannot be fitted properly with any reasonable theoretical curve depicted in Fig. 7.[83] (B) Linear dependence of the peak current on the scan rate for voltammograms of cytochrome c on PyC11/C10 SAM modified gold electrodes at scan rates of 10, 20, 40, 100, 200, 300, and 500 V/s, the temperature is 298 K.[78] The panels (A) and (B) show the data for both anodic and cathodic peaks. Rerpinted with permission from Ref. 78 (2006) American Chemical Society and Ref. 83 Copyright (2010) Proc. Natl. Acad. Sci.

tortion, Fig. 8A.[83] For a disc shaped electrode, the solution resistance can be calculated using Eq. (31):[222,223]

$$R_S = \frac{\arctan(r_r / r_w)}{2\pi \quad r_w \kappa} \tag{31}$$

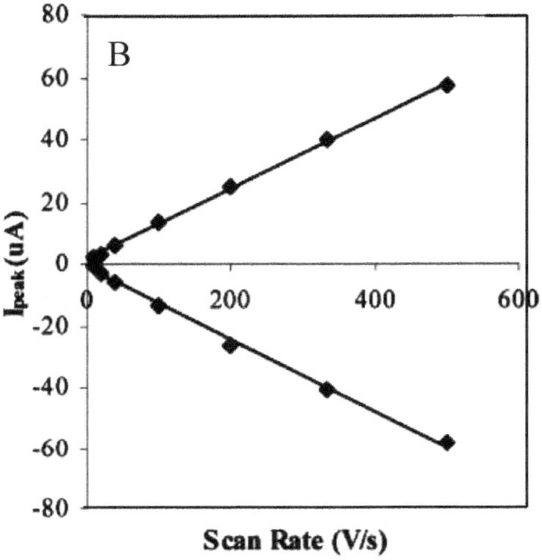

Figure 8. Continuation.

where r_w is the radius of the working electrode, κ is the conductivity of the medium and r_r is the distance between the working and reference electrodes. As one can see from the formula above, the resistance of the solution is mainly determined by the solution conductivity and the electrode area.

IV. MODEL REDOX-ACTIVE SPECIES AT Au/SAM JUNCTIONS

1. Overview: The Case of Immobilized Redox Species

Systematic studies of electron exchange between electrodes and redox-active metallocomplexes have deepened our understanding of ET with electrodes and provided us with insight into the operation of SAM/electrode assemblies, under the conditions where the elementary event is not complicated by a redox molecule's slow conformational degrees of freedom. Such understanding has provided a framework in which to interpret bioelectrochemical systems with their additional complexity. Results for different redox species, functionalized at electrode/SAM junctions in both the freely diffusing and the immobilized regimes are available. Electron exchange of redox species, in particular hexacyanoferrate, $Fe(CN)_6^{3-/4-}$, and hexaminoruthenate, $Ru(NH_3)_6^{3+/2+}$ redox couples, with metal electrodes coated with alkanethiol SAMs of variable thickness in the freely diffusing regime have been thoroughly investigated (see Refs. 47 and 224, respectively, and for the earlier work—literature cited therein). While such systems can be quite susceptible to film defects, they are uniquely important because they have also been studied at bare Au and Pt electrodes, see Refs. 135 and 136 and literature cited therein). Covalently tethering a redox moiety to the terminus of the SAM film, that is exposed (totally or partially, depending on the length of diluting SAM components) to the electrolyte simplifies the kinetic analysis, by eliminating diffusion and mitigating the importance of defects, and this approach has been widely used; however, the ferrocene (Fc), $Fe(cp_2)^{+/0}$ (cp = Cyclopentadiene), and ruthenium pyridine pentamine, (Ru), $RuPy(NH_3)_5^{3+/2+}$ (Ru) redox couples have received the most activity.[15,24-28,50-53]

As noted, the kinetic data obtained for SAM-immobilized redox species are simpler to analyze because the rate constants, k^o_{ET}, are not influenced by diffusional transport from the bulk solution and can be directly extracted from the voltammetric or other electrochemical data. Figure 9 depicts the structural arrangement of a metal/SAM/electrolyte junction with a redox active component, $Fe(cp_2)^{+/0}$ or $RuPy(NH_3)_5^{3+/2+}$, covalently attached to some portion of the thiol molecules comprising the SAM; i.e., the electroactive thiolates are diluted by an inert SAM constituent.[50-53]

While a number of workers have studied how the ET rate constant between ferrocene and an electrode changes with SAM thickness and composition, the work of Smalley and collaborators has been the most extensive (see Ref. 53 for a recent review). Figure 10 shows this dependence for SAMs composed of methylenes (from 5 to 16), oligophenylenevynylenes (OPV), and oligophenylenethynylenes (OPE). The plot of $\ln(k^o_{ET})$ vs. ET distance, calculated by using a model for the SAM thickness,[53,150] reveals a linear dependence with a slope of ca. 1 $Å^{-1}$ (typical for long-range tunneling, that is a nonadiabatic mechanism.)[1,5,18,23-46,51-53,57-63,66-79] The same figure presents data for ET through OPV and OPE bridges (π-conjugated SAMs). Rate constants in these latter sys-

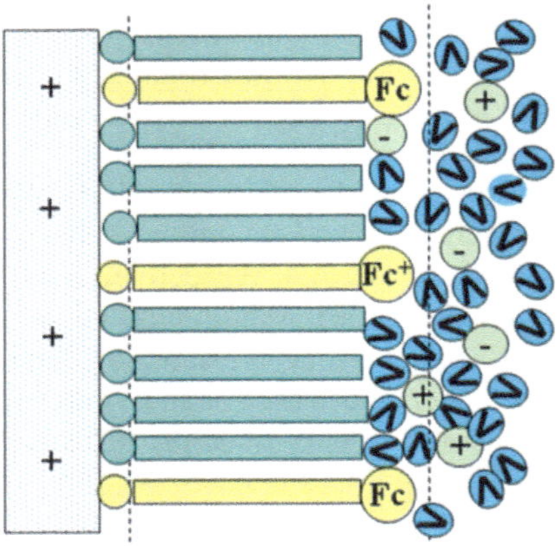

Figure 9. Schematic representation of the electrode/SAM/electrolyte junction including a mixed, electroactive SAM. The black dashed lines represent the inner (IHP) and outer (OHP) Helmholtz planes. The circles with the minus sign represent anions and the circles with the plus sign represent cations. The tethered ferrocene (ferrocenium) is labeled by the circle with Fc (Fc+). The water solvent is illustrated by a circle with a 'v' symbol. The plain circles are S atoms, and the alkane chains are represented by horizontally aligned rectangles.

tems show a slightly undulating distance dependence and are very nearly independent of distance for large thicknesses.[50,53]

Temperature dependent studies of the rate constant have provided the activation energies (enthalpies) through an Arrhenius analysis, hence allowing for a separation of exponential and preexponential factors.[50-53] Panel A of Fig. 11 shows how the activation energy changes with distance from the electrode surface. This distance dependence is explained by the change in reorganization energy that arises from image charge effects (see section 2.2) and a change in coupling with distance (Eqs. 1 and 6). Panel B of Fig. 11 plots the preexponential factors versus the ET distance and reveals a plateau-like region for ET distances within 6 to 10 Å (SAMs with n = 5 to 9) of the electrode, for the methylene SAMs. The conjugated films show a weak distance dependence over the entire range, albeit at a higher value for the preexponential term. The dependence in panel B might be interpreted in terms of a change in the intrinsic ET mechanisms; i.e., a transition from control of the rate by electron tunneling at large distance to control by the nuclear dynamics at short distance/higher couplings (see section 2). From the magnitude in the rate in the plateau region, Fig. 11B Smalley et al.[51,53] have estimated that the preexponential component A_ν for the cases of tethered Fc' and Ru' couples is of the order of 10^{10} to 10^{11} s^{-1}. The graph in panel B also demarks their estimated value of the aqueous longitudinal polarization *pre-exponential* term, ca. 5×10^{14} s^{-1};[51,53] however, the value of ν_{eff} for the aqueous longitudinal polarization time itself is ca. 10^{11} s^{-1}

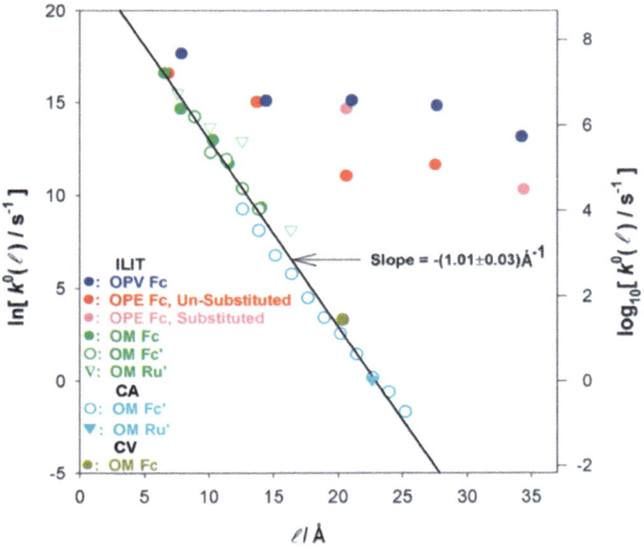

Figure 10. Standard rate constants (k^0_{ET}) versus ET distance (here l) at 298 K. ILIT data for directly-linked ferrocene (Fc), filled green circles; for ester-linked ferrocene (Fc'), open green circles; and for Ru', open green triangles; chronoamperometry (CA) data for Fc', open cyan circles, and for Ru' (n = 15), filled cyan triangle; cyclic voltammetry (CV) data for Fc (n = 16), the filled lime green circle. OPV bridges attached to Fc (all ILIT data): filled blue, red and pink circles.[53] Copyright (2007). Reproduced by permission of the Roytal Society of Chenmistry.

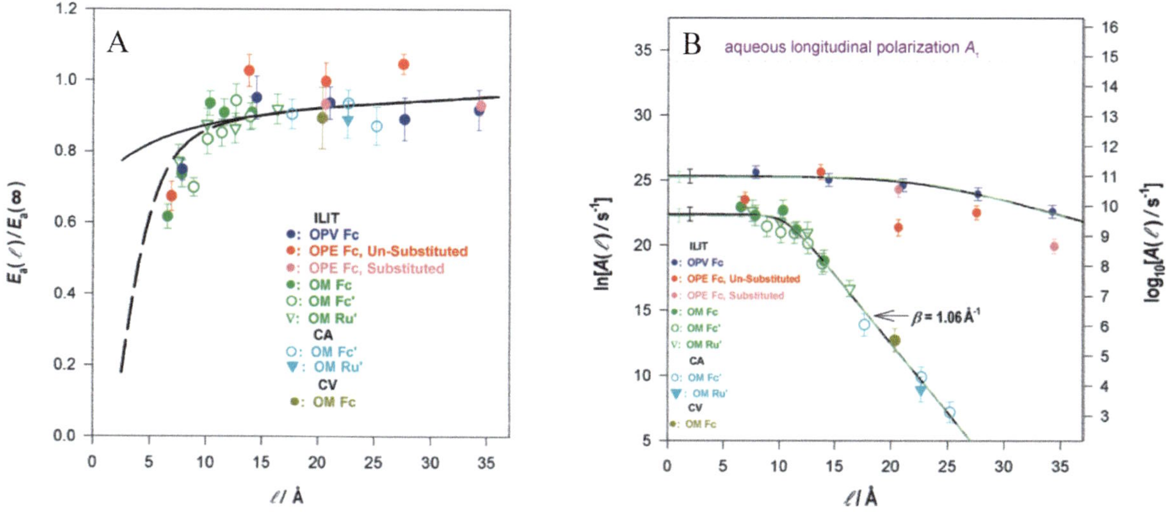

Figure 11. A) The value of reduced activation energy (enthalpy) determined through the Arrhenius analysis, as a function of ET distance (see the Fig. 10 caption for a complete key to the shapes and colors of the various points). The solid black curve is calculated according to Liu and Newton,[150] dashed black curve incorporates the effects of electronic coupling as well as image charges. B) Semilogarithmic plots of preexponential terms, $A(l)$, of rate constant expression versus ET distance (here l) for all of the bridges and redox couples depicted in Fig. 10 (see the Fig. 10 caption for a complete key to the shapes and colors of the various points).[53] Copyright (2007). Reproduced by permission of the Roytal Society of Chenmistry.

around room temperature.[128-130] Perhaps it is not surprising that ν_{eff} might be notably lower near the SAM/solution interface than the characteristic relaxation rate of bulk water electrolyte. For example, the redox species covalently attached to SAM terminal groups may be influenced by the polarization fluctuations of water molecules that are partially immersed in the *outer* SAM interior, forming a more slowly fluctuating *mixed phase*.[149] What remains unclear is what process(es) control the relaxation rate near the electrode surface; see Sections IV.3 and IV.4 below.

2. Overview: The Case of Freely Diffusing Redox Species

Systematic studies of electron exchange for freely diffusing redox couples with SAM-coated metal electrodes show a similar dependence of the rate constant on the SAM thickness. The most systematic studies of the couples $[Fe(CN)_6^{3-/4-}]$[47] and $[Ru(NH_3)_6^{3+/2+}]$[224] at Au/SAM junctions show that the nature of the redox couple including the charge type and distribution within the ligand sphere (*vide infra*) plays an important role in determining whether water and counter-ions can penetrate into the SAM interior (the case of $Ru(NH_3)_6^{3+/2+}$), or be almost ideally excluded from the SAM (the case of $Fe(CN)_6^{3-/4}$). Figure 12 plots data for the heterogeneous rate constant as a function of the number of methylene units for $Ru(NH_3)_6^{3+/2+}$ in panel A[224] and for $Fe(CN)_6^{3-/4-}$ in panel B.[47] Note the large difference in the extent of scatter between the rate constants in these two systems.

The results[225-228] indicate that the kinetically fast redox couple $[Ru(NH_3)_6]^{3+/2+}$ is better able to penetrate into unicomponent, CH_3-terminated SAMs of the general type: –S– $(CH_2)_n$–CH_3, with n running from 2 to 18, than is $Fe(CN)_6^{3-/4-}$, and thus it can detect small defects in the SAM. This interpretation has been substantiated by quantum-chemical calculations of the charge distribution within the complex ions, which demonstrate that hexamine metal complexes have most of the excess charge located on the metal ion core, whereas hexacyano metal complexes have it localized on the terminal nitrogen atoms of cyano ligands.[229] This key difference suggests that $[Ru(NH_3)_6]^{3+/2+}$ species are able to penetrate more readily into SAMs and diffuse along the SAM chains, in the presence (or even absence) of probable collapsed sites, as compared to $[Fe(CN)_6]^{3-/4-}$. This difference could cause

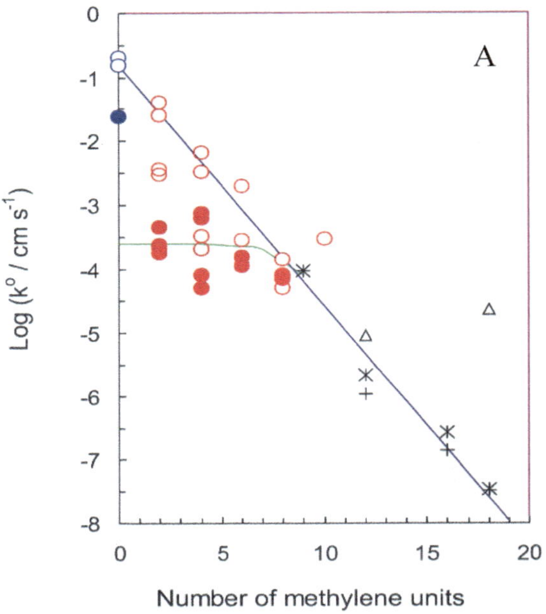

Figure 12. A) Logarithm of heterogeneous standard rate constants for the $Ru(NH_3)_6^{3+/2+}$ electron exchange at the SAM-coated electrodes with different methylene numbers. Data for the short-range ET in the absence (open circles) and presence (closed circles) of viscous additives (602 g/L glucose).[224] Asterisks represent data of Ref. 230 and triangles – of Ref. 231 and 232. Results for bare electrode (blue open and closed circles, respectively) are taken from Ref. 136. B) The logarithm of the experimental standard rate constant is plotted against the number of methylene units at different relative viscosities η_r. η_r = 1 are the filled diamonds (cyclic voltammetry method); η_r = 1.8 are the filled circles; η_r = 4.0 are the filled triangles; η_r = 11.25 are the filled squares;[47] The dashed line shows a linear extrapolation for the distance dependence of the n = 4, 6 and 8 rate constant data. Rerpinted with permission from Ref. 47, Copyright (2001) American Chemical Society.

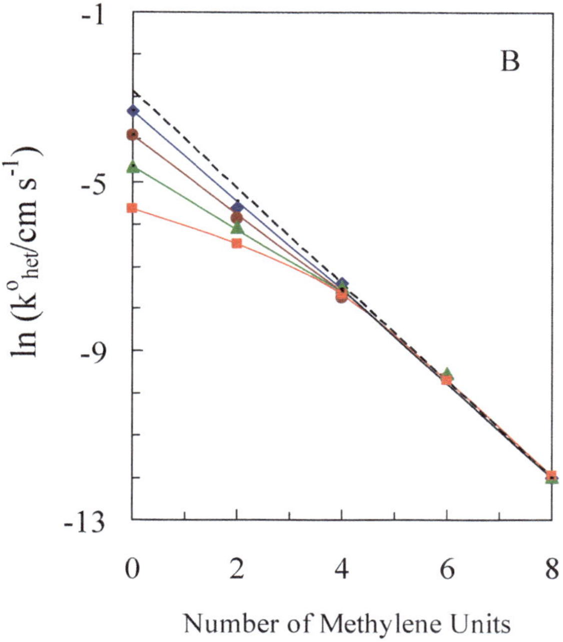

Figure 12. Continuation.

more variation in the effective charge-transfer distance, as compared to the *ideal* case with impermeable SAMs. In addition, we conjecture that the SAM defects can be roughly classified as *static* and *dynamic* ones, ascribing the former to different kinds of pinholes and collapsed sites, whereas the latter corresponds to increased mobility, and may or may not be associated with particular structural defects. We suppose that the *dynamic defects* have an essentially cooperative nature and contain many methylene chains. Consequently, the permeability of the CH_3-terminated alkanethiol SAMs with respect to the $Ru(NH_3)_6^{3+/2+}$ redox species (and, possibly a few molecules of solvating water and glucose), is believed to cause the remarkable scatter of experimental kinetic data.[224] In contrast, the data reproducibility was excellent for the $Fe(CN)_6^{3-/4-}$

couple with Au electrodes, both in the absence and presence of viscous additives.[47]

Figure 13 illustrates the arrangement of interfacial layers in the cases of the bare (panel A) and SAM-coated (panel B) Au electrodes with $Fe(CN)_6^{3-/4-}$ as the freely diffusing redox couple.[47] For Au electrodes coated by $-CH_3$ terminated *n*-alkanethiol SAMs with $n = 2, 4, 6$ and 8, the double-layer capacitance measurements under variable *n* indicated that, over the relevant potential range for the experiment, the electrolyte anions (Cl^-) are localized at the interface and these act to inhibit the penetration or adsorption by the $Fe(CN)_6^{3-/4-}$ to the surface, over the course of the experiment (*vide infra*). This feature leads to a specific variation of the effect-

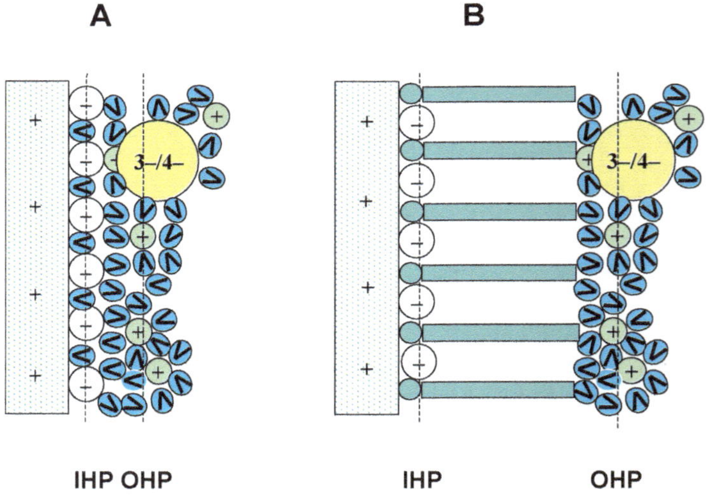

Figure 13. Interfacial region envisioned for the bare electrode (A) and the SAM-coated electrode (B), depicted as vertically aligned rectangles. The circles with the minus sign represent Cl^-; the circles with the plus sign represent K^+; the $Fe(CN)_6^{3-/4-}$ species are represented by a circle with '3-/4-' label; and the water solvent is illustrated by a circle with a 'v' symbol. The inner Helmholtz plane (IHP) and the outer Helmholtz plane (OHP) are indicated by the vertical dashed lines. The plain circles are S atoms and the alkane chains (represented by horizontally aligned rectangles). Rerpinted with permission from Ref. 47, Copyright (2001) American Chemical Society.

tive potential, Φ_r (Eq. 23) near the OHP, and a somewhat variable preequilibrium term, K_A, but whose change with n is estimable.[47]

The overall well behaved nature of this system allowed for the estimation of the relevant physical parameters and a more rigorous interpretation of the data, than is possible for $Ru(NH_3)_6^{3+/2+}$. In the following sub-sections we consider the electron exchange for freely diffusing $Fe(CN)_6^{3-/4-}$ (Ref. 47) in more detail and underscore its similarities to that of the findings for the immobilized model species.

3. Free $Fe(CN)_6^{3-/4-}$: Impact of SAM Thickness and Solution Viscosity

Cyclic voltammetry was used to assess the quality of the films, i.e., the defects and their development in time.[47] Figure 4a shows three voltammograms for SAM-coated Au electrodes (alkanethiols having methylene unit numbers of $n = 2$, 4, and 8) in contact with a KCl solution containing $Fe(CN)_6^{3-/4-}$. The change in the measured current (hence, the rate constants) and the shape of the voltammograms with the increase in the methylene number, n, indicates an increased hindrance to electron transfer. This observation is consistent with earlier data on SAM-coated electrodes with longer chains in which it was interpreted as a manifestation of the tunneling mechanism.[29-33,45,46] Table 1 presents the values of experimental (apparent uncorrected) standard rate constants, the corresponding nonadiabatic rate constants, the intrinsic charge transfer (corrected) rate constants, the double layer capacitance, and some calculated parameters (all in the absence of viscous additives). These values are the average rate constants obtained from a number of different electrodes and measurement sets.

The role of pinholes, collapsed sites and other possible defects (except for a pronounced transparency to Cl^- ions, see below) were deemed to be negligible for the following reasons:

a) all the current-potential curves, in all instrumental regimes applied, were well-behaved and repeatedly reproducible for 20–30 minutes, before the first distortions from SAM defects appeared;

Table 1

Experimental and Calculated Values of Different Kinetic and Intrinsic Energetic Parameters for the Electron Exchange Involving the Au(bare)/$Fe(CN)_6^{3-/4-}$ and Au(-S-$(CH_2)_{n-1}$-CH_3)/$Fe(CN)_6^{3-/4-}$ (n = 2, 4, 6, 8) Redox Systems.[47] Calculated Values are Given in Italics.

No.	Type of the electrode, mechanism	R_e, Å	$\ln(k^o_{ET})$, [cm s^{-1}]	λ, kJ mol^{-1}	H_{if}, cm^{-1}	ΔG_a^*, kJ mol^{-1}	$-\Phi_r$, V	$\ln(k^o_{ET})$, (corrected) [cm s^{-1}]	$-\ln(K_A)$, [cm]	$\ln(k^o_{ET})$, [s^{-1}]	C, μF cm^{-2}
1	Bare (n = 0) SF (observed)	*7.9*	–3.30	33.5	*~ 350*	*≈ 4.2*	0.125	–	*28.10*	*24.80*	35
2	Bare (n = 0) NA (hypoth.)	*4.3*	*1.40*	*57.3*	*1150*	*≈ 1.7*	0.125	*1.40*	*28.10*	*29.50*	35
3	Bare (n = 0) NA (hypoth.)	*7.9*	*–2.60*	*74.5*	*~ 350*	*14.6*	0.125	*–2.60*	*28.10*	*25.50*	35
4	SAM (n = 2) Med (observed)	*8.5*	–5.60	*76.1*	*63*	*18.4*	*0.065*	*–3.20*	*23.44*	*17.84*	33
5	SAM (n=4) NA (observed)	*10.7*	–7.40	*81.2*	*18*	*20.5*	*0.051*	*–5.70*	*22.36*	*15.00*	30
6	SAM (n=6) NA (observed)	*12.9*	–9.60	*84.5*	*5.0*	*21.3*	*0.043*	*–8.30*	*21.74*	*12.14*	25
7	SAM (n=8) NA (observed)	*15.2*	–12.0	*87.0*	*1.4*	*21.8*	*0.037*	*–10.80*	*21.28*	*9.30*	16

b) for the case of the shortest alkanethiol chain (*viz.*, $n = 2$), the voltammograms were peak-shaped, as in the case of the bare electrode. No sigmoidal curves, which are indicative of kinetically active pinhole-like defects,[28] were observed in this case (except for a few cases that were rejected); and

c) for SAMs with n equal to 6 and 8 no peak-shaped curves, which could be indicative of a large contribution from defect sites,[39] were observed (except for a few rejected cases).

For these reasons the electrochemical rate constants were deemed to be reliable and reproducible.

However, accessibility of the SAM-coated Au surface by electrolyte anions, such as Cl^-, does not necessarily imply the existence of pinholes or other structural defects. The adsorption may take place through well-oriented SAM domains because of dynamic fluctuations in the SAM. The adsorption is driven by the specific interaction of the Au surface with the electrolyte anions. Work of Porter et al.[23,28] indicates that easily desolvated ions like Cl^- can be adsorbed on a Au surface through alkanethiol SAMs with $n < 10$, while strongly solvated ions, like $F(H_2O)^-$ cannot (the latter is known also as a relatively poorly adsorbing ion on the bare metal surface.)[2,11-13] As a result, the compact part of the double layer for the bare and the SAM-coated ($n \leq 8$) Au electrodes may be comparable.[47]

Khoshtariya et al.[47] measured the heterogeneous standard rate constants as a function of solvent viscosity and the SAM film thickness by using both cyclic voltammetry (CV)[215-219] and steady-state,[217,218] including the rotating disk electrode (RDE), methods. Figure 14 shows how the apparent standard rate constant for electron exchange of the $Fe(CN)_6^{3-/4-}$ redox couple depends on the solution viscosity for bare and *n*-alkanethiol coated ($Au\text{-}S\text{-}(CH_2)_{n-1}\text{-}CH_3$; $n = 2, 4, 6, 8$ (see Ref. 47 for details of the experimental methods and analysis). From Fig. 14 it is evident that the two different methods give similar rate constants for the bare Au electrode. The rate constants show a strong dependence on the solution viscosity with a slope of $\delta = 0.96 \pm 0.04$, see Eq. (3), and this dependence becomes progressively weaker as the SAM thickness increases and the rate constant drops. The data for $n = 2$ have a slope of $\delta \approx 0.3$ and for the thicker SAMs ($n = 4, 6, 8$) no viscosity dependence is apparent; in these latter cases the nonadiabatic, tunneling, limit is reached. This limiting behavior is clear in Fig. 12B, in which the rate constants determined at all the viscosities are plotted versus the methylene number, *n*. Even without considering the real distance-dependent plot of the type presented above for immobilized species (Fig. 11B) and discussed below for free $Fe(CN)_6^{3-/4-}$, it is evident that a *plateau* like behavior occurs for these systems at an electrode-reactant (center) separation distance of ca. 8.5 Å. This result is qualitatively similar to the plateau-like behavior observed for the SAM-immobilized model redox species, see Figs. 10 and 11.

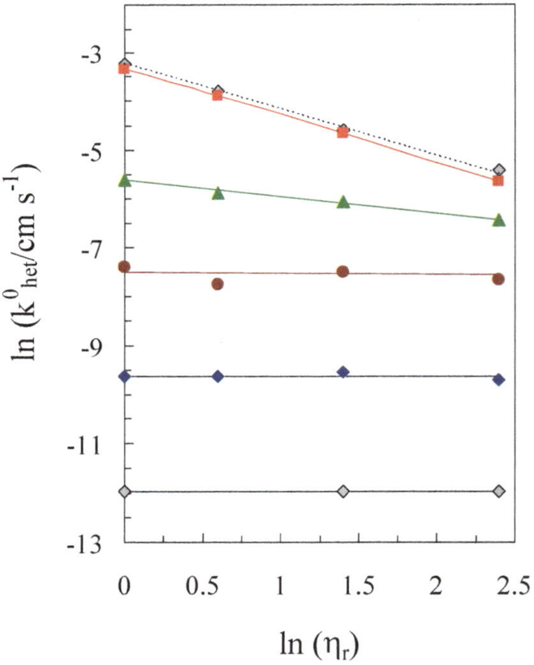

Figure 14. The viscosity dependence of the electron transfer rate constant is shown for the bare $Au/Fe(CN)_6^{3-/4-}$ (red squares circles, rotating disk electrode method and grey diamonds, cyclic voltammetry method) and SAM-coated electrodes ($Au[-S-(CH_2)_{n-1}-CH_3]$ $/Fe(CN)_6^{3-/4-}$ with $n = 2$ green triangles, $n = 4$ brown circles, $n = 6$ blue diamonds, $n = 8$ grey diamonds).[47] Rerpinted with permission from Ref. 47, Copyright (2001) American Chemical Society.

4. Free $Fe(CN)_6^{3-/4-}$: Impact of the Variable Reorganization Energy, Preequilibrium, and Electronic Coupling

A quantitative understanding of how the reorganization energy, λ, changes with SAM thickness is important for a quantitative understanding of the viscosity effect. As is the case for the ferrocene couple, the contribution of the inner-sphere reorganization energy $\lambda_{(IS)}$ to the total reorganization energy is small for the $Fe(CN)_6^{3-/4-}$ couple (ca. 5.9±0.8 kJ/mol[33,135]), so that $\lambda \sim \lambda_{(OS)}$. For thick SAMs the experimental reorganization energy is in reasonable semiquantitative agreement with the models of Marcus[1,2] and of Liu and Newton,[53,150] however for thin films they differ substantially. An experimental value of $\lambda \sim 96$ kJ mol^{-1} (corresponding to $\lambda_{(OS)} \sim 90.0$ kJ mol^{-1}) was extracted from the Tafel-like current-voltage curve obtained for Au electrodes that were coated with a -S-$(CH_2)_{16}$-OH SAM ($R_e \approx 20$ Å).[33] This experimental value should be compared to the value calculated from the model of Liu and Newton ($\lambda_{(OS)} \approx 73.2$ kJ mol^{-1}) and gives an error of 19%. Given the uncertainties in defining molecular radii, distances, and dielectric constants, this agreement is considered reasonable and illustrates the models predictive power. As discussed by Khoshtariya et al[47] for the ferri/ferrocyanide case, the image charge model causes the reorganization energy to decrease at the closest electrode-reactant approach by 20% to 30% with distance. For this system, one predicts a value of 57.3 kJ/mol whereas the measured value (for bare electrode, determined by three different procedures)[135,233,234] is 33.5 kJ/mol. This barrier lowering has been explained as a freezing out of the Marcus-like energy barrier by the friction controlled mechanism[47] in accord with the prediction by Zusman[89,94] for cases with sufficiently high H_{if}. This lowering of the activation free energy is in addition to the trivial one, predicted through Eq. (6).[1-7] The occurrence of a remarkably high H_{if} for $Fe(CN)_6^{3-/4-}$ at bare electrodes can be rationalized by an efficient superexchange mechanism involving specifically adsorbed Cl^- ions and π-conjugated CN^- ligands of the complex.[235-239]

Khoshtariya et al.[47] estimated the value of the effective potential, Φ_r, near the OHP and it is plotted as a function of the ET distance, R_e, in Fig. 15a. These estimates and the ansatz that $Z_{eff,(Ox)} \approx Z_{eff,(Red)} \approx -2$ (see Refs. 135 and 168 for the relevant argumentation)

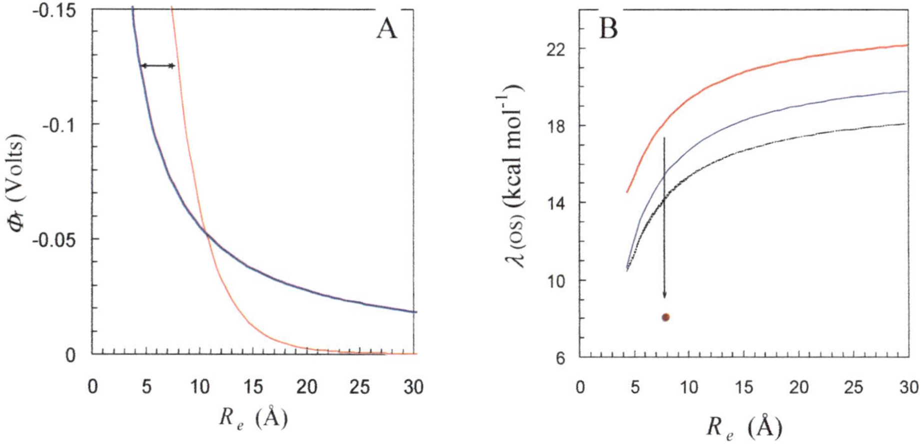

Figure 15. (A) The thick blue curve gives the value of the Φ_r potential at the OHP for the *n*-alkanethiol coated electrodes (according to Eq. 23). These values were used for the rate constant corrections given in Figure 16. For illustrative purposes, the calculated Φ_r potential is shown as a function of the electrode-reactant separation distance for the bare electrode (red line). The horizontal bar indicates the distance shift corresponding to the Cl^- diameter.[47] (B) The curves show the dependence of the outer-sphere reorganization free energy on the electrode-reactant distance: according to the model of Liu and Newton[150] (Eq. 16; black bottom curve), Marcus model[1] (Eq. 15, middle blue curve), and the rescaled curve with $\lambda_{(OS)} \sim 90.0$ kJ mol^{-1} at $R_e \approx 20$ Å[33] (top red curve). The arrow points to the experimental value for a bare electrode (filled brown circle).[47] Rerpinted with permission from Ref. 47, Copyright (2001) American Chemical Society.

allowed for the estimation of the variation of K_A with distance (according to Eq. 23). Together with the distance dependence of $\lambda_{(OS)}$ (estimates are shown in Fig. 15b) and of the experimental rates k^o_{het}, it is possible to estimate the variation of H_{if} with distance (including the parameters H^o_{if} and β (Eqs. 1 and 9).[47]

Figure 16 plots the data of Figure 12B (experimental points in filled symbols) against an electrode-reactant separation distance. The effective thickness of the *n*-alkanethiol chain length was determined according to Ref. 150 (see also Section VI) and a reactant radius of 4.3 Å. The entire range of experimental data can be fit by using the unified expression for the rate constant, Eq. (5), which interpolates smoothly between the two limiting regimes, tunneling control and friction control. As for $k^o_{ET(DC)}$, the experimental values obtained at different viscosities on the bare electrode were directly used for this plot; the ΔG^*_r and Φ_r corrections compensate each other and the preexponential term depends only weakly on H_{if} in the strong coupling regime.

The electronic coupling parameter H_{if} (the magnitude of which prevailingly determines the mechanism turnover) at different electrode-reactant separation distances was estimated from the rate constants obtained with SAM-coated electrodes at $n = 4$, 6 and 8, because they are viscosity independent and display the pure tunneling (nonadiabatic) mechanism. In Fig. 16, the lower three points (filled circles, thin line) correspond to these three experimental rate constants, corrected for the distance-dependent reactive site (the OHP) potential Φ_r.[47] The upper three points (filled circles, thin line) represent the same rate constants that would be obtained for a distance independent reorganization free energy of $\lambda_{(OS)} = 66.5$ kJ mol^{-1}, but not correcting for the distance dependence of the potential Φ_r. The three points given by crosses situated above the experimental points represent the rate constants that are corrected for the distance dependence of both the double layer potential and the reorganization energy; the straight line extrapolates these data toward zero separation distance. It is clear that these two factors give corrections to the experimental data that almost compensate one another. The straight line that connects the crosses and extrapolates to the minimum separation distance, $R_o \approx$ 4.3 Å, illustrates the variation of $(H_{if})^2$ (more precisely, the preex-

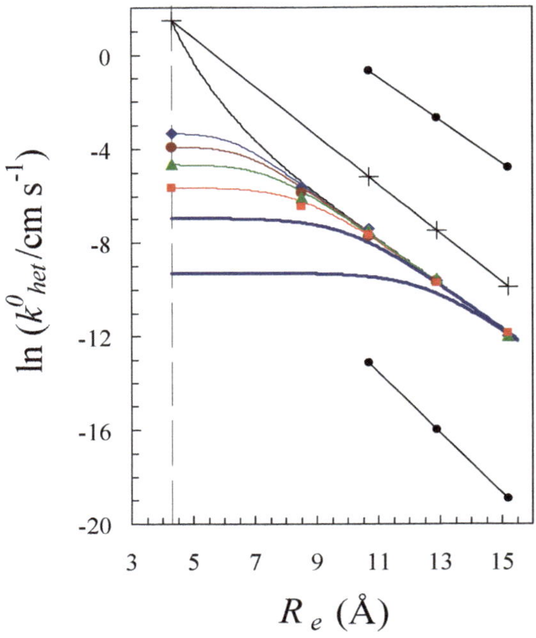

Figure 16. The data of Figure 12B plotted against the estimated electrode-reactant separation distance (see text for the estimation details). Simulated *resulting* curves for the nonadiabatic (Eq. 8) rate constants (thinner curve with the upward curvature), scaled preexponential factors (crosses and bold straight line), and hypothetical resultant curves for viscous media (50% and 100% glycerol) as the reaction medium (bold blue curves) are also drawn for comparison. Note the shift of transition regions for simulated (experiment) and hypothetical (50% and 100% glycerol) resultant curves to higher R_e with the increase of solution viscosity that is reminiscent of the distance dependence for protein ET (see Sections V–VIII below). Adapted with permission from Ref. 47, Copyright (2001) American Chemical Society.

ponential component of Eq. 8, scaled by K_A, Eq. 23, versus R_e). The distance dependence of $k^o_{ET(NA)}$ that is hypothetical (based solely on calculations) for thinner SAMs, is represented by the thin line curved upward.

The distance dependence extracted from the long straight line in Fig. 16, corresponding to $(H_{if})^2$, gives a value of $\beta = 1.04 \pm 0.05$ Å, which is in good agreement with theoretical estimates and most experimental values for methylene chains.[1,5,18,23-46,51-53,57-63,66-79] Table 1 reports the values of H_{if} at different electrode-reactant separation distances (at $n = 0, 2, 4, 6$, and 8). The value of $H_{if} = 0.017$ eV (about 140 cm^{-1}) for the $n = 2$ SAM is close to the value specified previously (≥ 0.030 eV or 245 cm^{-1}) for the onset of the solvent friction regime,[123,128,240] and is consistent with a turnover between the strong and weak coupling limits. At some minimum distance the extrapolated line should correspond to the rate constant of a hypothetical nonadiabatic electron transfer process at the close contact of the reactant ion and the bare Au electrode, with $R_e = 4.3$ Å. The extrapolated rate constant value of $k^{o}_{ET(NA)} = 4.3$ cm s^{-1} gives a value of $H^{0}_{if} = 0.152$ eV (1224 cm^{-1}) for values of $\lambda = 66.5$ kJ mol^{-1} (see above) and $\rho_m = 0.52$ meV.[47]

Figure 16 shows that the transition between the nonadiabatic and solvent friction regimes occurs between $R_e = 6$ and 9 Å. Taking this changeover to occur for $g = 1$ and using the values above, ν_{eff} can be calculated from Eq. 7; this calculation gives 2.4×10^{12} s^{-1} (if one uses a value of $\varepsilon_s^{W} \sim 20$ for the diffuse part of the double layer near the OHP at the bulk concentration of 1 M KCl and a Debye relaxation time of $\tau_D = 8.5$ ps, then one finds $\nu_{eff} = \nu_L = (\varepsilon_S/\varepsilon_\infty)\,\tau_D^{-1} = 4.7 \times 10^{11}\ s^{-1}$ (estimated for $\varepsilon_\infty = 5$, see Ref. 47 for the detailed discussion of this issue). The latter value is about four times smaller than that accepted for pure water ($1.9 \times 10^{12}\ s^{-1}$ at $\varepsilon_S = 78$),[128-130] however it falls within a reasonable range. Although the estimate used here is rough, the limiting value for ν_{eff} is plausible and about an order of magnitude larger than that deduced from the study of the tethered ferrocene redox couples discussed in Section 4.1. This difference could arise from the different influence of the SAM-adjacent solvating water/electrolyte in the two cases: much more hindered motion for more closely situated immobilized versus presumably solvent layer separated free redox species (partial immersion into SAMs, *vide supra*). The difference among the estimates made here and those made in Section 4.1 underscore the difficulty in quantifying the relaxation characteristics of these complex systems.

In summary, these systematic studies of electrochemical ET for SAM assemblies and small model redox-active species (metallocomplexes, in most cases) provide strong evidence for at least two extreme ET mechanisms, a nonadiabatic (tunneling) mechanism for thick films and a friction controlled (dynamically controlled) mechanism for thin films.[47,53] Unlike the nonadiabatic (tunneling) mechanism that displays a well recognized experimental signature, namely an exponential decay of the rate constant with distance, the friction-controlled mechanism is more difficult to recognize. To identify the friction-controlled regime requires experiments that test for a dependence on ν_{eff} (or η) (see Figs. 12, 16, and Refs. 47 and 224) and that explore whether the rate controlling step in the reaction is ET or not.[164-166,215-220] Figure 16 depicts a distance dependence for the k^o_{ET} of a freely diffusing redox couple which displays a viscosity dependent plateau region whose extent is dependent upon the solution's relaxation properties. These findings suggest that the dynamically controlled mechanism may be more pronounced in slowly relaxing media such as biomolecules, ionic liquids or liquid crystals. In the next Sections we review experimental results for proteins at the surface of electrode/SAM assemblies that display this same general behavior. Indeed, the interior and the SAM- and solvent-adjacent layers of proteins can be viewed as an extremely viscous media with characteristic dielectric relaxation times similar to pure glycerol.[106-111,155-163] Another interesting development of this issue was realized in electrochemical studies involving room temperature ionic liquids (RTIL) as ET media,[169,185-187] and while they are not reviewed here electrochemical studies in RTILs may offer new perspectives, especially in combination with biomolecules. The insights gained from the small model redox species studied above have proved extremely useful for developing a proper mechanistic understanding of more complex bioelectrochemical systems.

V. FREELY DIFFUSING CYTOCHROME *C* AT Au/SAM JUNCTIONS

1. The Overview

Despite its modest size and relatively *simple* function, the mechanistic understanding of the molecular recognition and charge-transfer functionality of cytochrome *c* (CytC) and the whole c-type cytochrome family remains incomplete.[241-243] While it is well known that CytC contains a covalently integrated iron heme moiety and shuttles electrons in cellular photosynthetic and/or respiratory systems of living cells,[241-243] the recently discovered role of CytC in cell apoptosis[244,245] indicates that it may have a much higher functional and mechanistic complexity than previously thought. In part, a full understanding of CytC's functional role remains elusive because of its highly diverse intermolecular interactions. Indeed, much evidence indicates that at least some well-studied cytochromes, including CytC (or its modified variants), operate either in a strongly (irreversibly) bound, loosely bound (almost freely diffusing), or in both of these regimes depending on the specific intracellular (*in vivo*) or experimental conditions.[246-258] Furthermore, these diverse interactions may control the proteins' essential properties (including structural, stability and dynamic characteristics) and, hence, its intrinsic redox mechanisms and their interplay.[250-258] The latter aspect, in turn, is closely connected with an old puzzle that concerns the thermodynamic stability of globular proteins,[260-272] as well as the more recent issue of the hierarchical coupling between a protein's external and internal dynamic properties, *dynamic slaving*.[155-161] These issues are synergistically connected and require a diversity of experimental studies and cross-testing strategies (e.g., combining kinetic and thermodynamic, and/or different kinetic, *viz*., high-pressure and solvent viscosity approaches) to elucidate.

Experimental information on the ET kinetics in liquid-phase systems involving CytC and its native or artificial redox partners (redox proteins, complex ions, etc.) is extensive (see e.g., Refs. 111-114 and 273-278). However, even for this well-characterized protein, systematic mechanistic studies in *homogeneous* systems are complicated by the impact of both molecular recognition and the intrinsic ET mechanism on the observations. In addition, an extra complexity (structural and dynamic) is introduced by the participating redox partner and its intrinsic ET characteristics.[112-114,273-278] While the strategy of covalent attachment of *small* complex ions as redox counterparts at different external sites of CytC[273-278] has generated significant insights, it is limited because it does not probe molecular recognition features and it does not provide a smooth variation of intrinsic parameters, such as the electronic coupling (because of the highly inhomogeneous nature of the protein.)[1-3] Meantime, as already mentioned in the Introduction, artificial bioelectrochemical devices that are composed of Au/SAM assemblies of variable composition and thickness, with redox proteins irreversibly attached or freely diffusing to the SAM/electrolyte interface,[16-22,54-83,279] are free of these limitations.

The work discussed in this section, and Sections VI and VII, illustrates the advantages of bioelectrochemical methods for understanding aspects of CytC's ET kinetics. This section discusses electrode/SAM assemblies in contact with an electrolyte solution that contains freely diffusing CytC. Sections VI and VII describe systematic kinetic studies for the irreversibly attached CytC. While some earlier electrochemical work on freely diffusing CytC[56,57,111] set the stage for more comprehensive bioelectrochemical studies, they did not exploit synergies between different kinetic approaches (e.g., combined variation of SAM thickness, hydrostatic pressure and solution viscosity) that provide complementary mechanistic information. More recent systematic bioelectrochemical kinetic studies of irreversibly adsorbed CytC varied the electronic coupling strength over a broad range and allowed for the observation of a mechanistic change from a non-adiabatic (weak electronic coupling, thick SAMs) limit to an apparent adiabatic (thin SAMs) limit (*vide infra*).[74,76,78,82] In this Section we discuss complementary kinetic studies of electrochemical ET for freely diffusing CytC. Through the combined variation of the SAM thickness, solution viscosity, and hydrostatic pressure[80,81] these studies offer new insights into the ET mechanism and the issue of dynamic slaving.[155-157] We note that the same work[81] used differential scanning calorimetry (DSC) to address the impact of stabilizing (e.g., glucose) and destabilizing (e.g., urea) additives on CytC and identify links between the protein's bioelectrochemical performance and its stability, conformational flexibility, and function (for related studies see also Refs. 279-284). However, these latter studies are outside the scope of the present Chapter.

2. Impact of the ET Distance (Electronic Coupling)

The voltammetry of cytochrome *c* at SAM-coated gold electrodes, for which the SAM presents a hydroxyl functionality to the solution, are well-behaved.[80,81] Figure 3B shows voltammograms for cytochrome *c* for different SAM thicknesses, whereas Fig. 17 show voltammograms for cytochrome *c* for different voltage scan rates (panel a), and different pressures (panel b). Panel a shows voltammograms for three different SAM thicknesses at the same scan rate and ambient pressure, revealing that the thicker SAMs have slower ET rates. In these data the peak current is proportional to the square root of the scan rate, as expected for a freely diffusing redox couple.[80,81] Other aspects of the voltammograms are consistent with a freely diffusing redox couple as well.

The dependence of the ET rate on the SAM thickness and the solution viscosity shows a behavior that is qualitatively similar to that observed for the metallocomplex redox couples described in Section IV, however with a more obvious plateau region. Figure 18 presents a logarithmic plot of the experimental standard heterogeneous rate constant, k^{0}_{het}, versus the number of SAM methylene units (see also Table 2) for three different solution viscosities. The upper plot, $\eta_r = 1$, presents the data obtained for a 500 mM Tris-HCl buffer (pH 7.4) solution, whereas the two lower curves have glucose added to the buffer to increase its viscosity (200 and 400 gL^{-1} glucose). For the thick SAMs with $n = 6$ and 11, the rate constants are the same (within experimental error) in each of the solutions, indicating that it is independent of the solution viscosity and that the glucose does not cause significant structural changes of the protein. In this regime the data can be described by an exponential distance decay (Eq. 1) with a slope of *ca.* 1 per CH_2 unit, in excellent agreement with other methylene SAM studies (see Sections II and IV of this Chapter). For thinner SAMs (methylene number 2 and 3), the rate constant data reveal a clear dependence on the solution viscosity and a deviation from the exponential distance dependence that is expected for an electron tunneling mechanism. Although this plateau-like region only spans two CH_2 units, it is well pronounced. We note that Miller et al. indicated essentially

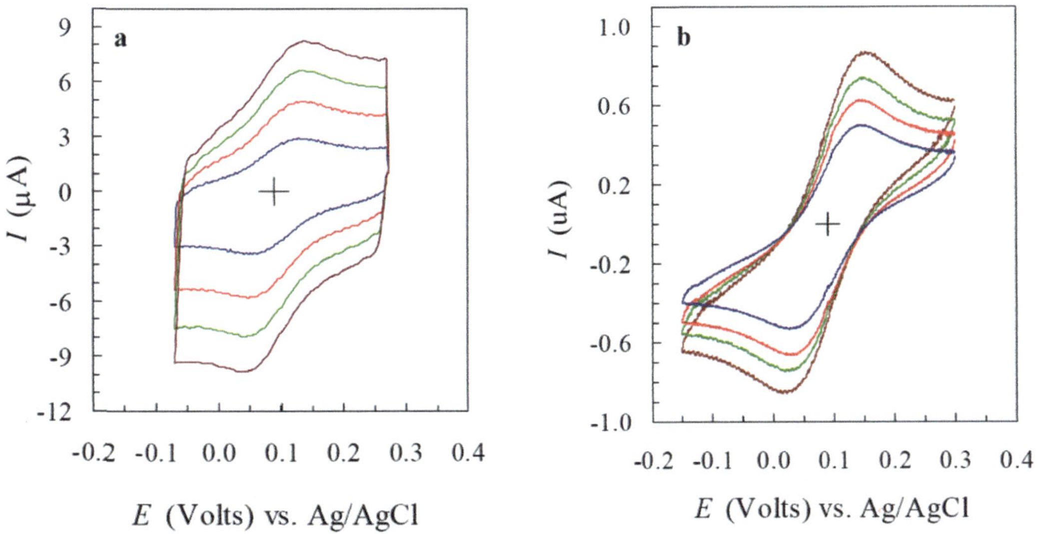

Figure 17. (a) Voltammograms for CytC electron exchange at Au electrodes modified by hydroxy-terminated n-alkanethiol SAM, $n = 2$, $P = 150$ MPa, scan rates 1, 2, 3 and 4 V s^{-1} (peak uprising); (b) Voltammograms for CytC electron exchange at Au electrodes modified by hydroxy-terminated *n*-alkanethiol SAM, $n = 6$, $P = 150$ MPa, scan rates 0.03, 0.05, 0.07 and 0.1 V s^{-1} (peak uprising). CytC: 5 mg ml^{-1}, 0.5 M Tris-HCl buffer (pH 7.4). Ref. 81, Copyright (2006) Wiley-VCH Verlag GmbH & Co KgaA. Reproduced with permission.

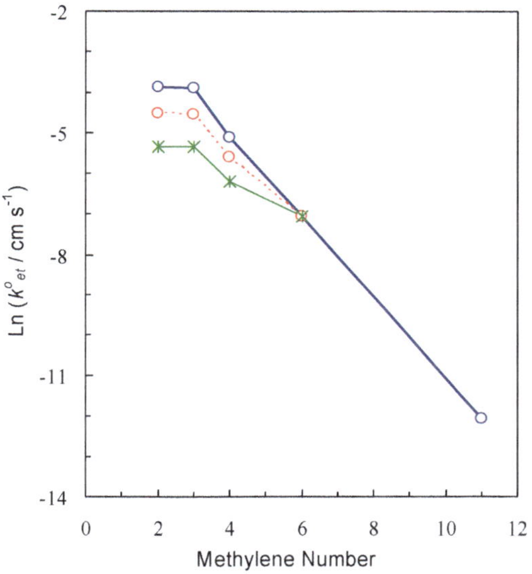

Figure 18. Logarithm of the heterogeneous standard rate constant for CytC electron exchange at Au electrodes modified by hydroxy-terminated *n*-alkanethiol SAMs ($n = 2,3,4,6,11$) vs. methylene unit number under variable solution viscosity. Upper curve (circles): no viscous additive, $\eta_r = 1$; middle curve (open circles), $\eta_r = 1.80$; lower curve (asterisks) $\eta_r = 3.96$ (note all three curves merge at $n = 6$ and 11, and form a single sloped line). Ref. 80 and 81, Copyright (2006) Wiley-VCH Verlag GmbH & Co KgaA. Reproduced with permission.

the same result, albeit incomplete, in their pioneering report.[56] The extent of the plateau region is similar to that observed for the ferricyanide/ferrocyanide redox couple; however, we will see that it is notably shorter than that observed for immobilized CytC (Sections 6 and 7). The points for n = 4 apparently fall in an intermediate region and is discussed more below.

Table 2

Values of Diffusion Coefficients and Heterogeneous Standard Rate Constants for Cytochrome *C* Electron Exchange at Au Electrodes Modified by Hydroxy-Terminated *n*-Alkanethiol SAMs (n = 2, 3, 4, 6, 11) Under Variable Solution Viscosity. The Values Are Averaged Over 3 to 5 Independent Experiments with a Maximum Error Within 5 %.[81] Ref. 81, Copyright (2006) Wiley-VCH Verlag GmbH & Co KgaA. Reproduced with permission.

[Glucose], gL^{-1} (M)	Relative Viscosity (η_r)	D_O x 10^7, $cm^2\ s^{-1}$	k^o_{ET}, cm s^{-1}					
			n = 2 (Tris)	n = 3 (Tris)	n = 4 (Tris)	n = 6 (Tris)	n = 11 (Tris)	n = 3 (Phosph.)
0 (0)	1	8.02	$2.065x10^{-2}$	$2.00x10^{-2}$	$6.0x10^{-3}$	$8.6x10^{-4}$	$5.76x10^{-6}$	$2.77x10^{-2}$
100 (0.56)	1.33	5.98	–	$1.45x10^{-2}$	$4.8x10^{-3}$	–	–	$1.90x10^{-2}$
200 (1.12)	1.80	4.91	$1.075x10^{-2}$	$1.00x10^{-2}$	$3.7x10^{-3}$	$8.4x10^{-4}$	–	$1.44x10^{-2}$
300 (1.67)	2.63	2.95	–	$6.99x10^{-3}$	$2.8x10^{-3}$	–	–	$1.10x10^{-2}$
400 (2.24)	3.96	1.84	$4.708x10^{-3}$	$4.77x10^{-3}$	$2.0x10^{-3}$	$8.6x10^{-4}$	$5.82x10^{-6}$	$6.79x10^{-3}$

It is natural to interpret the biphasic behavior of the ln(k^o_{het}) versus n plot (Fig. 9) in terms of a mechanism change; i.e., while electron tunneling is rate limiting for thick SAMs some other mechanism (or other process) becomes rate limiting for thin SAMs. The origin of the exponential distance dependent region at high values of n in terms of a nonadiabatic ET mechanism[1,5,18,23-46,51-53,57-63,66-79] appears to be generally accepted. In contrast, the mechanistic origin of the *plateau-like* region for the thin SAMs has had a number of different explanations offered. For example, the unified ET model[5,76,80-83,89-98] described in Section 2 identifies the plateau region with a rate limiting step that corresponds to the polarization relaxation rate of the surrounding medium; i.e. a friction-controlled mechanism. This model implies control of the ET by the protein and/or the SAM assemblies' nuclear rearrangements that are intrinsically coupled to the ET in one inseparable (bio)electrochemical act. An alternative interpretation, which is frequently discussed, considers a conformational rearrangement (which is viscosity-sensitive) that is distinct from the ET event but is required to precede it, and to involve an essentially different elementary barrier-crossing.[58,59,66,67,100-104] Experiments that explore the nature of the ET rate constant more rigorously have been performed and used to distinguish between these mechanisms.[74,76,78,80-83]

3. Impact of Hydrostatic Pressure

Semi-logarithmic plots of k^o_{het} versus the hydrostatic pressure (up to 150 MPa) for SAMs with $n = 3$, 4 and 6, are presented in Fig. 19. It is evident that the pressure dependence is qualitatively different for these three cases. For the SAM with $n = 3$, which is in the plateau region, the ln(k^o_{het}) decreases linearly with pressure, yielding a positive volume of activation (+6.7 ± 0.5 cm^3 mol^{-1} via Eqs. 3, 10, 11, and 22). This value is similar to the +6.1 ± 0.5 cm^3 mol^{-1} value that was reported for 4,4'-bipyridyl- and 4,4'-bipirydyl-disulfide modified Au electrodes (to be also considered as thin, although *primitive* SAMs).[111] In contrast, the plot for the SAM with $n = 6$, which is in the tunneling/nonadiabatic limit, has a negative volume of activation (−5.5 ± 0.5 cm^3 mol^{-1}); see Eqs. (1), (8), (9), and (21), and Fig. 19. In the case of a SAM with $n = 4$, an intermediate behavior is displayed with no effect of pressure on k^o_{het}, yielding $\Delta V_a \approx 0$, Fig. 19, Table 3. Thus, the same ET reaction, i.e., well-behaved oxidation/reduction of cytochrome *c*, displays an essentially different behavior as a result of the change in the electronic coupling parameter, which was varied by changing the SAM thickness.

In order to fully understand the pressure effect in the different ET regimes, corresponding to the different SAM thicknesses, it is important to consider the change in the reorganization energy with pressure and the possible change in the tunneling distance with pressure, in addition to the change in the intrinsic protein viscosity.

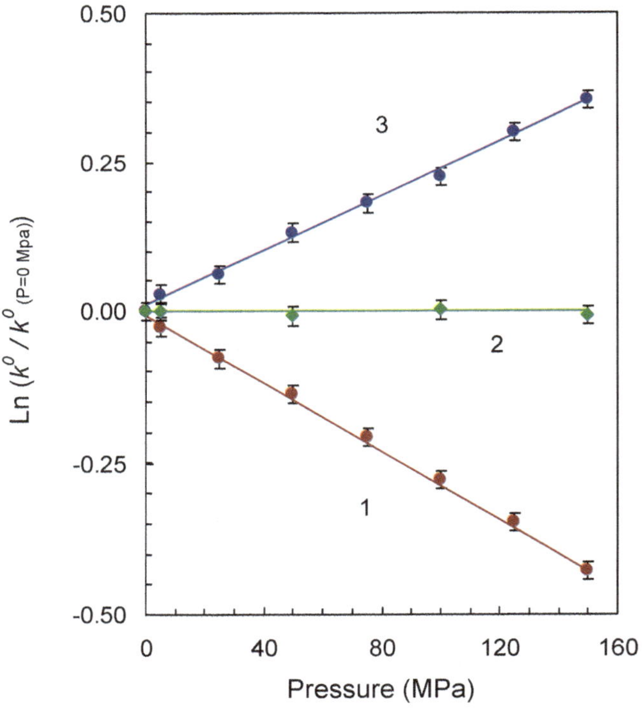

Figure 19. Logarithm of the rate constant (reduced to its standard value) for the CytC electron exchange at Au electrodes modified by hydroxy-terminated *n*-alkanethiol SAMs ($n = 3, 4, 6$) vs. hydrostatic pressure, (1) $n = 3$; (2) $n = 4$; (3) $n = 6$.[80,81] Refs. 80 and 81, Copyright (2006) Wiley-VCH Verlag GmbH & Co KgaA. Reproduced with permission.

Table 3
Experimental Volumes of Activation (± 0.5, in cm^3 mol^{-1}) for CytC Electron Exchange at Au Electrodes Modified by Hydroxy-Terminated *n*-Alkanethiol SAMs (*n* = 3, 4, 6) Under Variable Hydrostatic Pressure, and Estimated Values of Different Contributing Terms (in the Same Units).[80,81] Ref. 80 and 81, Copyright (2006) Wiley-VCH Verlag GmbH & Co KgaA. Reproduced with permission.

SAM	$\Delta V_{a(exp)}$	$1/4\ (\partial\lambda/\partial P)_T$	$RT\ (\partial \ln\eta/\partial P)_T$	$\beta RT\ (\partial R_e/\partial P)_T$
$n = 3$	+ 6.7	– (2 to 3)	+ (8 to 10)	(0)
$n = 6$	– 5.5	– (2 to 3)	(0)	– (2 to 3)
$n = 4$	~ 0	– (2 to 3)	+ (4 to 5)	– (1 to 2)

The outer-sphere reorganization energy changes with pressure because the bulk dielectric properties of the medium, especially the water, change with pressure. In the case of cytochrome *c,* the overall value of λ is about 0.6 eV, and it is roughly composed of *ca.* 50 % solution component that can contribute about half of the activation volume, and another *ca.* 50 % of the protein reorganization component (Franck-Condon term, see Eq. 18 in Section II) found to be negligibly altered by pressure, and therefore does not contribute significantly to the activation volume.[181] Note that the same can be concluded for the contribution from the SAM interior (λ_{SAM} ~ 0), since it can be assumed that $\varepsilon_s = \varepsilon_{\infty}$, *vide supra.* Thus, the most probable overall contribution of the second term of Eq. 21 would be -2 to -3 cm^3 mol^{-1}, arising from the bulk dielectric properties of the aqueous electrolyte. This value, together with the overall experimental value of +6.7 cm^3 mol^{-1}, yields a value for the frictional term of about +8 to +10 cm^3 mol^{-1} (Table 3). We stress again that this contribution originates from the direct influence of hydrostatic pressure on the intrinsic protein friction (viscosity).

By considering the rate data in the nonadiabatic limit ($n = 6$) and taking it to be viscosity independent (see Figs. 18 and 20), we can estimate how much the pressure modifies the electron transfer distance. Using Eq. (21) and the overall activation volume of –5.5 cm^3 mol^{-1}, gives an activation volume of ca. –2 to –3 cm^3 mol^{-1} for the SAM (see Table 3). This value corresponds to a change in dis-

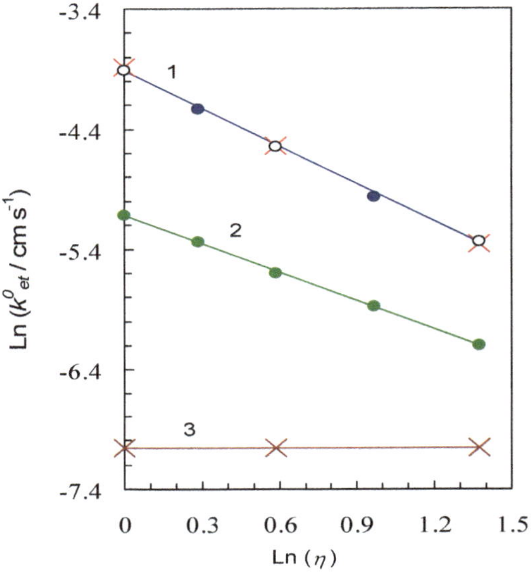

Figure 20. Logarithm of the heterogeneous standard rate constant for CytC electron exchange at Au electrodes modified by hydroxy-terminated *n*-alkanethiol SAMs *versus* logarithm of the solution viscosity. Plot 1: two coinciding plots for $n = 2$ (crosses) and $n = 3$ (open circles); plot 2: $n = 4$ (filled circles); plot 3: $n = 6$ (crosses).[80,81] Refd. 80 and 81, Copyright (2006) Wiley-VCH Verlag GmbH & Co KgaA. Reproduced with permission.

tance of a few hundredths of an angstrom. Table 3 summarizes the ranges of estimated values for the individual contributing factors to the experimental ΔV_a for the rate data in Fig. 19. These estimates are in good mutual agreement and consistent with the results obtained through other approaches.

4. Links between the Solution (External) vs. Intra-Globular Viscosity

It is useful at this point to consider the ET friction control in terms of a viscosity model. Figure 20 shows log-log plots of the rate data versus viscosity for three different hydroxyl terminated SAMs in contact with a cytochrome *c* solution. The slopes of the plots for $n = 2$ and 3 indicate that *full* frictional control is realized ($\delta \approx 1$, Eqs. 3, 10 and 11); whereas the plot for $n = 6$ is independent of the viscosity. In general, one finds that a description of the viscosity dependence by Eq. (3) usually gives $0 < \delta < 1$.[76,106-109,112-114,246,247] For example, a *maximum* value of $\delta \approx 0.6$ was reported for the case of specific adsorption of cytochrome *c* on pyridine-terminated SAMs (see section 7).[76] Somewhat larger slopes have been reported for the case of ET within the *homogeneous* (solution) system involving zinc-substituted CytC and cupriplastocyanins, ranging from 0.7 to 0.9.[112-114] Interestingly, full viscosity control ($\delta \approx 1$) has been observed for photo-induced ET from the artificial Ru-coordinated polypeptide electron donor to ferri-cytochrome *c*,[247] occurring through a loosely bound (encounter) complex, as opposed to a weaker dependence ($\delta \approx 0.6$) for a *preformed* (tight) complex.[247] These latter findings are similar to the overall pattern seen with cytochrome *c* in bioelectrochemical studies. In addition, recent experimental and computational results obtained for a related system, the *charge-modified* complex of the photosynthetic reaction center (PRC) and cytochrome c_2 [248,249] strongly suggest formation of a solvent-separated, softly stabilized complex as a reactive encounter pair.

Early work suggested that the overall *effective* friction additively originates from both solvent and protein components, and operates through Eq. (32),[200,201]

$$k_{ET} = \frac{C}{\sigma + \eta} exp\left(-\frac{\Delta G_a}{RT} \right) \tag{32}$$

in which C is a constant, and σ represents the protein's intrinsic viscosity; however, this form does not consistently account for the experimental observations. For cases where $\delta \approx 0.5$–0.8, the external (solution) and protein interior contribute comparably (the ratio σ / η amounts only to ca. 2 to 4), which is highly questionable because the protein interior is known to be much more viscous compared to bulk water, or even compared to aqueous solutions containing conventional viscous additives.[106-111,155-163,285-290] For the cases where $\delta \approx 1$ (as in the present case), this model suggests that $\sigma << \eta$, which contradicts the existing viewpoint about the protein's internal dynamic properties and does not agree with the high pressure results (discussed above) that convincingly reveal an essential role for the protein's intrinsic friction (for further criticism of Eq. 32, see Ref. 155). In this context, the earlier work that estimated the effective relaxation time, which controls the ET of cytochrome *c,* to be about 200 ps should be mentioned.[76,163] This rather slow relaxation time roughly corresponds to about a thousand-fold enhancement over the bulk aqueous environment.

The question emerges as to how a moderate variation of external viscosity can alter the much higher internal viscosity of the protein so effectively? The answer can be found in a recently introduced concept of dynamic slaving,[155-157] which implies that "solvent fluctuations dominate protein dynamics and function".[156] Indeed, for another well-studied *model* protein, myoglobin, it has been established that several of its dynamic features (including the rate constants of functional elementary processes, $k_F(T)$) change in parallel with the rate coefficient of the solution's dielectric fluctuations, $k_D(T)$, such that:[156]

$$k_D(T)/k_F(T) \approx m = Const \tag{33}$$

where, for different events, m varies over the range of 3×10^2 to 3×10^4. The origin of this effect may lie in a cooperative multi-site interaction of the solvent components (water, sugar, glycerol, etc.) with charged and polar groups on the protein surface.[158,159,201,206,272] Indeed, numerous studies have indicated the essential role of protein solvation in triggering and further tuning of their dynamic characteristics and functions.[155-161,200,201,204-206] Our finding that cytochrome *c* exhibits the high frictional coupling limit ($\delta = 1$) under the present conditions (weak interaction with SAM terminal groups) as compared to the case where it is (specifically) tightly adsorbed at the surface ($\delta \approx 0.6$, see Ref. 76) suggests that when the external solution is screened by a tight protein/SAM contact (that mimics the protein-membrane or protein-protein multi-site contact), the slaving effect is not complete because of the lack of the full protein-solution interaction.

The complementary calorimetric data[76] have revealed the thermodynamic stabilizing effect of glucose on the native conformation of cytochrome *c* and correlates it with the friction (viscosity) control in the dynamically controlled ET regime (manifested through Eq. 10). This observation is in remarkable agreement with the recent result of Chalikian et al. for the stabilizing effect of pressure on cytochrome *c*[289] and can also be linked to the dynamic intra-globular friction control in the same regime. Further analogy between the pressure and stabilizer dynamic effects comes from the *static* effect of the volume and compressibility decrease in both cases.[285-290] While the protein's internal viscosity (conformational flexibility that is coupled to ET) and its associated *solvent slaving* remain constant as the SAM becomes thicker ($n = 4, 6, 11$), the electronic coupling becomes weaker and the viscosity dependence weakens, according to Eq. (5), until the onset of a non-adiabatic regime in which Eq. (8) holds.

Overall, these data imply a change in the intrinsic ET mechanism with SAM thickness rather than an essential change in rate-determining step (see Section II). Indeed, it would be difficult to justify the existence of a rate-determining (and viscosity controlled, *vide supra*) large-scale rearrangement of the protein-SAM moiety (gating mechanism) for the case of freely diffusing cytochrome *c*. In the next sections, we draw comparisons with the kinetic results obtained for cytochrome *c* irreversibly bound to SAMs, for which such an alternative mechanism is more plausible. While it is difficult to compare directly the heterogeneous standard rate constants, k^o_{het} and k^o_{ET} when the actual value of the pre-equilibrium constant, K_A, in Eq. 23 is unknown (*vide supra*), it is possible to make a sound comparative analysis of the *effective* protein-to-terminal group distances for freely diffusing and adsorbed cytochrome *c* by the direct comparison of *apparent turnover distances*, i.e., the formal ET distances at which the mechanistic changeover occurs; see Section VII.

VI. SPECIFICALLY IMMOBILIZED CYTOCHROME *C* AT AU/SAM JUNCTIONS

1. The Overview

This Section describes a strategy for immobilizing cytochrome *c* on a chemically modified Au electrode through a specific ligand-receptor interaction that provides selectivity for the biomolecule's orientation on the surface.[73,75] This pioneering work demonstrates the ability to create mixed monolayer films that present nitrogen ligands to bind directly with the heme of cytochrome *c* (see Fig. 21 left). By creating mixed films of pyridinal alkanethiols, or other nitrogen receptors (see Fig. 21 right), in a diluent of shorter chain alkanethiols, cytochrome *c* has been immobilized through association of the functionalized longer chain thiols with the heme iron of the cytochrome. In particular, pyridine and imidazole were found to interact strongly, whereas nitrile-terminated chains interacted more weakly. This immobilization motif has been demonstrated by electrochemical voltammetry measurements (a large negative shift in the apparent redox potential), STM imaging, and surface-enhanced resonance Raman spectroscopy.[75,291] The voltammograms are nearly ideal, indicating well-defined sites on the surface of the electrode and allowing the electron-transfer kinetics to be quantified.

The quality of the voltammetry is revealed by the data in Figures 22 and 23. Figure 22 shows voltammograms of the imidazole and pyridine mixed films at some selected voltage scan rates. The voltammograms display well-defined peaks and a shift of the peak (oxidation)-to-peak (reduction) separation with the scan rate. Note that the traces are shown for different potentiostat time constants (see original work for details). Table 4 presents data for the different immobilizations and compares it to some nonadsorbing films. Note the large negative shift in the apparent redox potential for the mixed films with nitrogen ligands and its dependence on the nature of the nitrogen ligand. Spectroscopic studies have shown that the pyridine, imidazole, and nitrile functionalities can bind to the heme iron of cytochrome *c* in free solution,[292-297] and that the negative shift in redox potential indicates an interaction between the nitrogen ligand and the cytochrome's heme. The redox potential shifts that are observed on the SAMs correlate with those found in ho-

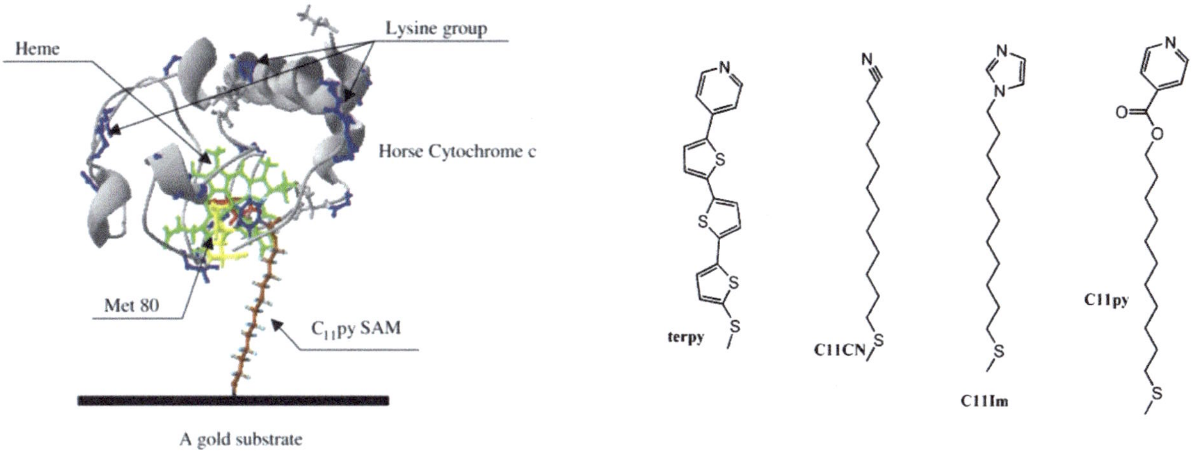

Figure 21. Left: Association of a pyridine tether to the heme of cytochrome *c*. Also shown are the lysine groups that are present on the protein surface.[73] Right: molecular structures for the different receptor-based tethering molecules.[75]
Adapted with permission from Ref. 75, Copyright (2002) American Chemical Society and from Ref. 73, Copyright (2001) by permission of the Reoyal Society of Chemistry.

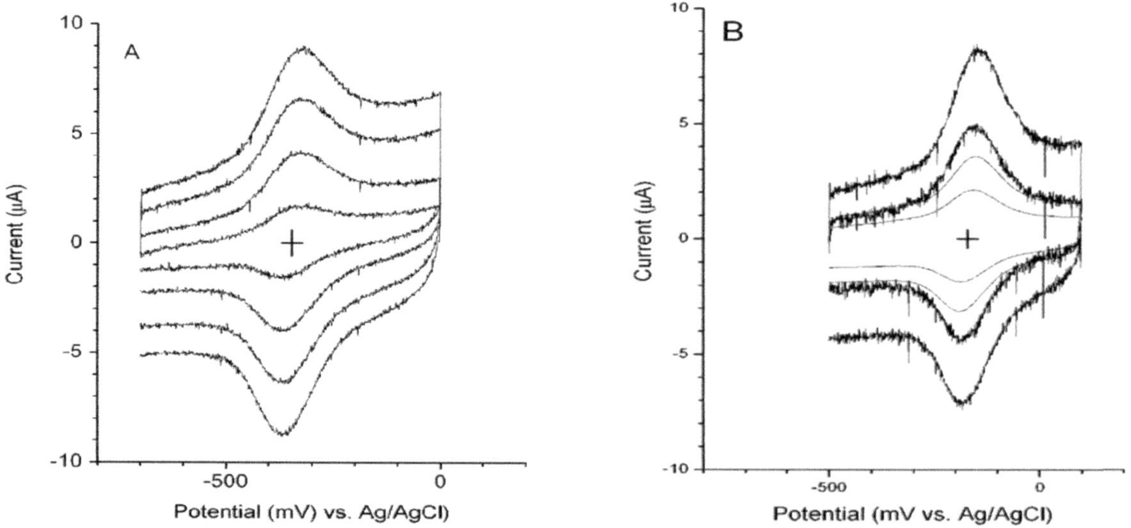

Figure 22. Voltammograms are shown for cytochrome *c* immobilized on the surface of mixed monolayer films containing imidazole functionalities (panel A) and pyridine functionalities (panel B). The scan rates for these voltammograms are 20 15, 10, and 6 V/s. Rerpinted with permission from Ref. 75, Copyright (2002) American Chemical Society.

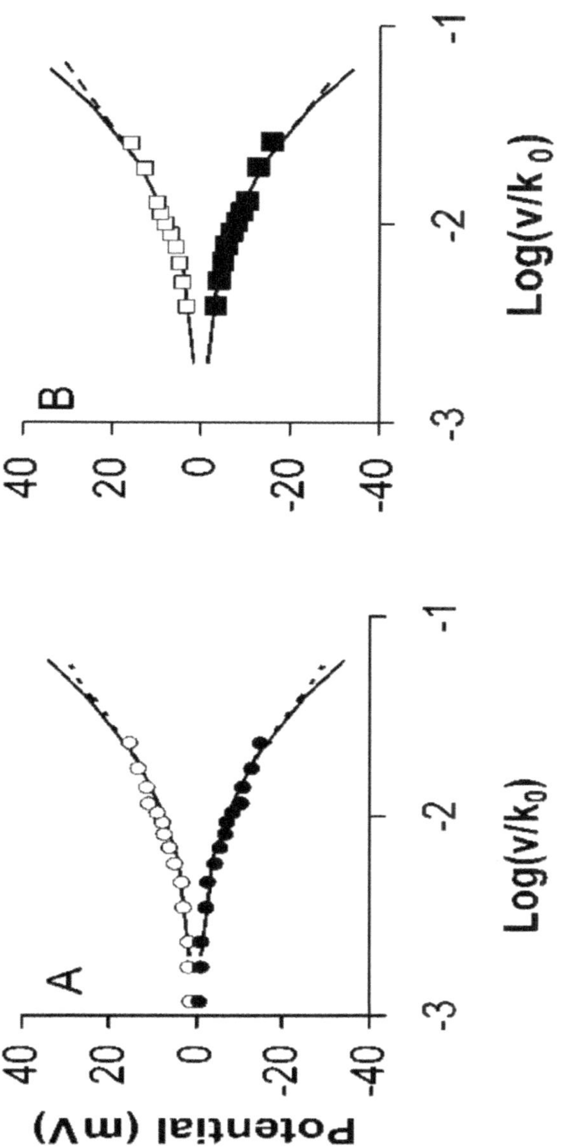

Figure 23. The dependence of the peak potential on the scan rate is shown for the imidazole system (Panel A) and the pyridine system (Panel B). Fits of the data to Marcus theory predictions are also shown for two different reorganization energies (0.8 eV is the solid line and 0.9 eV is the dashed line). Rerpinted with permission from Ref. 75, Copyright (2002) American Chemical Society.

Table 4
Electrochemical Parameters for Different Electrode/Cytochrome Systems. Rerpinted with permission from Ref. 75, Copyright (2002) American Chemical Society.

System	$E^{0'}$ (mV)	ΔE (mV)	Scan Rate (V/s)
$HOC(CH_2)_6S$ [a]	44 ± 2	58	0.2
$PyCO_2(CH_2)_2S$ [a]	5	56	0.2
$HOOC(CH_2)_{10}S$	12 ± 3	99	0.6
$PyCO_2(CH_2)_{12}S/C_{11}H_{21}S$	−172 ± 10	108	1.0
$Im(CH_2)_{11}S/C_8H_{15}S$	−346 ± 20	117	1.0
$NC(CH_2)_{11}S/C_8H_{15}S$	−415 ± 20	132	8.0

[a] In this system the cytochrome *c* is freely diffusing in solution at a concentration of 50 μM.

mogeneous solution studies of cytochrome *c*; however, they are typically somewhat smaller in magnitude.[18,298-300] The redox potential shift that is observed upon addition of an exogenous ligand to solution can arise from a conformational change (even denaturation) of the protein, as well as ligation to the heme iron. A study by Fan et al.[298] examined the pyridine binding to cytochrome *c* in solution extensively and found that the heme bound pyridine had a redox potential of -161 mV (close to the value of −172 mV observed on the SAM, see Table 4), but that higher concentrations of pyridine in solution shifted the redox potential to −294 mV because it denatured the protein and exposed the heme iron to the solvent (see also Ref. 81). Fan's findings substantiate the view of cytochrome *c* immobilization that is indicated by the cartoon in Fig 21. At the slowest scan rates, the voltammetric peak widths of these assemblies are near the ideal limit of 91 mV and indicate a high degree of homogeneity. In contrast, an earlier study had used pure layers of pyridine-terminated alkanethiols and reported broadened voltammograms and a large asymmetry between the oxidation and reduction responses.[73]

The dependence of the peak separation on scan rate can be analyzed to obtain the electron transfer rate constant according to the procedure described in Section III.[48,164,165, 301] Figure 23 shows plots of the peak shift (E_p-$E^{0'}$) versus the voltage scan rate for the pyridine and imidazole ligands, along with a best fit by the classical Marcus theory for k^0. The two theoretical curves correspond to two different reorganization energies, 0.8 and 0.9 eV, and reveal the sensitivity of the analysis to the choice of the reorganization energy.[56] The best fit standard rate constants (k^0) were 780 s^{-1} for the pyridine-terminated layer (electron transfer through a C12 chain) and 850 s^{-1} for the imidazole-terminated layer (electron transfer through a C11 chain), with a symmetry factor of 0.5 for the reaction.[217]

2. Pyridine-Terminated SAMs: Impact of the Electron Transfer Distance

Figure 24 shows a plot of the measured heterogeneous rate constant as a function of the methylene number in the tethering chain for some different pyridine-terminated SAMs[76] and –COOH terminated SAMs from the work of Niki et al.[58,59] and Bowden et al.[54,55] (also see Section 7). For thick SAMs, the pyridine terminated alkanes and the COOH terminated SAMs display an exponential dependence on the charge-transfer distance, see Eqs. (1) and (3), with a decay constant of about one per methylene. This decay length is similar to that found in other tunneling studies with saturated hydrocarbons and is a signature for nonadiabatic electron transfer. Note that the thickness value at which the rate constant starts to fall off exponentially is shifted to higher methylene numbers for the pyridine ligated data, as compared to the –COOH data. Both data sets show a weak distance dependence at short donor-acceptor separations (the *plateau* region); the maximum rate constants differ by about a factor of two and the rate constants in the pyridine-bound systems are consistently higher than those for the electrostatically bound system.

Figure 24 and Table 5 also report rate data for cytochrome *c* adsorbed through three other tethers of a similar length to the C11py. In two of these systems the C11 tether is retained, but the receptor group has been modified from a pyridine to an imidazole and/or to a nitrile unit. Although these ligands cause a significant shift in the apparent redox potential, they have a minor effect on k^o_{ET}. This result demonstrates that the methylene chain length controls the rate constant. In contrast, the conjugated terthiophene tether with a pyridinal head group has an apparent redox potential

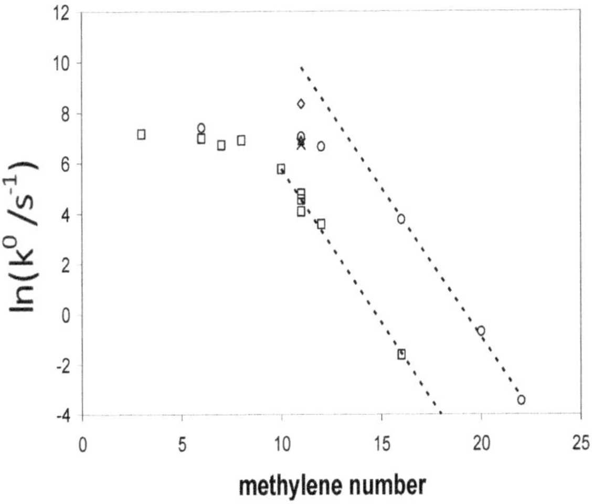

Figure 24. This diagram plots the apparent standard electron transfer rate constants for the different systems. The data for systems bound through coordination with the heme are represented by circles for pyridine, × for imidazole, triangle for CN, and diamond for terthiophene. The squares are the data for electrostatic adsorption on COOH. The dashed lines are fits to the nonadiabatic model at large layer thickness. Ref.74 Copyright (2002) Wiley-VCH Verlag GmbH & Co KgaA. Reproduced with permission.

like that of the alkylpyridine but has a k^{o}_{ET} that is a factor of three to four larger than the C11 tether. The electronic coupling via the conjugated linker is expected to be higher than that of the C11 chain and cause an increase in the rate constant (see Section 2).

The data in Fig. 24 show that the onset of the exponential decline (see Eq. 1) occurs at larger film thicknesses (ca. 12 methylenes) for the pyridine-terminated SAMs than for the COOH-terminated SAMs. Once in the exponential regime, however, both films display a tunneling behavior like that observed for hydrocarbon chains, 1.19 per CH_2 for the pyridinal SAMs and 1.22 per CH_2 for the COOH SAMs.[15,302] Wei et al.[77] showed that the shift in the onset distance for the exponential region arises from the fact that

Table 5
Rate Constant Data for Cytochrome *c* Immobilized on Different Mixed SAMs. Rerpinted with permission from Ref. 76, Copyright (2003) American Chemical Society.

System[a]	k^o_{ET} (s^{-1})	E^0 (mV)	#Trials	System	k^o_{ET} (s^{-1})	E^0 (mV)	#Trials
C6py/C5	1580	−175	6	C11CN/C8	1000	−415	2
C11py/C10	1150	−168	14	C11Im/C8	860	−346	4
C12py/C11	785	−172	4	Terthiophene py/C7	4200	−188	2
C16py/C15	52	−158	12	C11py/C10	1150	−168	14
C20py/C19	0.50	−156	3				
C22py/C20	0.032	−145	2				

[a] The numbers indicate the number of methylene units in the chain and py indicates the pyridinal receptor. For example, C6py/C5 denotes ω-pyridinalhexanethiol tethers in a pentanethiol diluents SAM.

the two different immobilization motifs have two different dominant tunneling pathways. The COOH terminated groups electrostatically bind the cytochrome by its lysine groups[60,303,304] and the pyridine-terminated alkanethiols bind through ligation with the heme iron.[73,75,291]

Wei's study was based on an earlier study by Niki et al.[60], which examined a number of different mutants of cytochrome *c* that were electrostatically adsorbed to –COOH SAMs. Niki replaced the lysine units on the surface of the cytochrome *c* and studied how it changed the electron transfer rate for the electrostatically bound protein. They showed that replacement of the lysine-13 with an alanine (mutant RC9-K13A) changed the electron transfer rate by five orders of magnitude. Even when they swapped the lysine 13 with a glutamic acid (at position 90 and adjacent on the protein surface), they observed a similar decrease in rate constant. The large sensitivity of the rate constant to the lysine 13 and its relative insensitivity to the movement or elimination of other surface lysines, led them to conclude that the dominant tunneling pathway in electrostatically adsorbed cytochrome *c* proceeds through lysine 13. Using the cytochrome *c* crystal structure, one can estimate a physical, *through-space*, distance of 5.8 Å from the lysine 13 to the heme and a *through-bond* distance of about 20 Å. Wei et al.[77] measured k^0 for the RC9-K13A mutant and the native rat cytochrome *c* on COOH-terminated SAMs and on the mixed pyridine-terminated SAMs (see Table 6 and Figure 25). Wei's rate data in the electrostatic assemblies show excellent agreement with that of Niki et al.[60] revealing a nearly 5000-fold decrease in the rate constant for the mutant form, as compared to the native form of the protein. In contrast, the pyridine immobilized protein has a similar k^o_{ET} for both the mutant and native cytochrome *c*.

Figure 25 shows how the measured k^o_{ET} changes with the distance from the electrode surface for the native rat cytochrome *c*, horse heart cytochrome *c*, and K13A rat mutant immobilized on the COOH terminated SAMs and the pyridine terminated SAMs. For the pyridine-terminated films (open symbols), the native rat cytochrome *c* and the K13A cytochrome *c* rate constants are nearly the same; they display a weak distance dependence for thin SAMs and an exponential distance dependence for thick SAMs. In contrast to the pyridine terminated SAMs, the carboxyl-terminated

Table 6
Electron Transfer Rate Constant of Rat Heart Cytochrome *c* and the Mutant K13A Adsorbed on Different Electrodes. Rerpinted with permission from Ref. 77, Copyright (2004) American Chemical Society.

Systems	Native Cytochrome *c*		Mutant K13A	
	k^o_{ET} (s^{-1})	# Trials	k^o_{ET} (s^{-1})	# Trials
C6Py/C5			789 ± 155	3
C11Py/C10	903 ± 130	3	816 ± 122	4
C12Py/C11	770 ± 42	3	737 ± 103	3
C16Py/C15	55.3 ± 2.1	3	80 ± 16	5
C20Py/C19	0.62 ± 0.03	3	0.73 ± 0.12	4
C3COOH	920 ± 60	2	0.13 ± 0.04	2
$(C_6H_4)COOH$	570 ± 45	5	0.20 ± 0.06	2
C5COOH	680 ± 68	5	0.0035 ± 0.001	4
C10COOH	19 ± 7.2	4		

Note: Errors indicated for the rate constant represent one standard deviation from the average value found for the different trials.

SAMs have different rate constants for the native and mutant forms of cytochrome *c*. At short film thickness the native form of cytochrome *c* approaches the limiting (plateau) value that is found for the pyridine tethered protein, but the mutant cytochrome *c* never reaches this value; in fact, it is orders of magnitude lower than the native cytochrome *c*.

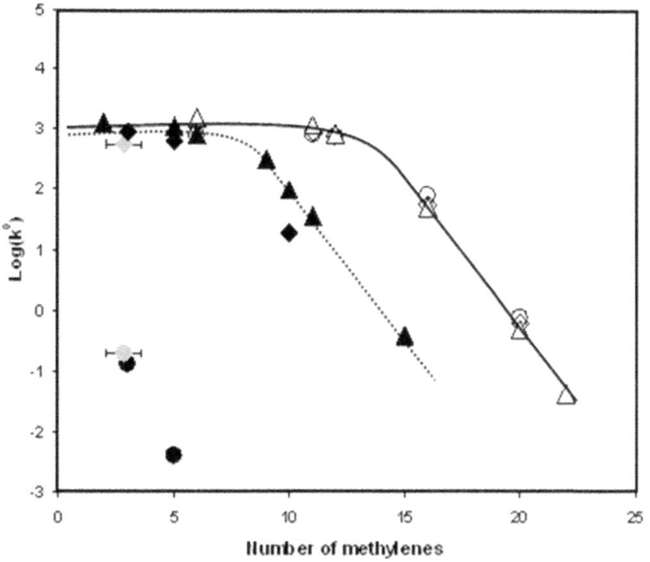

Figure 25. The measured electron transfer rate constant of surface immobilized rat heart cytochrome *c* and its K13A mutant is plotted as a function of SAM thickness. The unfilled symbols represent pyridine immobilized cytochrome *c*: the triangle for native horse heart cytochrome *c*, the circle for mutant K13A, and the diamond for native rat cytochrome *c*. The filled symbols represent electrostatic adsorption by carboxylic acid films: the black diamond is native rat cytochrome *c*, the black circle is the K13A mutant, and the black triangle is horse heart cytochrome *c*. The grey symbols are for a $S(C_6H_4)COOH$ monolayer and the bar shows the uncertainty in assigning it a length equivalent to some number of methylenes. The solid curve and the dashed curve represent the distance dependence of cytochrome c with the pyridine and carboxylic acid system, respectively. Rerpinted with permission from Ref. 77, Copyright (2004) American Chemical Society.

Because the electron transfer rate depends strongly on the reorganization energy, Wei et al.[77] measured the reorganization energy of both the mutant and native forms of the protein. To achieve reasonable accuracy on the reorganization energy determination, they performed voltammetry studies for cytochrome *c* immobilized on C20Py/C19 mixed films, for which the rate constant is very slow. This choice of system allowed them to scan to high overpotentials and determine the reorganization energy to an accuracy of 0.1 to 0.2 eV.[28] The data for the rate constants and reorganization energies are reported in Table 7.

These data make it clear that the binding mode of the cytochrome *c* to the SAM film changes the dominant electron tunneling pathway and thus the point at which the transition between the short distance/plateau region and the long distance tunneling regime is observed in SAM/protein assemblies. The cartoon in Figure 26 illustrates the two binding modes considered here. For the electrostatic binding (Case 1), the k^o_{ET} of the native cytochrome *c* is 5000 times larger than that found for the K13A mutant because the electrons must tunnel through carboxylate:lysine contacts on the protein surface. By removing the lysine 13 contact on the SAM surface that tunneling pathway is removed. In contrast, the similar k^o_{ET}'s observed for the pyridine-terminated SAMs result because

Table 7
Summary of Reorganization Energy Measurements of Rat Heart Cytochrome *c* and Mutant K13A Obtained from Immobilization on C20Py/C19 Mixed Monolayer Films. Rerpinted with permission from Ref. 77, Copyright (2004) American Chemical Society.

	Mutant K13A Cytochrome *c*			Native Cytochrome *c*		
Trial	k^o_{ET} (s^{-1})	λ (eV)	$E^{0'}$(mV)	k^o_{ET} (s^{-1})	λ (eV)	$E^{0'}$(mV)
1	0.82	0.50	−150	0.60	0.62	−141
2	0.85	0.45	−157	0.65	0.55	−153
3	0.66	0.70	−132	0.62	0.58	−148
4	0.60	0.65	−145			
Avg.	0.73±0.12	0.58±0.12	−146±11	0.62±0.03	0.58±0.04	−147±6

Note: Errors indicated for the rate constant represents one standard deviation from the average value found for the different trials.

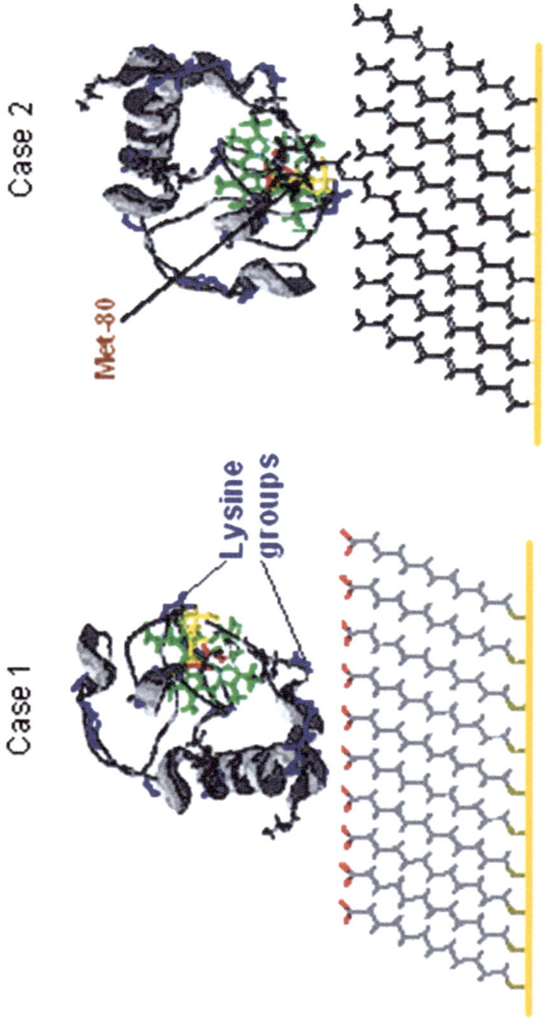

Figure 26. Cytochrome *c* adsorption on self-assembled monolayers. Case 1: electrostatic adsorption on carboxylic acid SAM. Case 2: Ligand immobilized cytochrome *c* on pyridine terminated mixed SAM. Rerpinted with permission from Ref. 77, Copyright (2004) American Chemical Society.

the dominant electron tunneling pathway does not change, along the alkane tether of the pyridine ligand to the heme iron, when the lysine is replaced. This finding substantiates the conclusion that the adsorption of the protein to the pyridine terminated SAMs proceeds by axial coordination between the pyridine receptor and the protein's heme, and the primary difference between the two cases is the nature of the link between the protein and the electrode.

Given this view of the tunneling pathways, it is possible to estimate the physical distance between the electrode surface and the heme unit of the protein for the two cases. If one considers that the pyridine unit coordinates to the heme iron and that the pyridine unit has little effect on the charge-transfer distance because of its π-conjugation, then the charge transfer distance d in the pyridine-terminated SAMs can be written as:[150]

$$d = 1.90 + 1.12\, n\ (\text{Å}) \tag{34}$$

where n is the number of methylenes in the alkane chain and 1.90 Å accounts for the S atom of the thiol. For the electrostatically adsorbed cytochrome c the charge transfer distance has contributions from the SAM and from the peptide between the SAM surface and the heme unit. By assuming that the tunneling proceeds via the lysine 13 [60] and using the cytochrome c crystal structure, *vide supra*, one can estimate a physical distance. Figure 39 shows plots of the k^0 for both of these data sets versus distance, which is calculated from these models. The good agreement between the two data sets suggests that the difference in k^0 values for the two immobilizations can be *corrected* by accounting for differences in the electron transfer distance.

3. Pyridine-Terminated SAMs: Impact of the Solution Viscosity and Heavy Water

For the thin SAMs, the rate constant changes with solution properties, such as viscosity and D_2O versus H_2O, but for thick films it does not.[74,76] Figure 27 shows plots of k^o_{ET} measured on three different pyridine-terminated SAMs versus solution viscosity (data is in Table 8), which was varied by addition of glucose. Fits of the data by Eq. (3) give δ values of 0.58 for C6Py tethers, 0.28 for C11Py tethers, and ≈ 0 for C16Py tethers. Thus, k^o_{ET} displays a viscosity dependence for thin films, the plateau region of Fig. 24, but it is independent of the viscosity for thick films, the tunneling region. The lack of a viscosity dependence for the C16Py tether indicates that the tunneling probability controls the rate and demonstrates that the method used to change the viscosity does not cause some other change in the protein or its adsorbed state. Although the k^o_{ET} values for C11Py and C6Py are similar, their viscosity dependence is different; the significantly larger δ for the shorter tether indicates a stronger frictional coupling and is consistent with a stronger electronic coupling. The importance of viscosity for the thin films demonstrates a change in the mechanism of the electron transfer reaction with distance.[98]

Table 9 shows data that display a dependence of k^o_{ET} on the water isotope used in the experiments for thin SAMs. For thick SAMs, the C16Py/C15 assemblies, no dependence of the rate constant on the water isotope can be observed. This control experiment demonstrates that changing the *normal* buffer solution to a D_2O buffer solution does not impact the adsorbed state of the protein. These experiments show a modest change in k^o_{ET} with water isotope if it is exposed for a long time (*ca* 30 minutes or more) and the assemblies are on thin SAMs, the plateau region of Figure 24. These results suggest that water present in the protein, or exchangeable protons, modulate k^o_{ET}. Murgida and Hildebrandt have reported a much stronger isotope dependence, however their SAM films were thinner than those studied here.[61] According to the detailed studies by Davis and Waldeck,[305] a significant portion of the

Table 8
Rate Constants of Immobilized Cytochrome *c* for Different Solution Viscosities. Rerpinted with permission from Ref. 76, Copyright (2003) American Chemical Society.

	η=0.98 cP		η=1.76 cP		η=3.88 cP	
System	#Runs	k^o_{ET} (s^{-1})	#Runs	k^o_{ET} (s^{-1})	#Runs	k^o_{ET} (s^{-1})
C6py/C5	3	1512	4	1050	3	670
C11py/C10	2	1155	5	990	4	780
C16py/C15	2	60	2	60	2	61

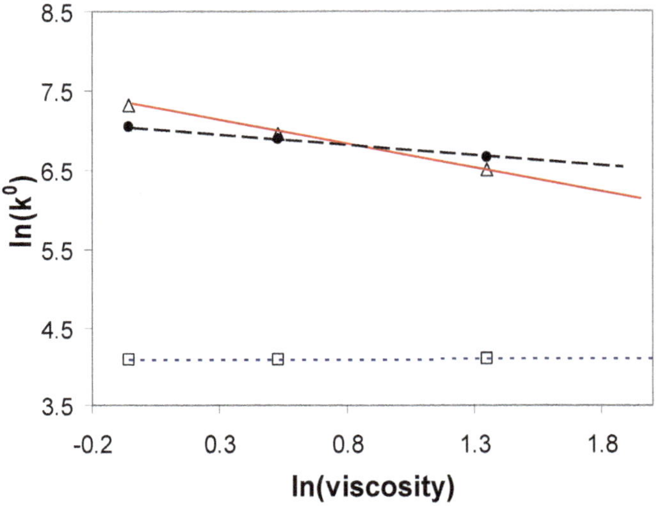

Figure 27. The viscosity dependences of the observed electron transfer rate constant are shown for three different alkanethiol chain lengths: the triangles are C6, the circles are C11, and the squares are C16. The dashed line has zero slope. Rerpinted with permission from Ref. 76, Copyright (2003) American Chemical Society.

Table 9
D_2O Dependence of the Rate Constant Data for Immobilized Cytochrome *c*. Rerpinted with permission from Ref. 76, Copyright (2003) American Chemical Society.

C11Py/C10			C16Py/C15		
Cell	Incubant	k^o_{ET} (s^{-1})	Cell	Incubant	k^o_{ET} (s^{-1})
H_2O	H_2O	1140	H_2O	H_2O	58
D_2O	H_2O	1100	D_2O	H_2O	
H_2O	D_2O	890	H_2O	D_2O	
D_2O	D_2O	879	D_2O	D_2O	55

change in rate constant with the water isotope arises from a local viscosity effect (see Section 7.3).

These results require a change in the electron transfer mechanism as one proceeds from thick SAMs to thin SAMs. For thick films (> C12), k^o_{ET} decreases exponentially with increasing alkane chain length and does not depend on the solution viscosity or the isotope of the buffer. This behavior is consistent with a rate constant that is controlled by the tunneling probability. For thin films the reaction mechanism changes. It no longer decreases exponentially with distance; it displays a viscosity dependence; and it depends on the use of D_2O versus H_2O. Early workers[59,306] who examined COOH-terminated SAMs, explained this mechanism change by hypothesizing that a conformational rearrangement to a precursor state acts to gate the electron transfer. For example, the cytochrome *c* might rotationally diffuse at the SAM/electrolyte interface until the heme position becomes close to the electrode, at which time the electron transfer occurs rapidly. Such a scenario is not consistent with the data for the pyridine terminated chains, which show a similar distance dependence but should not involve reorientation of the protein on the SAM surface. Nevertheless, a more subtle conformational change might become rate controlling; e.g. a change in the nitrogen-iron distance in the protein/SAM complex. To distinguish gating from polarization dynamics, Yue and coworkers[78] measured the dependence of the electron transfer rate constant on the overpotential, and their findings are discussed in the next section. The gated mechanism should not depend on the overpotential, because the rate is limited by a nuclear rearrangement rather than electron transfer, whereas the friction controlled mechanism should change with overpotential.

4. Evidence for Dynamic Control rather than Gating at Short Distances

Yue et al.[78] used cyclic voltammetry and time-resolved surface-enhanced resonance Raman spectroelectrochemistry to measure the temperature, distance, and overpotential dependences of the electron transfer rates for pyridine-terminated SAMs. They concluded that the electron transfer mechanism changes from a tunneling controlled reaction at long distances (thick films) to a friction-controlled reaction at short distances (thin films). They measured the k^o_{ET} on pyridine-terminated Au electrodes by cyclic voltammetry, and for each SAM thickness ($n = 6$, 11, 12 and 16) the temperature was varied from 2°C to 40°C.[307-309] Figure 28 shows the temperature and distance dependence of k^o_{ET} for the four different SAM assemblies. The rate data are qualitatively similar to that in Figs. 24 and 25; namely, the rate increases strongly from $n = 16$ to $n = 12$, consistent with a nonadiabatic reaction, and it increases weakly for shorter chains, $n < 12$. The activation energies that were found for each of the SAM thicknesses show a monotonic increase with decreasing film thickness (see Table 9). Yue et al.[78] showed that the change in the activation energy could be explained by considering it to be a sum of two terms; one that arises from the solution viscosity and one that arises from the Marcus reorganization energy. By empirically determining and accounting for the viscosity contribution, they were able to extract a reorganization energy for the electron transfer (0.4 to 0.6 eV) that changed only modestly with the SAM thickness; see the last column of Table 10.

Surface-enhanced resonance Raman (SERR) spectroscopy was used to measure the rate constant for cytochrome *c* immobilized by pyridine-terminated SAMs on Ag electrodes. Stationary potential-dependent SERR[291] has been used to show that the immobilized protein exists in equilibrium between two forms; both of which lacked axial ligation with the native ligand Met 80. The dominant form was observed to have a six coordinate, low spin heme (6cLS), in which the sixth ligand was the pyridine group, and the minor form was a five coordinate, high spin heme (5cHS); presumably it binds to the mixed monolayer through hydrophobic interactions. The equilibrium was shown to be potential dependent with the ferric form displaced towards the 6cLS state, reflecting the larger affinity of Fe^{3+} for pyridine as compared to Fe^{2+}. By performing time-resolved SERR spectroscopy, Yue et al.[78] measured the electron transfer rate at different overpotentials by performing potential jump experiments.

Figure 29 shows a plot of the measured electron transfer rate constants for the two different forms of the immobilized protein on PyC_6/C_5 coated Ag electrodes. The two forms of the protein can be treated as independent on the times scale of the experiment (i.e., they do not interconvert) and both respond to the applied overpotential. The overpotential dependence for the pyridine-coordinated

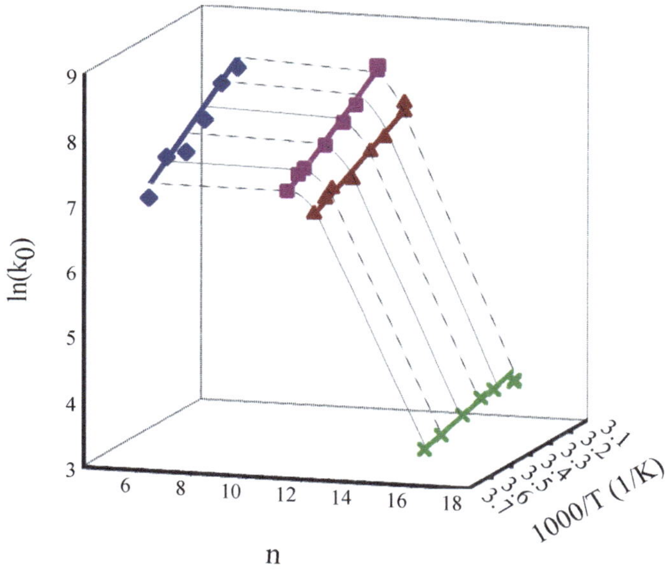

Figure 28. The graph shows the distance and temperature dependence of k^o_{ET} for cytochrome c on SAMs of different composition on Au. Green crosses: PyC_{16}/C_{15}. Purple triangles: PyC_{12}/C_{11}.Pink squares: PyC_{11}/C_{10}. Blue diamonds: PyC_6/C_5. Rerpinted with permission from Ref. 78, Copyright (2006) American Chemical Society.

Table 10
Activation Energies, Reorganization Energies, and Viscosity Parameters for Cytochrome *c* on the Different SAM/Au Assemblies. Rerpinted with permission from Ref. 78, Copyright (2006) American Chemical Society.

	Arrhenius	Classical analysis	Viscosity corrected
SAM	ΔG_a^* (eV)	λ_C (eV)	λ (eV)
PyC_6/C_5	0.225	0.84	0.48
PyC_{11}/C_{10}	0.205	0.76	0.56
PyC_{12}/C_{11}	0.15	0.56	0.44
PyC_{16}/C_{15}	0.082	0.38	0.38

form (6cLS) gives equally good fits to either a nonadiabatic model with a λ = 0.3 eV or to a friction-controlled model with a λ = 0.58 eV. It is important to note that the value of k^{o}_{ET} for the PyC_6/C_5 SAMs is substantially smaller on Ag as compared to Au (146 s^{-1} *vs*. 2400 s^{-1}). Thus, these experiments provide clear evidence for a dependence of the rate constant on the overpotential for thin SAMs on silver electrodes.

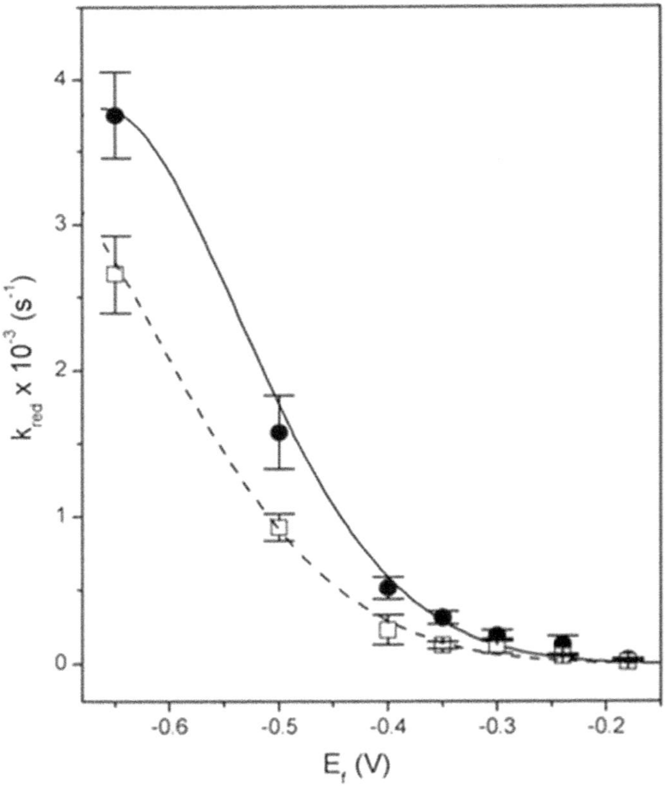

Figure 29. Rate constant of reduction of cytochrome *c* on PyC_6/C_5 SAM modified silver electrodes as a function of the final potential in the TR-SERR experiments (see text for details). Circles: 6cLS redox couple. Squares: 5cHS redox couple. Rerpinted with permission from Ref. 78, Copyright (2006) American Chemical Society.

5. Comparison with Homogeneous Electron Transfer Involving Cytochrome *C*

Figure 30 plots electron transfer rate constants as a function of the reaction free energy, ΔG, for many different systems involving cytochrome *c*; including the electrochemical value (k^o_{ET} for thin SAMs), *unimolecular* systems, and bimolecular systems. The solid curve is generated by fitting the unimolecular rate data (G), a series of ruthenium-modified cytochrome *c*'s,[310-314] to a Gaussian-shaped curve, as predicted by the classical Marcus model. The dashed line shows the same curve shifted down by one order of magnitude. The electrochemical rate constant (filled circle) appears to follow the Marcus curve. This analysis assumes that the electron transfer rate is determined primarily by the Franck-Condon factor (free energy and reorganization energy, rather than the electronic coupling) and that the reorganization energy does not change too significantly between the different systems. The observed free energy dependence and the viscosity-sensitive behavior for some of the data (see figure caption) suggest that the electron transfer belongs to the friction-controlled regime, rather than to a conformationally-gated mechanism.[101]

The general correspondence of the data in Fig. 30 with the reaction free energy supports the assignment of the rate limiting step to an electron transfer process. Although the data display a large amount of scatter, the diversity of the systems used to obtain these rates is large and scatter should be expected. The peak of the curve corresponds to the condition $\Delta G_o = \lambda$, so that λ is 0.7 to 0.8 eV. This reorganization energy value is somewhat higher than that found through the electrochemical studies of the immobilized cytochrome *c*, which ranges from 0.4 to 0.6 eV. A number of experimental and theoretical studies[56,153,275,321-324] report λ values in the range of 0.8 eV to 0.4 eV for the protein in solution. Theoretical studies[153,322-324] have tried to separate the reorganization energy into an intrinsic protein component, an outer sphere (solution) component, and a contribution from the redox partner (see also Eq. 18, Section II above). Although the reorganization energy depends on both partners in a redox reaction, these data suggest that the protein dominates the contribution and is fairly consistent between systems. It is plausible that this difference arises from the change in the solvation environment of the protein in the two cases. For the electrochemical studies, the kinetic data probe the reorganization energy through the dependence of k^o_{ET} on ΔG_o; i.e., the overpotential.

A number of the observations are in conflict with conformational gating as the rate limiting step, but support frictional coupling as the rate controlling step for thin SAMs.[79] First, the electrochemical data, ac impedance and cyclic voltammetry, indicate a simple charge-transfer step; i.e., there is no evidence for a pre-equilibrium. The larger rate constant that is found for the conju-

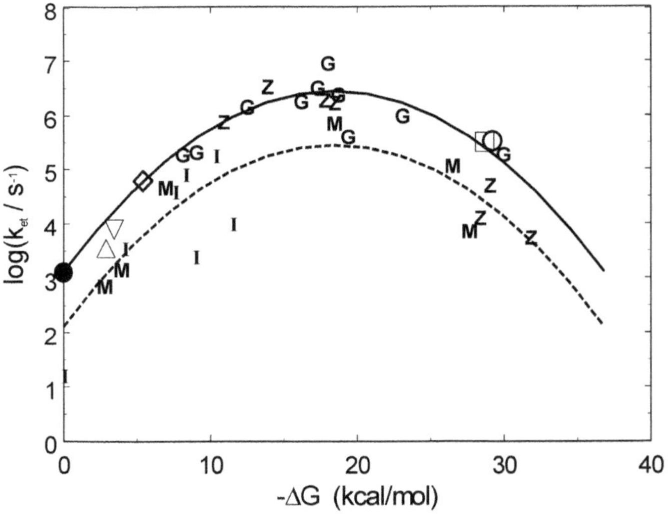

Figure 30. This Marcus plot shows the free energy dependence of cytochrome *c*'s electron transfer rate constant from a number of different studies, mostly homogeneous solution -the data are from Gray et al.[310-313] (G) for Ru-modified cytochrome *c*; Zhou et al.[315] (Z) for cytochrome *c*/uroporphyrin complexes; McLendon for interprotein system cytochrome *c*/cytochrome b_5[316] (M); and Isied for Ru- modified cytochrome *c*[314] (I). The open symbols (◊,[317]; ∇,[318]; □,[319]; Δ,[320]; ○,[246]) correspond to rate constants that exhibit a dependence on the external solution viscosity. The filled circle shows the electrochemical k^o_{ET} at short distances (plateau region), which also displays a viscosity dependence. The solid curve shows the free energy dependence expected from the Marcus model, and the dashed curve is the same model shifted down by a factor of ten. Rerpinted with permission from Ref. 76, Copyright (2003) American Chemical Society.

gated terthiophene tether, as opposed to the methylene tether, cannot be explained by a conformational gating mechanism but can be rationalized through the effect of the electronic coupling on the activation barrier for a direct electron transfer reaction. The temperature dependence of the measured rate constant gives a consistent activation energy for a direct electron transfer mechanism, as long as the temperature dependence of the viscosity is analyzed; note that the temperature dependence would require a barrier that changes with SAM thickness for a gating mechanism. The overpotential dependence that is assessed through potential jump experiments (Fig. 29) and the reaction free energy analysis (Fig. 30), indicates a direct electron transfer for thin SAMs. The change from tunneling control to friction control can be understood to arise from an increase in electronic coupling and the slowing of the polarization relaxation in SAM/protein assemblies, as opposed to free solution.

Next we proceed to the case of electrostatically immobilized cytochrome *c* (Section VII) and consider the polarization relaxation mechanism for all of the cytochrome assemblies in Section VII.5.

VII. ELECTROSTATICALLY AND COVALENTLY IMMOBILIZED CYTOCHROME *C* AT AU/SAM JUNCTIONS

1. The Overview

Electrostatic immobilization of cytochrome *c* onto negatively charged surfaces has been reported by numerous workers. While some workers have reported that cytochrome *c* may adsorb on unicomponent SAMs with hydroxyl terminal groups under exceptional conditions,[64,65] in the presence of electrolyte such SAM/cytochrome contacts operate in a freely diffusing regime.[56,57,82] In contrast, the immobilization of cytochrome *c* at neat carboxyl terminated SAMs under conventional electrochemical conditions (up to 60 mM electrolyte) normally is strong and irreversible.[54,55,325-327] Moreover, mixed SAMs of COOH and OH terminated methylene chains have proven useful for optimizing the ideality of the electrochemical response (voltammetry signal), because they provide a way to chemically tune the strength of the protein/SAM surface interaction and minimize the inhomogeneity of the adsorption sites; i.e., they act to narrow the distribution of distinguishable redox species at the SAM interface.[328-331] Figure 31 (panels A and B) illustrates these two different model SAM assemblies. The mixed ω–COOH/ω–OH composite SAMs have been well-characterized by a variety of physicochemical methods, including X-ray photoelectron spectroscopy[327,330] and ellipsometry,[332] and have been shown to be well-organized and well-behaved regarding their structure (presenting a mostly randomly textured molecular surface) and electrochemical performance (nearly ideal current waves).[326,327,329] Importantly, mixed SAMs with longer active chains (here –COOH terminated) and relatively shorter diluent (here –OH terminated) components provide an extended manifold of selectively variable protein/SAM interactions[82,330] and may serve as more advanced models for biomimetic studies.

Recently, Yue et al.[329] compared the electrochemical response for four different cytochrome *c* immobilizations and its effect on the electron transfer rate constant. These binding motifs are shown in Figure 31 and include electrostatic binding to both pure and mixed films, pyridine ligation, and covalent binding on mixed films. The preparation of covalently bound cytochrome *c* electron transfer complexes using carbodiimide cross-linking reagents is a well-established technique. The cross-linking occurs by formation of one or more amide bonds between lysine residues on the cytochrome *c* surface and a complementary carboxyl group on the SAM (or other partner), as shown in Figure 31. The resultant complexes are similar in structure to their electrostatic counterparts and appear to be functional and stable.[325, 328-330]

Figure 32 shows how the standard electron exchange rate constant k^o_{ET} of cytochrome *c* changes with ionic strength for a SAM that is thick enough to ensure that electron transfer occurs in the nonadiabatic/tunneling limit. All of the experiments were performed at pH 7, for which the protein is stable; denaturation does not occur until the pH has decreased to between 4 and 5.[333,334] For the electrostatically immobilized protein, they found that the average standard electron exchange rate constant k^o_{ET} decreased and

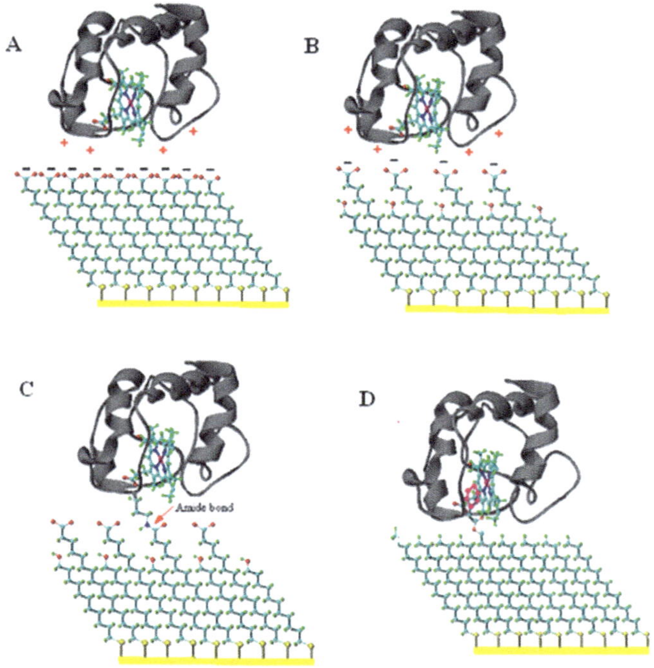

Figure 31 Schematic diagram depicting four different cytochrome *c*/SAM assemblies. A) Cytochrome *c* adsorbed electrostatically to a pure SAM composed of carboxylic acid-terminated alkanethiol molecules. B) Cytochrome *c* adsorbed electrostatically to a mixed SAM composed of hydroxyl-terminated and carboxylic acid terminated thiols (C15COOH/C11OH). C) Cytochrome *c* is immobilized through the formation of one or more amide bonds between a protein lysine and the carboxylic groups of the SAM. D) A pyridine moiety is used to ligate to the protein's heme iron; a motif discussed in Section 6. Rerpinted with permission from Ref. 329, Copyright (2006) American Chemical Society.

the inhomogeneity of the voltammetry increased with the increase of the solution ionic strength. For ionic strength values higher than 80 mM, the cytochrome *c* desorbs from the carboxylic-terminated SAM. For the pure C15COOH SAM (panel A) and the mixed C15COOH/C11OH SAM (panel B), the rate decreases systematically as the ionic strength increases from 10 mM to 80 mM, for

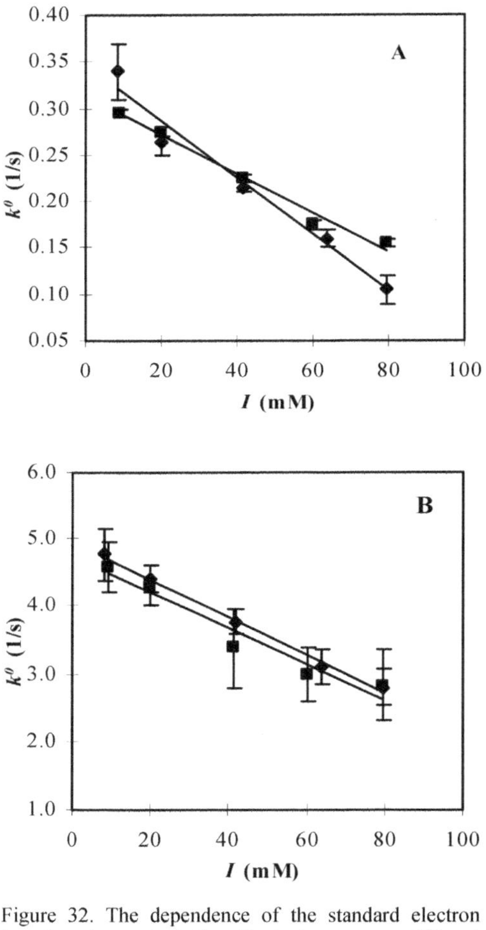

Figure 32. The dependence of the standard electron transfer rate constant k^o_{ET} of cytochrome *c* on different SAMs. Black diamonds (♦) are data from phosphate buffer solutions at pH 7, and black squares (■) are data from tris-acetate buffer solutions at pH 7. The solid curves are a linear fitting of experimental data. A) cytochrome *c* on pure C15COOH SAM; B) on C15COOH/C11OH SAM; C) covalently attached to C15COOH/C11OH SAM; D) tethered to PyC16/C15 SAM.Rerpinted with permission from Ref. 329, Copyright (2006) American Chemical Society.

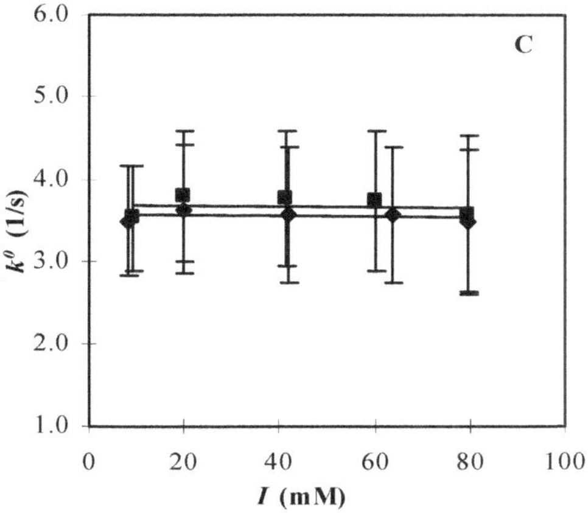

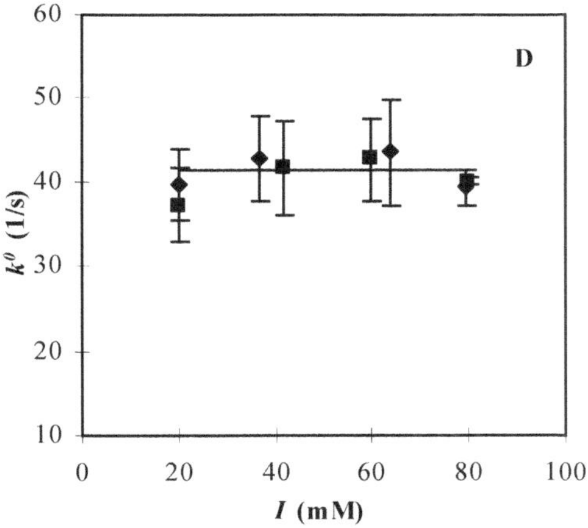

Figure 32. Continuation.

both of the buffer systems that were studied. For the covalent immobilization (panel C) and the ligated protein (panel D), the rate constant did not change with ionic strength. Thus, the decrease in k^o_{ET} with the increase of ionic strength can be understood to arise from an increase in the effective distance between the cytochrome *c* and the film.[329]

The standard rate constants for the different immobilization strategies are quite different and summarized in Table 11. For the pure COOH terminated films, the standard rate constant is about ten times smaller than that observed in the mixed films (Fig 32B). In part, this difference can be explained by a distribution of adsorbate orientations (inhomogeneity) that include unfavorable sites; a finding that is corroborated by analyzing the widths of the current waves in the voltammograms.[329] In addition, it may be possible for the heme of cytochrome *c* to approach closer to the electrode in the case of mixed SAMs, because of the more flexible nature of the solvent-exposed fragments of SAM *active* chains.[82,330] In contrast, the mixed assemblies (Fig. 32B) and the covalently attached SAM assemblies (Fig. 32C) have the same electron transfer rate; a fact that likely originates from the similarity of the SAMs textured surface. Despite the similarity in their average rate constants, the covalently immobilized cytochrome *c* does not change with ionic strength. This finding indicates that the electron tunneling is not affected by ions that may be present in the solution or solvated near the SAM/electrolyte boundary. The high rate constant for the pyridine immobilized protein was explained in Section VI.2 as arising from a change in the dominant electron tunneling pathway. Comparison of the data in panels A and B of Fig 32 with those in panels C and D show that the solution properties (in this case, io- nic strength) can change the protein's immobilization geometry and thus modulate its electron transfer rate.

Table 11
Rate Constant k^o_{ET} (s^{-1}) of Cytochrome *c* at I = 40 mM.

	Pure C15COOH	Mixed C15COOH/C11OH	Covalent attachment	PyC16/C15
Phosphate buffer	0.22 ± 0.01	3.6 ± 0.4	3.4 ± 0.8	43 ± 5
Tris/Acetic	0.23 ± 0.01	3.7 ± 0.6	3.5 ± 0.8	42 ± 4

2. Results for Cytochrome *c* Immobilized on Mixed SAMs with Large Chain Length Difference

The dominant electron tunneling pathway(s) between an electrode and the heme unit of the cytochrome *c* depends on the composition of the mixed SAM. Yue et al.[330] studied mixed C15-carboxylic acid/hydroxyl-terminated SAMs as a function of the SAM stoichiometry and the length of the hydroxyl-terminated alkanethiol. It was found that by increasing the concentration of shorter-chain hydroxyl diluent thiol in the SAM, the rate constant increased in a systematic way. When the diluent composition was high, the k^o_{ET} increased exponentially with decreasing chain length of the diluent molecules until the chain length of the diluent molecule reached a length of about eight methylene groups. Their findings showed that the diluent molecules can mediate the electronic interaction and change the effective tunneling distance, if they are short enough.

From earlier studies it was not clear how the increased flexibility, which is more or less evident for mixed SAMs might affect the bioelectrochemical electron transfer properties in the friction-controlled regime. Specifically, the increased flexibility of the Au/SAM/cytochrome *c* electrostatic complex may affect the frictional coupling; e.g., the value of δ. Khoshtariya et al.[82] analyzed how the electron transfer rate depends on the composition of assemblies that are prepared from 50:50 mol percent solutions of C_5COOH/C_2OH, $C_{10}COOH/C_6OH$ and $C_{15}COOH/C_{11}OH$. These three systems span a broad range of electron transfer distances for mixed SAMs and include the critical short-range electron transfer region.

Figure 33 presents some typical cyclic voltammograms recorded for selected Au/SAM/Cytochrome *c* assemblies, and Table 12 gives rate parameters extracted from the analysis. Note that the difference in rate constants for the $(CH_2)_{15}$-COOH/$(CH_2)_{11}$-OH mixed SAMs, which are reported in Tables 11 and 12, likely result from a difference in the percentage of hydroxyl terminated chains in the SAM assemblies that were prepared in these two studies (see Fig. 5 of Ref. 330).

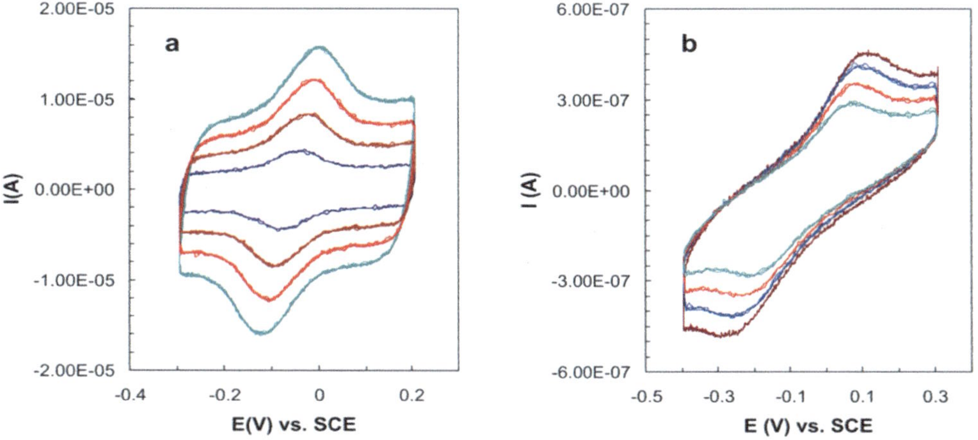

Figure 33. (A) Voltammograms for cytochrome *c* at HS-$(CH_2)_5$-COOH + HS-$(CH_2)_2$-OH, 4.4 mM phosphate buffer, pH 7.0, scan rates: 10, 20, 30 and 40 V/s; (B) Voltammograms for cytochrome *c* at HS-$(CH_2)_{15}$-COOH + HS-$(CH_2)_{11}$-OH, 4.4 mM phosphate buffer, pH 7.0, scan rates: 1, 1.5, 2 and 2.5 V/s. Ref. 82, Copyright (2007) Wiley-VCH Verlag GmbH & Co KgaA. Reproduced with permission.

Table 12
The Values of Standard Rate Constants, and the Value of the Reorganization Energy Obtained Using Phosphate and Tris-HCl (Both in 4.4 mM, pH 7, in Each Case) Buffers through the Established Procedure. Ref. 82, Copyright (2007) Wiley-VCH Verlag GmbH & Co KgaA. Reproduced with permission.

No	The SAM Composition	k^o_{ET} (s^{-1})		λ, eV
		Phosphate buffer (4.4 mM)	Tris-HCl buffer (4.4 mM)	
1	$HS\text{-}(CH_2)_5\text{-}COOH + HS\text{-}(CH_2)_2\text{-}OH$	330	155	0.65±0.1
2	$HS\text{-}(CH_2)_{10}\text{-}COOH + HS\text{-}(CH_2)_6\text{-}OH$	440	205	0.65±0.1
3	$HS\text{-}(CH_2)_{15}\text{-}COOH + HS\text{-}(CH_2)_{11}\text{-}OH$	2.5	–	0.7 ±0.2

Figure 34 shows an example of a *trumpet* plot, i.e., plot of the anodic and cathodic peak potentials versus the log of the potential scan rate, and a fit of these data by the Marcus theory. The value of the reorganization energy parameter (see Table 12) in this analysis was essentially the same for cases of long-range (n = 11, 16) and short-range (n = 6) electron transfer, despite a clear mechanism changeover. Figure 34B shows plots of the logarithmic dependence of the standard rate constant on the methylene number of the *active* (-COOH) components of the mixed SAMs and compares it to rate data for unicomponent COOH-terminated SAMs and pyridine-terminated mixed SAMs. Comparison of the n = 11 and n = 16 kinetic data reveal that the rate constants for mixed SAMs are higher than for their unicomponent analogues, however, they each display similar tunneling decay lengths of about one per methylene unit. In contrast, for the case of n = 6 the rate constant is higher for the unicomponent SAM, as compared to the mixed SAM assemblies (Figure 34B). Note that the rate data are averaged over a number of measurements (symbols on Figure 34B are larger than actual error bars), however the data of different workers agree reasonably well. Lastly, parallel kinetic experiments that were performed using Tris-HCl buffer gave relatively lower rate constants, but revealed a similar plateau region (Table 12 and Fig. 34B).

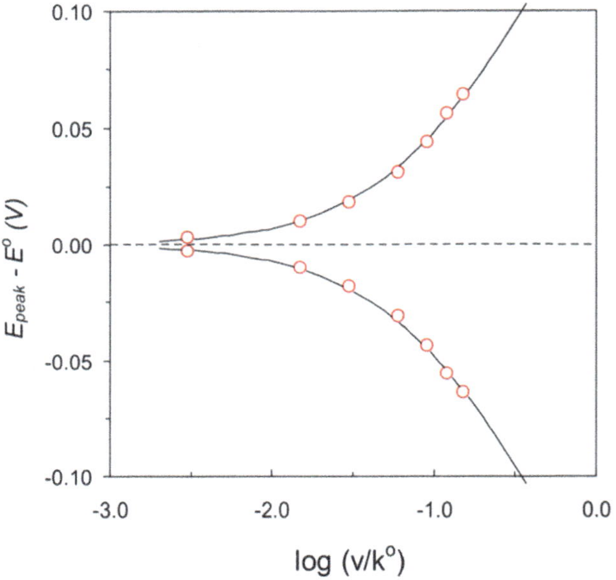

Figure 34. (A) Fitting of the experimental peak separation as a function of scan rate – the procedure that allows simultaneous determination of both the rate constant and reorganization free energy; the case of *thin* SAM (HS-$(CH_2)_5$-COOH + HS-$(CH_2)_2$-OH), 4.4 mM phosphate buffer (see Table 11). (B) Logarithmic dependence of the rate constant on SAM carbon number of *active* (longer) components of different SAMs. Closed red circles and blue crosses are results obtained in phosphate and tris buffers, respectively; open red squares: data of Bowden et al.[326,327] for comparable mixed SAMs; open violet circles: data of Niki et al.[58,59] and Bowden et al.[54,55] for unicomponent carboxyl terminated SAMs; closed green squares: data of Waldeck et al.[74] for pyridine terminated diluted SAMs. Ref. 82, Copyright (2007) Wiley-VCH Verlag GmbH & Co KgaA. Reproduced with permission.

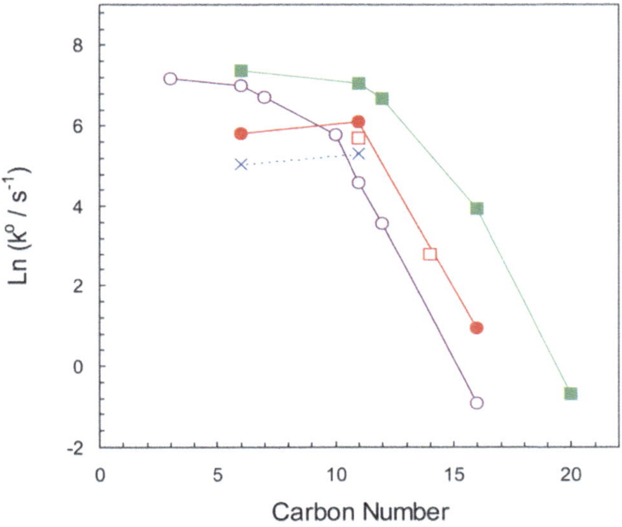

Figure 34. Continuation

3. Kinetic Isotope and Viscosity Effects for Systems with Coulomb and Covalent Binding

Davis et al.[305] evaluated the dependence of the electron transfer rate on the presence of D_2O versus H_2O for electrostatic and covalent assemblies, in both the tunneling and plateau regions. In the thick film (tunneling regime), they found no change in the rate constant as the solvent was changed from H_2O to D_2O; i.e., the kinetic isotope effect was unity. For thin SAMs, they observed a dependence of the rate on whether H_2O or D_2O was used in the assembly preparation and electrolyte solution. While the dependence was significant, a total kinetic isotope effect of about 1.4, it was moderate. They showed that at least half of the isotope effect can be explained in terms of the change in the solution viscosity.

For thin SAMs, they identified two separate isotope effects: a cell isotope effect that is manifest at short timescales (< 30 s) and arises from the viscosity difference between H_2O and D_2O, and an incubant isotope effect that is manifest at long timescales (> 2 h) and results from H-D (or H_2O/D_2O) exchange. These two categories have their origin in the manner by which the assemblies are prepared. Once a SAM is assembled on the electrode, it is immersed into a solution containing the protein (the incubant solution) for a period of about 30 minutes to an hour. After incubation, the electrode assembly is rinsed with a buffer solution and then placed in an electrochemical cell that contains only a buffer solution (the cell solution), no protein. Davis and Waldeck[305] performed experiments that independently controlled whether D_2O or H_2O were used in these solutions, incubant (inc) or cell. They defined the cell isotope effect, KIE_{cell}, by the ratio k^o_{ET} (H_2O, H_2O): k^o_{ET} (H_2O, D_2O), in which the first term in parentheses defines the incubant solution's isotope and the second term defines the electrochemical cells isotope. They defined the incubant isotope effect, KIE_{inc}, by the ratio k^o_{ET} (H_2O, H_2O): k^o_{ET} (D_2O, H_2O); this definition removes the contribution of the increased viscosity of D_2O. The data in Table 13 report KIE_{inc} for both the electrostatic and covalent systems, and show that its magnitude is comparable to that of KIE_{cell}. Furthermore, they found that these two ratios are similar and that the total isotope effect, which is defined as KIE_{total} = k^o_{ET} (H_2O, H_2O): k^o_{ET} (D_2O, D_2O) could be constructed from a product of KIE_{cell} and KIE_{inc}. The electron transfer rate is slower in D_2O buffer than in H_2O buffer, and this slowing of the rate occurs for both the electrostatic and covalent assemblies. The degree of slowing in the rate is similar to that reported by Murgida and Hildebrandt for electrostatic assemblies of cytochrome *c* on SAM coated Ag electrodes.[19,61]

Table 13
Evaluation of the Kinetic Isotope Effects Observed for Short Chain Cytochrome *c*/SAM Assemblies. Rerpinted with permission from Ref. 305, Copyright (2008) American Chemical Society.

	Electrostatic	Covalent
KIE_{cell}	1.2 ± 0.2	1.2 ± 0.2
KIE_{inc}	1.3 ± 0.3	1.1 ± 0.2
$KIE_{total,}$	1.6 ± 0.3	1.4 ± 0.3
$KIE_{total,\ pred}$‡	1.6 ± 0.4	1.4 ± 0.4

‡$KIE_{total,\ pred.}$ = $KIE_{cell} \times KIE_{inc}$. Error obtained by propagating the standard error in KIE_{cell} and KIE_{inc}.

Davis[305] examined the viscosity dependence of the electron transfer rate in these assemblies and contrasted it with the cell isotope effect that they observed. While a solution viscosity dependence had been reported on carboxylate terminated films previously, it had not been reported for covalent assemblies or mixed electrostatic assemblies $-S(CH_2)_mCOOH/-S(CH_2)_mOH$ systems. The viscosity of the aqueous electrolyte was changed over the range of 1.0 to 2.5 cP by the addition of dextrose to the solution. Their k^o_{ET} data are shown in Fig. 35 (filled points show the viscosity dependence data of electrostatic (panel A, squares) and covalent (panel B, triangles) assemblies). The shaded area shows the acceptable best-fit region, as determined by χ^2 values using Eq. (3), and a best fit gave a value of $\delta = 1.1$ for the electrostatic assemblies and $\delta = 1.0$ for the covalent assemblies; i.e., the reaction appears to be fully in the friction controlled regime.

Figure 35 also shows data from the isotope studies and reveals that the cell isotope effect arises from a viscosity change. The open diamonds correspond to k^o_{ET} (H_2O, D_2O), and the open circles to k^o_{ET} (D_2O, D_2O) respectively, for each assembly type. For the samples incubated in H_2O and measured in a D_2O buffer, k^o_{ET} (H_2O, D_2O), the measured rate constants lie near the boundary of the best-fit viscosity region; in contrast, the k^o_{ET} (D_2O, D_2O) rate constants lie clearly below the best fit viscosity region, indicating that H/D exchange affects the rate constant at longer exposure time. To further test this conclusion, Davis performed switching experiments between a normal water buffer solution and a buffer solution whose viscosity was adjusted to 1.09 cP (that of D_2O). The results are shown in Table 14, and they support their conclusion that the observed change in rate constant by the cell solution arises from viscosity rather than H/D exchange.

From the k^o_{ET} (D_2O, D_2O) data in Table 13 and Fig. 35, it is clear that, on long timescales some other factor decreases the rate constant, presumably H/D exchange of some protein residues or of one or more water molecule(s).[335-337] This KIE_{inc} contribution is similar in magnitude to that of KIE_{cell}. However, KIE_{inc} is somewhat larger in electrostatic assemblies (1.3) than in covalent assemblies (1.1). While this difference may indicate that specific bonded contacts between cytochrome *c* and the SAM are less important for the covalent assemblies than for the electrostatic com-

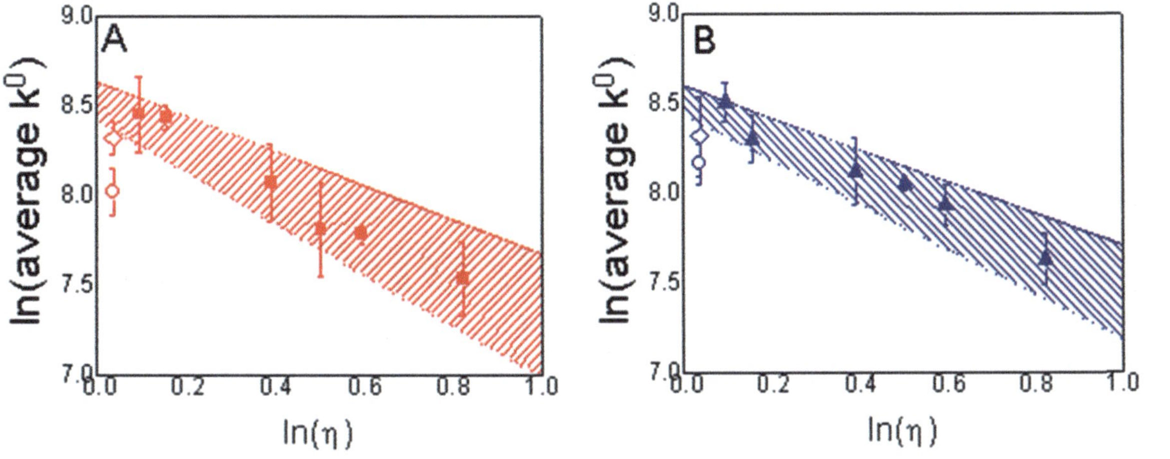

Figure 35. This figure plots the electron transfer rate constant versus the logarithm of solution viscosity for electrostatic (Panel A) and covalent (Panel B) assemblies. Open diamonds correspond to k^{o}_{ET} (H_2O, D_2O), while open circles correspond to k^{o}_{ET} (D_2O, D_2O). The hashed regions show the best-fit region as a result of χ^2 goodness-of-fit analysis of the weighted linear regression of k^{o}_{ET} (H_2O, dextrose solution), which is plotted as closed squares for electrostatic systems and closed triangles for covalent systems. Rerpinted with permission from Ref. 305, Copyright (2008) American Chemical Society.

Table 14
Summary of Cell and Incubant Effects and Comparison to Viscosity Effect. Rerpinted with permission from Ref. 305, Copyright (2008) American Chemical Society.

	Incubant solvent, cell solvent	k^0(incubant, cell): $k^0(H_2O, H_2O)$
Electrostatic	H_2O, D_2O[a]	1.2 ± 0.2
	H_2O, viscous H_2O	1.1 ± 0.3
Covalent	H_2O, D_2O[a]	1.2 ± 0.3
	H_2O, viscous H_2O	1.2 ± 0.2

plexes, it is difficult to say. Although they examined a number of possibilities for this latter effect, such as proton-coupled electron transfer, no firm conclusions could be drawn. One plausible explanation that is consistent with the friction-controlled limit of the electron transfer would be that the H/D and/or H_2O/D_2O exchange cause a change in the internal viscosity of the protein, beyond that arising from the buffer solution, and this depresses the rate. From the plots in Fig. 35, it is evident that the magnitude of the shift, the effective viscosity, lies in a plausible range of values.

4. Comparison of Different Systems Involving Coulomb, Covalent and Through-Heme Interactions

The distance dependence of the electron transfer rate constant for the electrostatically adsorbed protein, the covalently linked protein, and the pyridinally ligated protein assemblies reveal a tunneling behavior for thick SAMs and a *plateau* region for thin SAMs. Sections VI.4 and VI.5 discuss the mechanism for the electron transfer rate in the plateau region for the pyridine assemblies and contrast it with that for electrostatic assemblies. The two major mechanisms that have been used are the friction-controlled electron transfer and conformational gating (i.e., a pre-equilibrium that becomes rate limiting). The assemblies for which cytochrome *c* is covalently bound to the –COOH termini of the SAMs offer another way to test whether the conformational gating model can explain the distance dependence. Because the covalent assembly restricts the rotational diffusion/rearrangement of cytochrome *c* at the SAM/solution interface, it can assess the importance of large-

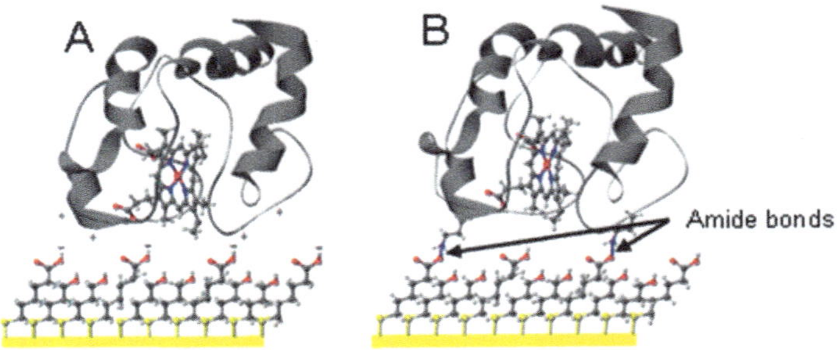

Figure 36. Panel A: cytochrome *c* is adsorbed to the carboxylate moieties on a mixed 2:3 $-S(CH_2)_5COOH/-S(CH_2)_4OH$ SAM via electrostatic interactions between lysine residues and carboxylate groups. Panel B: cytochrome *c* is tethered to the same monolayer as in A via two covalent bonds between lysine residues and terminal carboxylate groups. More than two bonds are possible. Rerpinted with permission from Ref. 328, Copyright (2008) American Chemical Society.

amplitude rotational diffusion of the cytochrome *c* on the SAM surface in determining the rate for thin films.

The cartoon in Figure 36 shows a view for the electrostatic and covalent assemblies on mixed films. The similarity of the surface environment is reflected in the similar redox potentials for the cytochrome *c*, when it is adsorbed on these films. Petrović et al.[338] investigated how the formal potential of cytochrome *c* changes for the freely diffusing protein and the protein immobilized in different manners onto a SAM (Fig. 37). From this graph it is evident that the formal potential for the pyridine immobilized protein is shifted strongly negative of that for the protein in solution, but that they have a similar dependence on the ionic strength. For the electrostatic assemblies and the covalent linkage on mixed –COOH/-OH films the formal potentials are very similar and intermediate between the two extremes of free in solution and pyridine ligated. It is also evident that the ionic strength dependence is similar for these assemblies but different than that observed for the pyridinal and freely diffusing cases. The different ionic strength dependences, and their fit to a model represented by the lines in Figure 37, were discussed at length by Petrović et al.[338]

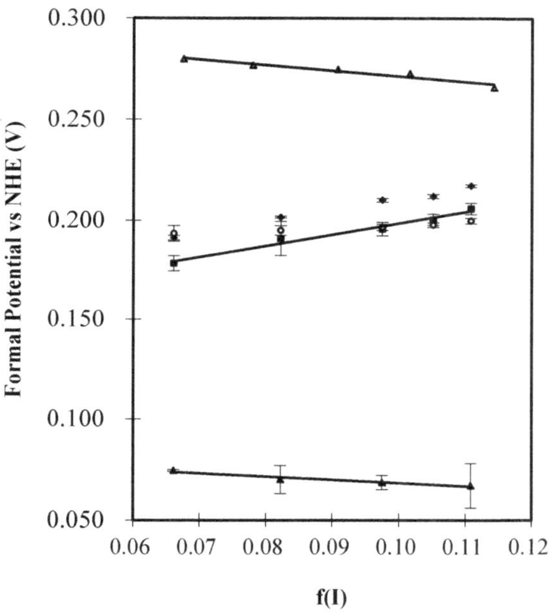

Figure 37. Example Tris/Acetate data for $E^{o'}$ as a function of ionic strength, f(I), for cyt *c* adsorbed on pure carboxylic acid SAMs (solid diamonds), mixed carboxylic acid SAMs (solid squares), covalently attached to mixed monolayers (open circles), datively bound to the electrode (solid triangles), and freely diffusing in solution (open triangles). Reproduced from Reference 338, with permission. Rerpinted with permission from Ref. 338, Copyright (2005) American Chemical Society.

Figure 38 shows a plot of the standard electrochemical rate constant, which was measured by quantifying how the faradaic peak position shifts with scan rate, versus the thickness of the SAM in methylene units, for the three different assemblies. The data show that the covalently-bound protein assemblies have electron transfer rate constants that are very similar to those found for the electrostatically adsorbed cytochrome *c* at a given methylene number. Despite the similarity, a number of differences can be identified and are discussed at length below. In general, these data robustly support the conclusion that rotational or large-amplitude rearrangement of the protein on the surface of the SAM is not responsible for the plateau region, and a different mechanism should be considered.

(*i*) *The Distance for a Transition from Nonadiabatic to Solvent Control*

As shown in Fig. 38, the electron transfer rate in covalent cytochrome *c*/SAM assemblies (red symbols) depends on SAM thickness in a manner similar to the electrostatic adsorption (blue circles) and pyridine-ligated (green stars) cases. Distinct tunneling and plateau regions are evident; however, the turnover between

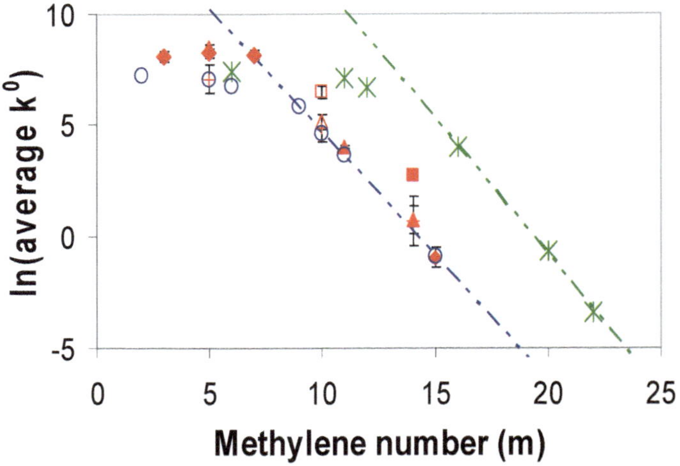

Figure 38. Distance dependence of the rate constant in covalently attached cytochrome *c*/SAM assemblies (red symbols) and comparison to the electrostatic (blue circles) and pyridine-ligated (green stars) cases. ♦: $(CH_2)_mCOOH/(CH_2)_{m-1}OH$ SAMs, +: pure $(CH_2)_mCOOH$ film, Δ: $(CH_2)_mCOOH/(CH_2)_{m+1}OH$, ▲: $(CH_2)_mCOOH/(CH_2)_mOH$, □: $(CH_2)_{10}COOH/(CH_2)_8OH$, ■: $(CH_2)_{14}COOH/(CH_2)_{11}OH$. *m* denotes the number of methylene groups in the carboxyl-terminated SAM component. Electrostatic values for pure –COOH terminated films were reported by Niki et al.[58] and Bowden et al.[55]; Py-ligated values are from Refs.74 and 76 The dashed lines correspond to the rate constant predicted for a tunneling mechanism. Rerpinted with permission from Ref. 328, Copyright (2008) American Chemical Society.

regions appears at $n < 7$ methylene groups rather than $n < 6$ for electrostatic assemblies and $n < 12$ for pyridine-ligated systems. This shift of the turnover point between the covalent data and the electrostatic data could be caused by stronger coupling of the iron center to the electrode through covalent bonds; for a given methylene chain length k^{o}_{ET} is slightly higher for the covalent system than for the electrostatically adsorbed cytochrome *c*, provided there is no diluent hydroxyl-terminated thiol or that it is very close in length to the –COOH-terminated thiol. The large shift in the onset of the tunneling regime between the pyridine-ligated system and the electrostatic and covalent systems results from a change in the dominant tunneling pathway; see Section VI.2 and Ref. 77. The shift in this transition region with the nature of the immobilization can be consistently explained in terms of the electronic coupling strength and charge transfer distance differences between the immobilizations.

(*ii*) *Effect of SAM Composition on the ET Rate in the Tunneling Regime*

A number of workers have examined how the concentration and chain length of the diluent thiol in a mixed SAM affects the electron transfer rate in the electrostatic assemblies,[82,325,328,330] and the findings of these different groups are in basic agreement with each other; also see Section VII.2. Each of these studies concludes that when the hydroxyl-terminated diluent thiol is much shorter than the COOH terminated chain, the rate is enhanced. The data in Figure 38 show this trend; namely, when two components of the mixed $(CH_2)_mCOOH/(CH_2)_nOH$ film have a comparable number of methylene groups ($m - 1 \leq n \leq m + 1$), the rate is similar to that of a pure $(CH_2)_mCOOH$ film; however, as n becomes significantly smaller than m ($n \leq m - 2$) the rate becomes significantly faster. When the diluent thiol is near the length of the COOH-terminated thiol ($n = m \pm 1$), no significant difference in the rate is found, while shorter diluent thiols ($n \leq m - 2$) result in at least a three-fold rate enhancement. As discussed by Yue et al.[330], the electron tunneling can proceed through multiple pathways, rather than exclusively through –COOH chains or covalent linkages, and the relevant importance of the pathways depends on the disparity in chain lengths.

This conclusion adds an important constraint to any attempts to quantify the dependence of the electron transfer rate on the distance in the nonadiabatic limit. As reflected by Eq. (1), the exponential decay of the rate constant with chain length can be quantified as $\exp(-\beta m)$, where $1/\beta$ is the characteristic tunneling decay length and m is the methylene number in the –COOH chain. Table 15 reports values of β, determined from the rate data in Fig. 38, for SAMs with different disparities between the active chain length (-COOH, m) and the diluents chain length (–OH, n). This comparison indicates that the tunneling decay parameter in mixed films is similar to that reported for pure electrostatic systems where $\beta \approx 1.1$ per methylene, and pyridine-ligated systems where $\beta \approx 1.2$ per methylene, as long as the length difference between the diluents and *active* chains is not too significant.

(*iii*) *Electron Transfer Mechanism in the Plateau Region*

The early explanation that was offered by Avila et al.[59] for the weak distance dependence of thin films invoked rotational diffusion of cytochrome *c* to a favorable electron transfer configuration that becomes the rate-limiting step. Although this explanation was consistent with the known dependence of the rate on pH, ionic strength, and viscosity, it is not consistent with the more recent data found for covalent assemblies, in which the conformational rearrangement of the protein on the surface is expected to be highly restricted. In contrast, the ionic strength, viscosity dependence, and the similarity in rates for the covalent and electrostatic assemblies can be explained by a rate-limiting friction-controlled mechanism at short distances.[76,328]

Table 15
Determining the Tunneling Decay Coefficient. Adapted with permission from Ref. 328, Copyright (2008) American Chemical Society.

Methylenes in diluent thiol (n)	Number of points	β
$m - 1 < n < m + 1$	4	1.19 ± 0.05
$n = 0$; $m - 1 < n < m + 1$	6	1.13 ± 0.05

5. Electron Transfer in the Friction-Controlled Regime

Figure 39 shows a plot of the electron transfer rate data for the pyridinal and electrostatic immobilizations on Au electrodes and for the unimolecular studies of Gray et al.[310-312] versus the charge transfer distance. In order to compare these different data sets, Khoshtariya et al.[76] converted the observed electron transfer rate constants to their maximum (optimal) values k_{max} by rearrangement of Eq. (5),

$$k_{\max} \equiv k_{et}^{o} \exp\left(\Delta G_a / RT \right) = \frac{|H_{if}|^2}{\hbar} \frac{\rho_m}{1+g} \sqrt{\frac{\pi^3 RT}{\lambda_o}} \quad (34)$$

This transformation removes the activation barrier from the considerations and allows the dynamical part (pre-exponential factor) of the rate constant to be studied. They used a reorganization energy of 0.8 eV (taken from Fig. 30) and generated the plot shown in Fig. 39. While quantitative aspects of the analysis are very sensitive to the choice of reorganization energy and act to rescale the maximum rate, they do not change the qualitative behavior that is observed in the graph. By way of example, changing the reorganization energy to 0.6 eV reduces the value of k_{max} by a factor of seven. Given this assumption about the reorganization energy, Fig. 39 shows a plot of k_{max} versus the charge transfer distance for the two electrochemical systems and the homogeneous studies as a function of the distance between the redox active heme of the cytochrome and the electron donor, gold electrode and ruthenium moiety. The •'s correspond to the rate constants of the pyridine terminated SAMs and the G's correspond to the unimolecular rate constant data of Gray et al.[310-312] The charge transfer distance for the COOH terminated SAMs (×, *, +) was computed by modeling the charge transfer distance by the length of the methylene chain and the through space distance from lysine 13 to the heme iron. The solid black curve shows a fit to Eq. (34), which describes the transition between electron transfer regimes. The dashed line corresponds to an extrapolation of the thick film (nonadiabatic) rate constant back toward short distances.

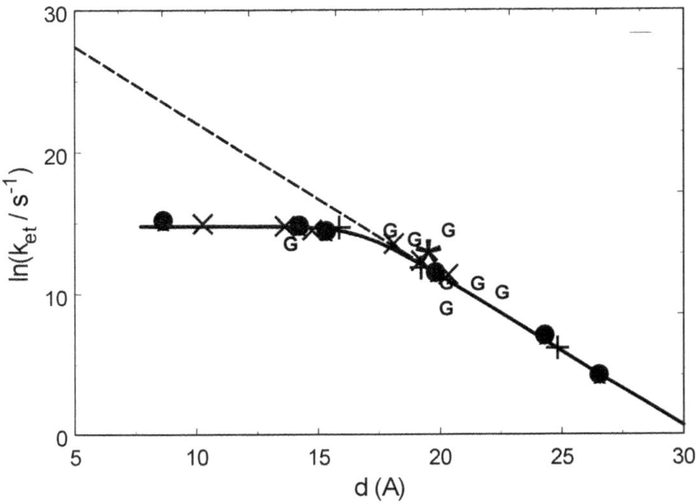

Figure 39. The maximum electron transfer rate constants (Eq. 34) for cytochrome *c* from Fig. 24 are plotted as a function of the electron transfer distance. A constant distance of 5 Å has been added to the electrochemical data on the carboxylic acid terminated films (x -Niki et al.[58,59]; + -Bowden et al.[54,55]; * - Khoshtariya et al.[76]) so that they coincide with the data on pyridine terminated layers (• - Khoshtariya et al.[76]) and the data of Gray et al.[310-312] (G). The solid black curves are fits to Eq. (34), and the dashed line shows the predicted nonadiabatic electron transfer rate constant at shorter distance. Rerpinted with permission from Ref. 76, Copyright (2003) American Chemical Society.

Fitting of the rate constant data in the different regimes allows the adiabaticity parameter *g* to be evaluated. By fitting the electron transfer rate constants at large distance (thick films) to the nonadiabatic model (Eq. 8), one can define the parameters that describe the nonadiabatic rate. Using a reorganization energy of 0.8 eV and a density of states for the Au electrode[15,47,339] of 0.28 eV^{-1}, one finds an electronic coupling between the Au electrode and cytochrome *c* of 0.17cm^{-1} at 17 Å. Using these parameters and fitting the whole data set by Eq. 34, one finds a characteristic relaxation time for the protein's polarization response to be $\tau_{eff} \sim$ 188 ns. While this characteristic time seems too long for a pure liquid solvent response, the highly structured SAM/protein assembly could plausibly have a slower polarization relaxation. This analysis requires that the electron transfer mechanism for cytochrome *c* lie in the strong to intermediate regimes at distances up to 17 Å.

VIII. AZURIN AT Au/SAM JUNCTIONS IMMOBILIZED THROUGH THE HYDROPHOBIC PATCH

1. The Overview

The copper-containing redox protein azurin, which is associated with oxidative stress responses in bacteria (e.g., *Pseudomonas aeruginosa*),[340] provides a useful model system for exploring electron tunneling through protein molecules.[274,341-344] Azurin exists in the Cu(II) or Cu(I) redox states and exchanges electrons with other redox proteins, such as cytochrome c551 or nitrite reductase.[345-346] The type 1 Cu ion binding site (Fig. 40, panel A) is characterized by a bright blue color, a narrow hyperfine splitting in its electron paramagnetic resonance (EPR) spectra, and a high reduction potential.[347-349] In addition, both its oxidation states have the copper ion coordinated with a cysteine (Cys) thiolate group and two histidine (His) nitrogen atoms in a trigonal planar conformation.[347-351] The coordination of the ion is completed by two axial ligands, typically one from a methionine (Met) thioether group and a second from the amide oxygen of a glycine (Gly).[352] The similar structures of the oxidation states is reflected in a relatively low reorganization free energy (λ) for the solution phase (homogeneous) redox process (0.6–0.8 eV,[351,353,363] *vide infra*) and a high ET rate.[1,273,274] Such structural similarity is unusual for copper ions, which typically have large structural changes between the I and II redox states.[355]

The origin of the small λ and the structural similarity of the azurin oxidation states remains a topic of debate. Early workers hypothesized that the protein forces the Cu(II) to be bound in a geometry closer to that preferred by Cu(I).[356,357] However, quantum chemistry calculations of Cu cluster model systems[358-361] indicate that the binding-site geometry is stable for both oxidation states and any strain imposed by the protein is weak and comparable to other metalloenzymes, e.g., alcohol dehydrogenase.[355,359] Furthermore, these studies claim that contributions to λ arise mostly from the active-site and that the outer-sphere reorganization energy (λ_{OS}), arising from contributions of the protein and the solvent, are not so important.[355,358] However, more recent estimates of azurin–azurin self ET,[362] based on a dielectric-embedded cluster model, suggest that λ_{OS} should be comparable to λ_{IS} in contributing to the total λ. Second, theoretical evidence suggests that the active site might be rather flexible and allows fluctuations of the ligands;[355,359] recent work points out that thermal fluctuations can play an extremely important role in the ET process.[369,370,389] Moreover, electron spin echo envelope modulation (ESEEM) measurements,[364] which probe the electron spin-density, indicate that the cluster calculations are too limited to accurately describe the redox site and larger models are needed. Electrochemistry measurements[365-367] find that the redox potential of unfolded azurin is 0.1–0.2 eV higher than that of the folded protein, implying that the protein environment affects the electronic energy of the copper-binding site. A direct coupling between the protein's dipolar field and its redox behavior has been proposed by Warshel and coworkers[368] for plastocyanin and rusticyanin, and we have found a direct effect of the protein electrostatic potential on the optical spectrum of oxidized azurin (data not shown).

Because of its high stability and excellent redox properties, azurin provides a nice model system for studying fundamental aspects of protein/electrode coupling.[372-374] The location of a surface disulfide Cys 3 Cys 26 group on one end of the protein and a hydrophobic patch around the copper center on the other end of the protein can be used to immobilize azurin under different, but controlled, orientations on electrodes. On bare Au surfaces the disulfide group orients the protein molecule with its copper center away from the electrode surface, about 26 Å.[375,376] An ET rate constant of $\approx 30\ s^{-1}$, which is consistent with intramolecular ET between the copper center and the Cys 3 Cys 26 site in homogeneous solution ($44\ s^{-1}$), has been reported.[342] Alternatively, azurin can be immobilized by noncovalent interactions between its hydrophobic patch and methyl terminated alkanethiols.[66;68]

2. Rate Constants and Reorganization Free Energies

Recently, workers[66-71] have immobilized azurin by the interaction of its hydrophobic patch with the hydrophobic CH_3-terminal groups of a SAM composed of alkanethiols: [Au–S–$(CH_2)_n$–CH_3]

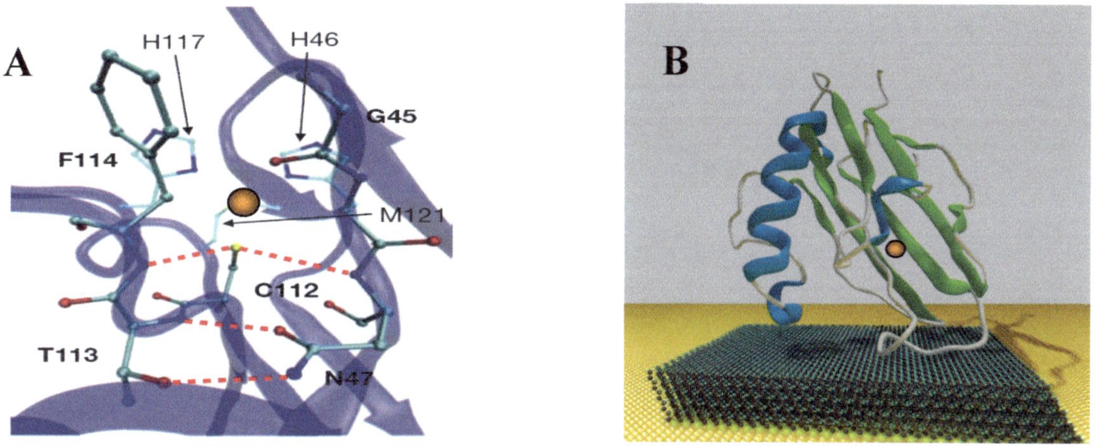

Figure 40. (A) The structure of the active site of azurin. Reproduced from Ref. 376 with permission. (B) Schematic illustration for the placement of azurin on $-CH_3$ terminated SAM through the hydrophobic patch. In both panels copper is shown as a brown sphere. Rerpinted with permission from Ref. 376, Copyright (2000) American Chemical Society.

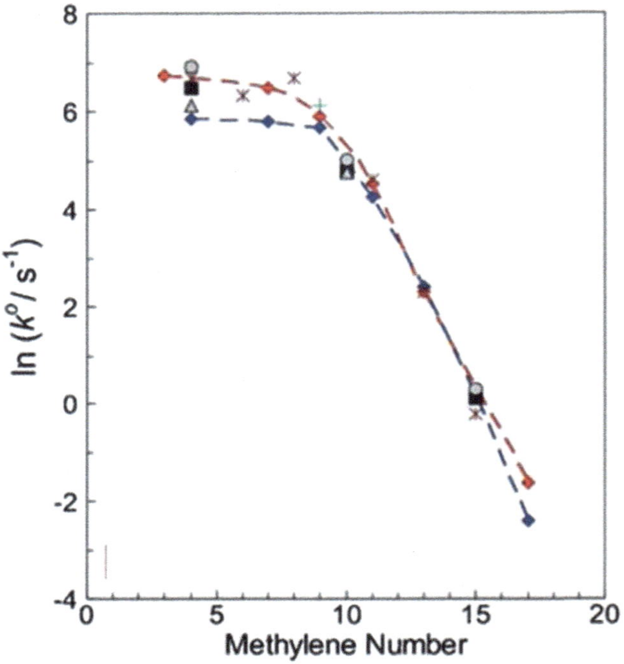

Figure 41. Logarithmic plots of unimolecular standard rate constants for Az electron exchange at Au electrodes modified by CH_3-terminated alkanethiol SAMs with variable methylene number. Red diamonds – data of Guo et al.[71]; blue diamonds – Chi et al.[69]; asterisks – Yokoyama et al.[70]; cross – Jeuken et al.[67]; black squares – Khoshtariya et al.[83], 2 °C (Az from Sigma), gray circles – the same, 25 °C; gray triangles – Khoshtariya et al.[83], 2 °C (Az from Fluka). The broken curves are for eye guidance only. Reproduced from Ref. 83, Copyright (2010) with permission from Proc. Natl. Acad. Sci.

(n = 2 to 17). Figure 40, panel B, illustrates this binding motif. Similar to cytochrome *c*, a plot of ln(k^o_{ET}) versus n (values from 2 to 17) exhibits biphasic behavior; *viz.*, an exponential dependence for thick films, the range of n = 11 to 17, and a plateau region, the range of n = 3 to 9; see Fig. 41. Khoshtariya et al.[83] examined the three systems with n = 4, 10 and 15 in depth: one in the plateau region (n=4), one in an intermediate region (n = 10), and one in the limiting tunneling region (n=15). In all cases, fast scan cyclic voltammetry gave well-defined anodic and cathodic peaks (see Fig. 42), so that measurements as a function of scan rate could be used to determine rate constants. The common method of Creager,[164] and Murray et al.,[165] which is based on the Marcus relation, was generalized to account for the impact of temperature over the range of conventional bioelectrochemical studies (0 to 60 °C), and used a global fitting procedure (Section III) to improve the accuracy of the λ_o determination, see Fig. 43. This procedure allows for the simultaneous fitting of all the data corresponding to ca. 100-fold variation of the potential scan rate and ca. 1000-fold variation of the rate constant throughout the experiments.[83]

First we consider and discuss the values of the reorganization free energies, λ_o, from this analysis.[83] Figure 43 depicts the temperature- and pressure-related data sets for the global fitting procedure. Both the traditional fitting method and the global fitting give a common value of $\lambda_o = 0.3$ eV, for all three types of SAMs. The value from the global fit has an accuracy of $\pm$ 0.03 eV, whereas the same analysis for the thinner SAM assemblies with $n = 10$ and 4 (with data points falling mostly in the middle and bottom parts of the global curves, Figures 43a and 43b) has an accuracy of $\pm$ 0.1 eV. The globally fitted value of $\lambda_o = 0.3 \pm 0.03$ eV seems especially reliable and agrees satisfactorily with the average of values determined for thicker SAMs in earlier work.[66,67,71] The rather small value of $\lambda_o \approx 0.3$ eV found for azurin in these bioelectrochemical systems[66,67,71,83] is considerably smaller than that one finds from various homogeneous experiments[351,353] or theoretical calculations.[154] This difference may result from partial burial of the protein's active site into the SAM's outer layer, hence its essential isolation from nearby water (See Ref. 83 for more discussion). As discussed in Section 3, the parameter λ_o explicitly depends on the electron donor-acceptor separation distance, R_e,[1,150] however, this dependence weakens at moderately large distances, say $R_e \geq 10$ Å (Section IV, Figs. 11A and 15B). Estimates of R_e range from 11 Å for $n = 4$ to 24 Å for $n = 15$, so that the dependence of λ_o on R_e should be weak over this range. The lack of any trend for λ_o within the series of SAMs with variable thickness (Table 16), along with the simple and well-defined voltammetry, indicates that the ET occurs by a single barrier-crossing event throughout the series

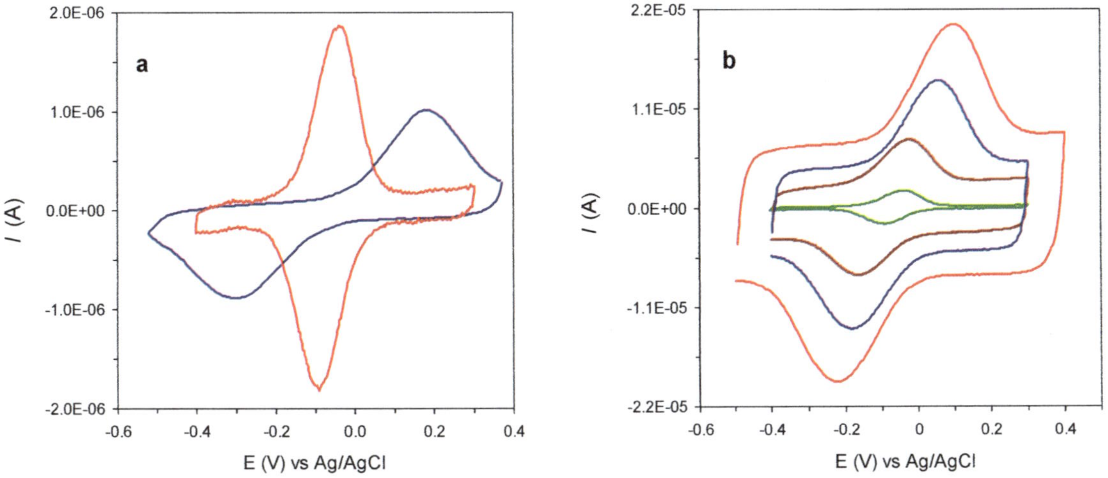

Figure 42. Experimental cyclic voltammograms for azurin electron exchange at Au electrodes modified by CH_3-terminated alkanethiol SAMs with variable *n*. (a) Red curve – 1-pentanethiol ($n = 4$); blue curve – *n*-hexadecanethiol ($n = 15$), scan rate: 5 V s^{-1}. (b) Representative voltammograms for azurin electron exchange at Au electrodes for the case of 1-undecanethiol ($n = 10$) scan rates: 5, 20, 40 and 60 V s^{-1} (respective peak current increase). Reproduced from Ref. 83, Copyright (2010) with permission from Proc. Natl. Acad. Sci.

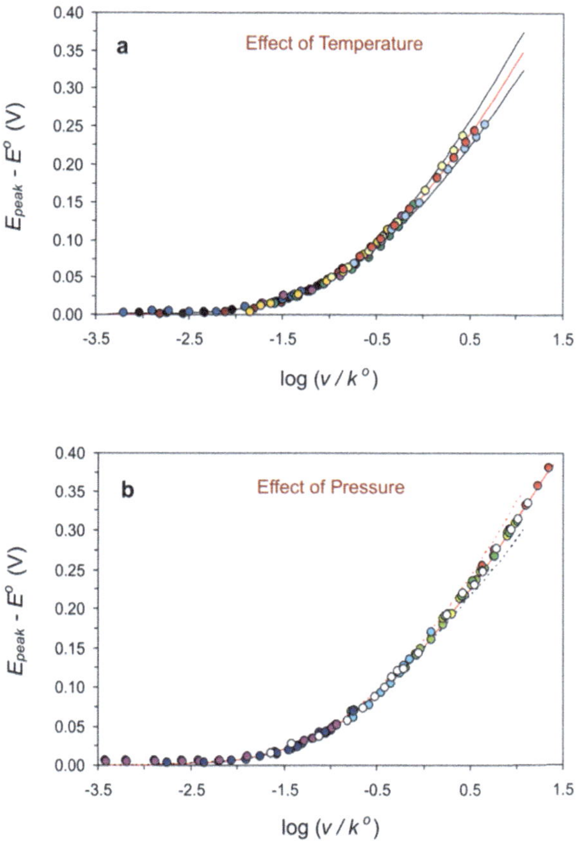

Figure 43. Master plots that demonstrate global fitting of the experimental data for the voltammetric peak shift versus potential scan rates, v, reduced to standard rate constants, k^o_{ET}. Standard errors for experimental points directly match the symbol size. (a) Data collection and global fitting for rate constants at n = 4, 10, 15, and temperatures of 2, 25 and 50 °C; theoretical curves generated for λ_o = 0.3 eV at temperatures of 2, 25 and 60 °C (from bottom to top). (b) Data collection and global fitting for the data obtained at n = 4, 10, 15, temperature 2 to 3°C, and pressure 5, 50, 100 and 150 MPa; theoretical curves generated for λ_o = 0.3 eV, 2°C (middle curve), λ_o = 0.3 eV, 25°C (upper curve), λ_o = 0.6 eV, 2°C (lower curve). See text for further details. Reproduced from Ref. 83, Copyright (2010) with permission from Proc. Natl. Acad. Sci.

Table 16
Standard Rate Constants, Intrinsic Free Energy of Activation (Derived from the Fitted Value of Reorganization Energy, Second Column), and Separately Derived *Effective* Activation Parameters for ET of Az at Au Electrodes Modified by Alkanethiol SAMs [–S–$(CH_2)_n$–CH_3] of Different Thickness (n = 4, 10, 15).[83] Reproduced from Ref. 83, Copyright (2010) with permission from Proc. Natl. Acad. Sci.

SAM (n)	k^o_{ET}, s^{-1} (25 °C)	$\Delta G_a^*(\lambda_o/4)$, kJ mol^{-1}	$\Delta H_{a(EXP)}$, kJ mol^{-1}	$\Delta H_{a(\eta)}$, kJ mol^{-1}	$\Delta V_{a(EXP)}$, cm^3 mol^{-1}	$\Delta V_{a(\eta)}$, cm^3 mol^{-1}
4	1098	7.2 ± 0.7	15.4 ± 2	7.8 ± 3.0	+1.7 ± 0.3	+4.2 ± 0.6
10	171	7.2 ± 0.7	11.7± 1.3	4.1 ± 1.5	−1.9 ± 0.3	+2.1 ± 0.6
15	1.42	**7.2** ± 0.7	**7.6 ± 0.6**	≈ 0	−5.1 ± 0.6	≈ 0

The values depicted in bold (second and third columns) may be considered as equal on theoretical grounds (see text for details).

covering long-range and short-range ET distances (*vide supra*). Thus, an explanation for the biphasic shape of the plot in Fig. 41 should be sought within fundamental ET mechanisms, rather than in specific gating patterns or other mechanistic complexity (see Sections II and IV to VII, above).

3. Activation Enthalpies from Arrhenius Analysis

Let us now consider the results from an Arrhenius analysis of the temperature dependent rate constant, according to Eq. (19). Figure 44 shows Arrhenius-like plots for azurin's k^o_{ET} as a function of T for the three different alkanethiol SAMs of different thickness (n = 4, 10, 15). The respective values of $\Delta H_{a(EXP)}$ are collected in Table 16. Using the Marcus model and the definition of k^o_{ET}, the activation free energy of the ET process may be related to the reorganization free energy through Eqs. (6) and (19) (see Section II). For the three cases encountered here, the values of R_e and H_{if} are such that $H_{if} << \lambda_o/4$ (see e.g., Refs.1,47,51,53, and 76 for representative examples, *vide supra*). Thus, for the long-range (nonadiabatic) ET we find that $\Delta G_a^* \approx \lambda_o/4$, implying that ΔG_a^* is determined solely by the Franck-Condon factor and does not have an appreciable contribution from the pre-exponential factor, see Eqs. (4) and (8).

For thinner films, which occur in the friction controlled regime (Eqs. 2, 3, 10,11), the thermal activation of ν_{eff} must be included in the analysis. In this case the value of $\Delta G_{a(EXP)}$ (or $\Delta H_{a(EXP)}$, as its constituent part) should contain an additional component, *viz.*, $\Delta G_{a(\eta)}$ (or, respectively $\Delta H_{a(\eta)}$), related to the *protein friction* (see Section II). For short-range ET (n = 4), the value of $\Delta H_{a(EXP)}$ is twice that of the long-range (n = 15) ET; see

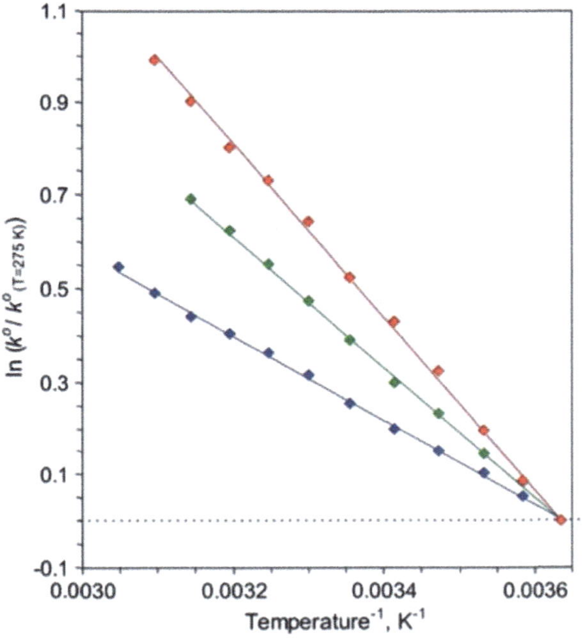

Figure 44. Arrhenius plots for unimolecular standard rate constants reduced to the standard value at 2°C for electron exchange of Azurin hydrophobically immobilized on SAM coated Au electrodes; the CH_3-terminated n-alkanethiol SAMs are n = 4 (red), 10 (green), 15 (blue). Standard errors for experimental points directly match the symbol size.[83] Reproduced from Ref. 83, Copyright (2010) with permission from Proc. Natl. Acad. Sci.

Table 16 and it is natural to propose that this increase comes from an additional component arising from the slow relaxation (viscosity-related, or frictional) term, such that $\Delta H_{a(EXP)} = \Delta H_a^* + H_{a(\eta)}$. Assuming that $\Delta H_a^* = 7.2$ kJ mol^{-1} (because $H_{if} << \lambda_o/4$, *vide supra*) one obtains $H_{a(\eta)} = 15.4 - 7.2 = 8.2$ (kJ mol^{-1}), a value that can be attributed to the accompanying relaxation process(es) of the slowly fluctuating protein/water/SAM environment [83] (also Refs.76 and 81). Interestingly, the value of $\Delta H_{a(EXP)}$ for $n = 10$ (the transition point between two regimes, Fig. 41) is intermediate to the other two cases, see Table 16.

4. Activation Volumes from High-Pressure Kinetic Studies

High-pressure kinetic studies provide activation volumes, obtained according to Eq. (20) (Section II), for the ET. Figure 45 shows logarithmic plots for k^o_{ET} as a function of pressure. Experimental values for $\Delta V_{a(EXP)}$ obtained for the cases of $n = 4$, 10 and 15 are presented in Table 16. One can see a dramatic change in the value and even the sign of $\Delta V_{a(EXP)}$ as the system changes from the long-range ET ($n = 15$) to short range ET ($n = 4$). From Eq. (21) for the nonadiabatic limit, it follows that $\Delta V_{a(EXP)}$ originates from the effect of pressure on the ET distance and/or the medium (SAM/protein/solvent) reorganization energy (Franck-Condon factor, Section II). The value of $\Delta V_{a(EXP)} = -5.1 \pm 0.6$ cm^3 mol^{-1} ($n = 15$, Table 16) is very similar to that reported earlier for the ET of cytochrome *c* freely diffusing to [Au–S–$(CH_2)_6$–OH] junctions, *viz.*, -5.5 ± 0.5 cm^3 mol^{-1} (Section V).[80,81] Assuming that the contribution to λ_o from the protein interior is small and hardly affected by pressure,[153] the contribution of the first term of Eq. (21) can be ascribed to reorganization energy changes of the remote bulk water amounting to ca. $-(2$ to $3)$ cm^3 mol^{-1},[81,111] and the remainder, $-(2$ to $3)$ cm^3 mol^{-1}, can be ascribed to the contribution from the system's compression in the transition state versus the initial state.[81,111]

For the friction controlled limit, Eq. (22) applies (see Section II). Using $\Delta V_{a(EXP)} = +1.7 \pm 0.3$ cm^3 mol^{-1} ($n = 4$, Table 16) and assuming that the second term of Eq. 22 is similar to that for the nonadiabatic case, the contribution of the first viscosity-related term, $\Delta V_{a(\eta)}$ (caused by *protein friction*) is found to be $+(4$ to $5)$

cm^3 mol^{-1} (Table 16).[83] According to Refs. 285 and 286, the native structure of azurin is preserved for pressures up to 300 MPa, but its internal flexibility decreases.[285] This finding substantiates the increase in internal friction that is deduced from these results. This value is about half of the analogous estimate for free cytochrome *c*, *viz*., +(8 to 10) cm^3 mol^{-1}.[81] Assuming that the friction con-

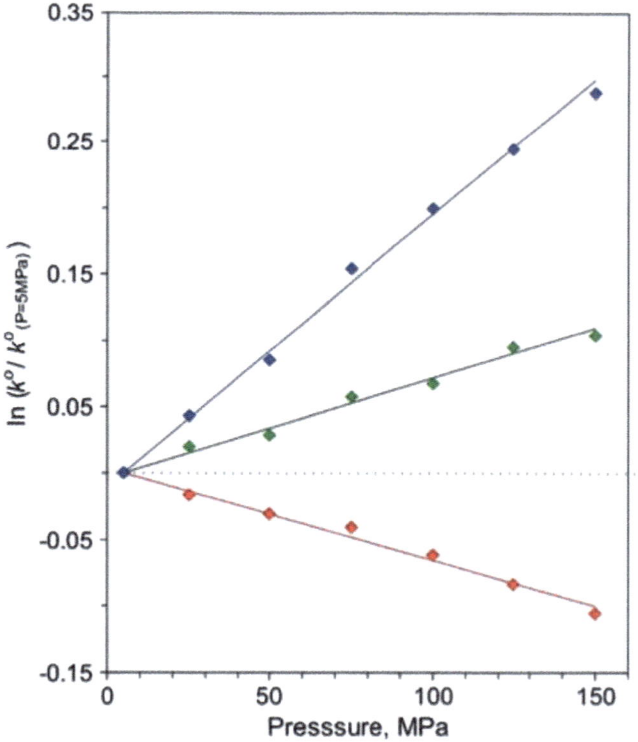

Figure 45. Logarithmic plots for unimolecular standard rate constants reduced to standard pressure at 5 MPa versus hydrostatic pressure for the electron exchange of Azurin hydrophobically attached to Au electrodes modified by CH_3-terminated *n*-alkanethiol SAMs, *n* = 4 (red), *n* = 10 (green), *n* = 15 (blue). Standard errors for experimental points are two times larger than the symbol size.[83] Reproduced from Ref. 83, Copyright (2010) with permission from Proc. Natl. Acad. Sci.

trolled mechanism operates for both proteins, this difference must reflect a different flexibility (mobility) of the cytochrome *c* and azurin interiors. The freely diffusing regime for cytochrome *c* versus the *rigidified*, immobilized azurin (because of its multipoint interaction with the SAM terminal groups) presumably allows for the stronger impact of pressure on its flexibility.

The agreement of the temperature and pressure kinetic studies and the reorganization free energy ($\lambda_o = 0.3 \pm 0.03$ eV) provide a self-consistent picture for the ET reaction. Namely, the long- and short-range ET reactions display distinctive signatures of non-adiabatic and friction controlled ET mechanisms and the indication of a smooth changeover between them through the mixed ET regime, notably for $n = 10$. This finding is in remarkable agreement with matching results for cytochrome *c* operating at Au/SAM junctions through both irreversibly adsorbed[74,76,82] and freely diffusing[80,81] modes (see Sections V to VII of this Chapter).

IX. CONCLUDING REMARKS

The studies discussed here demonstrate that electrochemical, particularly bioelectrochemical devices comprising metal/SAM/ reactant assemblies with nanoscopically tunable physical properties, provide a uniquely powerful system for fundamental electron transfer studies and nanotechnological applications. These electrode assemblies provide an avenue to investigate how the electron transfer mechanism changes with the electronic coupling and the nuclear relaxation characteristics of the assembly. In particular, studies with small redox molecules and with redox proteins were investigated and found to show two extreme regimes for their electron transfer: a nonadiabatic (or tunneling) regime for thick monolayer films (long range ET/ weak electronic coupling) and a friction controlled (dynamically controlled) regime for thin monolayer films (short range ET/ strong electronic coupling). Although this latter regime is now well established, it is less well appreciated and might be more pervasive and important a mechanism than currently believed. Furthermore, these studies reveal that electron transfer between electrodes and proteins in SAM/protein assemblies are in accord with basic theoretical notions concerning the intrinsic ET mechanisms. This basic mechanistic understanding of electrode/SAM/protein assemblies will allow for further progress in understanding more complex processes such as, proton-coupled electron transfer, and electrocatalysis, among others. In addition to these fundamental mechanistic insights, the studies on the protein systems reveal how the binding motif of the protein to the electrode can be changed to manipulate its behavior. For example, we showed that it was possible to create SAMs in which one (or a few) electron tunneling pathways dominate the ET and that it was possible to create SAMs that displayed multi-point tunneling pathways. These few highlights of the work illustrate the fundamental questions and issues that remain to be asked and make it clear that in depth fundamental studies are necessary for the further development and design of new bio-nanotechnologies, including biomimetic ones.

ACKNOWLEDGEMENTS

DHW and DEK are grateful to NRC for financial support in the framework of the Twinning Program Grant (1999-2000), and to NATO for a Collaborative Linkage Grant PST.CLG 975701 (2000-2001). DHW acknowledges financial support from the US-National Science Foundation (CHE-0718755). DEK acknowledges numerous resumptions of his Fellowship from the Alexander von Humboldt Foundation, Germany, during 2002-2005, the Fellowship from Dr. Hertha and Helmut Schmauser Foundation, Germany (2006), and the Cariplo Fellowship from the Landau Network – Centro Volta, Italy (2002, 2005, 2009). Current Research Grants from the Volkswagen Foundation, Germany (I/83 395), and the Georgian National Science Foundation (GNSF /ST08/2-374) for DEK are kindly acknowledged. We thank our coauthors, especially R. van Eldik, T.D Dolidze, S. Rondinini, K. Davis, J. Petrovic, H. Yue, J. Wei, H. Liu, H. Yamamoto, and R. Clark for a long-standing cooperation, and colleagues: J. Ulstrup, W. Schmickler, P. Hildebrandt, D. Beratan, L. Zusman and D. Murgida for past stimulating discussions, especially concerning the problem of a dynamical control in biomolecular electron transfer.

REFERENCES

[1] R. A. Marcus, and N. Sutin, *Biochim. Biohphys. Acta* **811** (1985) 265.

[2] L. I. Krishtalik, *Charge Transfer Reactions in Electrochemical and Chemical Processes*, Plenum, New York, 1986.

[3] A. M. Kuznetsov, *Charge Transfer in Physics, Chemistry and Biology*, Gordon and Beach: Reading, MA, 1995.

[4] P. F. Barbara, T. J. Meyer, and M. A. Ratner, *J. Phys. Chem.* **100** (1996) 13148.

[5] *Electron Transfer from Isolated Molecules to Biomolecules.* (*Advances in Chemical Physics*), J. Jortner, and M. Bixon, eds., Wiley, NY, 1999, Vols. 106 and 107.

[6] J.-M. Savéant, *Elements of Molecular and Biomolecular Electrochemistry: An Electrochemical Approach to Electron Transfer Chemistry*, John Wiley, Hoboken, New Jersey, 2006.

[7] A. M. Kuznetsov, and J. Ulstrup, *Electron Transfer in Chemistry and Biology*, Wiley, Chichester, 1999.

[8] A. Nitzan, and M. A. Ratner, *Science* **300** (2003) 1384.

[9] *Nano-Physics and Bio-Electronics: A New Odyssey*, T. Chakraborty, F. Peeters, and U. Sivan, eds., Elsevier, Amsterdam, 2003.

[10] F. A. Armstrong, H. A. Hill, and N. J. Walton, *Acc. Chem. Res.* **21** (1988) 407.

[11] B. E. Conway, *Theory and Principles of Electrode Processes*, Ronald Press, New York, 1965.

[12] J. O'M. Bockris, and A. K. N. Reddy, *Modern Electrochemistry*, vol. 2; Plenum Press, New York, 1970.

[13] W. R. Fawcett, *Liquids, Solutions and Interfaces*; Oxford University Press, 2004.

[14] A. Ulman, *An Introduction to Ultrathin Organic Films: From Langmuir-Blodgett to Self-Assembly*, Academic Press, New York, 1991.

[15] C. E. D. Chidsey, *Science* **251** (1991) 919.

[16] A. M. Bond, *Inorg. Chim. Acta* **226** (1994) 293.

[17] F. A. Armstrong, and G. S. Wilson, *Electrochim. Acta* **45** (2000) 2623.

[18] M. Fedurco, M. *Coord. Chem. Rev.* **209** (2000) 263.

[19] D. H. Murgida, and P. Hildebrandt, *Acc. Chem. Res.* **37** (2004) 854.

[20] C. Leger, and P. Bertrand, *Chem Rev.* **108** (2008) 2379.

[21] J. Zhang, A. M. Kuznetsov, I. G. Medvedev, Q. Chi, T. Albrecht, P. S. Jensen, and J. Ulstrup, *Chem. Rev.* **108** (2008) 2737.

[22] A. L. Eckermann, D. J. Feld, J. A. Shaw, and T. J. Meade, *Coord. Chem. Rev.* **254** (2010) in press (doi:10.1016/j.ccr.2009.12.023).

[23] M. D. Porter, T. B. Bright, D. L. Allara, and C. E. D. Chidsey, *J. Am. Chem. Soc.* **109** (1987) 3559.

[24] M. R. Linford, and C. E. D. Chidsey, *J. Am. Chem. Soc.* **115** (1993) 12631.

[25] J. F. Smalley, S. W. Feldberg, C. E. D. Chidsey, M. R. Linford, M. D. Newton, and Y.-P. Liu, *J. Phys. Chem.* **99** (1995) 13141.

[26] S. B. Sachs, S. P. Dudek, R. P. Hsung, L. R. Sita, J. F. Smalley, M. D. Newton, S. W. Feldberg, and C. E. D. Chidsey, *J. Am. Chem. Soc.* **117** (1997) 10563.

[27] H. O. Finklea, D. A. Snider, J. Fedyk, E. Sabatani, Y. Gafni, and I. Rubinstein, *Langmuir* **9** (1993) 3660.

[28] H. O. Finklea, in: *Electroanalytical Chemistry*, A. J. Bard, and I. Rubinstein, eds., Marcel Dekker, New York, 1996, vol. 19, p.109.

[29] C. Miller, P. Cuendet, and M. Grätzel, *J. Phys. Chem.* **95** (1991) 877.

[30] C. Miller, and M. Grätzel, *J. Phys. Chem.* **95** (1991) 5225.

[31] A. M. Becka, and C. Miller, *J. Phys. Chem.* **96** (1992) 2657.

[32] A. M. Becka, and C. Miller, *J. Phys. Chem.* **97** (1993) 6233.

[33] S. Terrettaz, A. M. Becka, M. J. Traub, J. C. Fettinge, and C. Miller, *J. Phys. Chem.* **99** (1995) 11216.

[34] L.-H. Guo, J. S. Facci, and G. McLendon, *J. Phys. Chem.* **99** (1995) 8458.

[35] L. A. Curtiss, and J. R. Miller, J.R. *J. Phys. Chem. A* **102** (1998) 160.

[36] J. Xu, H.-L. Li, and Y. Zhang, *J. Phys. Chem.* **97** (1993) 11497.

[37] M. French, and S. Creager, *Langmuir* **14** (1998) 2129.

[38] S. Creager, C. J. Yu, C. Bamdad, S. O'Connor, T. MacLean, E. Lam, Y. Chong, G. T. Olsen, J. Luo, J. Gozin, and J. F. Kayyem, J. F. *J. Am. Chem. Soc.* **121** (1999) 1059.

[39] P. Diao, D. Jiang, X. Cui, D. Gu, R. Tong, and B. Zhong, *J. Electroanal. Chem.* **464** (1999) 61.

[40] K. Slowinski, R. V. Chamberlain, R. Bilewicz, and M. Majda, *J. Am. Chem. Soc.* **118** (1996) 4709.

[41] K. Slowinski, R. V. Chamberlain, C. J. Miller, and M. Majda, *J. Am. Chem. Soc.* **119** (1997) 11910.

[42] K. Slowinski, K. U. Slowinska, and M. Majda, *J. Phys. Chem. B* **103** (1999) 8544.

[43] C. W. Sheen, J. X. Shi, J. Martensson, A. N. Parikh, and D. L. Allara, *J. Am. Chem. Soc.* **114** (1992) 1514.

[44] Y. Gu, Z. Lin, R. A. Butera, V. S. Smentkowski, and D. H. Waldeck, *Langmuir* **11** (1995) 1849.

[45] Y. Gu, and D. H. Waldeck, *J. Phys. Chem.* **100** (1996) 9537.

[46] Y. Gu, and D. H. Waldeck, *J. Phys. Chem. B* **102** (1998) 9015.

[47] D. E. Khoshtariya, T. D. Dolidze, L. D. Zusman, and D. H. Waldeck, *J. Phys. Chem. A* **105** (2001) 1818.

[48] A. M. Napper, H. Liu, and D. H. Waldeck, *J. Phys. Chem. B* **105** (2001) 7699.

[49] R. Bilewicz, S. Sek, and I. Zawisza, *Russ. J. Electrochem.* **38** (2002) 29.

[50] H. D. Sikes, J. F. Smalley, S. P. Dudek, A. R. Cook, M. D. Newton,C. E. D. Chidsey, and S. W. Feldberg, *Science* **291** (2001) 1519.

[51] J. F. Smalley, H. O. Finklea, C. E. D. Chidsey, M. R. Linford, S. E. Creager, J. P. Ferraris, K.Chalfant, T. Zawodzinsk, S. W. Feldberg, and M. D. Newton, *J. Am. Chem. Soc.* **125** (2003) 2004.

[52] J. F. Smalley, S. B. Sachs, C. E. D. Chidsey, S. P. Dudek, H. D. Sikes, S. E. Creager, C. J. Yu, S. W. Feldberg, and M. D. Newton, *J. Am. Chem. Soc.* **126** (2004)14620.

[53] M. D. Newton, and J. F. Smalley, *Phys. Chem. Chem. Phys.* **9** (2007) 555.

[54] M. J. Tarlov, and E. F. Bowden, *J. Am. Chem. Soc.,* **113** (1991) 1847.

[55] S. Song, R. A. Clark, E. F. Bowden, and M. J. Tarlov, *J. Phys. Chem.* **97** (1993) 6564.

[56] S. Terrettaz, J. Cheng, C.J. Miller, and R. D. Guiles, *J. Am. Chem. Soc.,* **118** (1996)7857.

[57] J. Cheng, S. Terrettaz, C. J. Blankman, C. J. Miller, R. D. Dangi, and R. D. Guiles, *Isr. J. Chem.*, **37** (1997) 259.

[58] Z. Q. Feng, S. Imabayashi, T. Kakiuchi, and K. Niki, *J Chem. Soc. Faraday. Trans.* **93** (1997) 1367.

[59] A. Avila, B. W. Gregory, K. Niki, and T. M. Cotton, *J. Phys. Chem. B* **104** (2000) 2759.

[60] K. Niki, W. R. Hardi, M. G. Hill, H. Li, J. R. Sprinkle, E. Margoliash, K. Fujita, R. Tanimura, N. Nakamura, H. Ohno, J. H. Richards, and H. B. Gray, *J. Phys. Chem. B* **107** (2003) 9947.

[61] D. H. Murgida, and P. Hildebrandt, *J. Am. Chem. Soc.* **123** (2001) 4062.

[62] D. H. Murgida, and P. Hildebrandt, *Phys. Chem. Chem. Phys.* **7** (2005) 3773.

[63] D. H. Murgida, and P. Hildebrandt, *Chem. Soc. Rev.* **37** (2008) 937.

[64] K. Ataka, and J. Heberle, *J. Am. Chem. Soc.* **125** (2003) 4986.

[65] K. Ataka, and J. Heberle, *J. Am. Chem. Soc.* **126** (2004) 9445.

[66] Q. Chi, J. Zhang, J. E. T. Andersen, and J. Ulstrup, *J. Phys. Chem. B* **105** (2001) 4669.

[67] L. J. C. Jeuken, J. P. McEvoy, and F. A. Armstrong, *J. Phys. Chem. B* **106** (2002) 2304.

[68] K. Fujita, N. Nakamura, H. Ohno, B. S. Leigh, K. Niki, H. B. Gray, and Richards, *J. Am. Chem. Soc.* **126** (2004) 13954.

[69] Q. Chi, O. Farver, and J. Ulstrup, *Proc. Natl. Acad. Sci. USA* **102** (2005) 16203.

[70] K. Yokoyama, B. S. Leigh, Y. Sheng, K. Niki, N. Nakamura, H. Ohno, J. R. Winkler, H. B. Gray, and J. H. Richards, *Inorg. Chim. Acta* **361** (2008) 1095.

[71] Y. Guo, J. Zhao, X. Yin, X. Gao, and Y. Tian, *J. Phys. Chem. C* **112** (2008) 6013.

[72] B. D. Fleming, S. Praporski, A. M. Bond, and L. L. Martin, *Langmuir* **24** (2008) 323.

[73] H. Yamamoto, H. Liu, and D. H. Waldeck, *Chem. Commun.* (2001) 1032.

[74] J. Wei, H. Liu, D. E. Khoshtariya, H. Yamamoto, A. Dick, D. H. Waldeck, *Angew. Chem. Int. Ed.*, **41** (2002) 4700.

[75] J. Wei, H. Liu, A. Dick, H. Yamamoto, Y. He, and D. H. Waldeck, *J. Am. Chem. Soc.* **124** (2002) 9591.

[76] D. E. Khoshtariya, J. Wei, H. Liu, H. Yue, and D. H. Waldeck, *J. Am. Chem. Soc.* **125** (2003) 7704.

[77] J. J. Wei, H. Liu, K. Niki, E. Margoliash, and D. H. Waldeck, *J. Phys. Chem.B* **108** (2004) 16912.

[78] H. Yue, D. E. Khoshtariya, D. H. Waldeck, J. Grochol, P. Hildebrandt, and D. H. Murgida, *J. Phys. Chem. B* **110** (2006) 19906.

[79] H. Yue, and D. H. Waldeck DH *Curr. Opinion Solid State Mat. Sci.* **9** (2006) 28.

[80] D. E. Khoshtariya, T. D. Dolidze, D. Sarauli, and R. van Eldik, *Angew. Chem. Int. Ed.* **45** (2006) 277.

[81] D. E. Khoshtariya, T. D. Dolidze, S. Seyfert, D. Sarauli, G. Lee, and R. van Eldik, *Chem. Eur. J.* **12** (2006) 7041.

[82] T. D. Dolidze, S. Rondinini, A. Vertova, D. H. Waldeck, and D. E. Khoshtariya, *Biopolymers* **87** (2007) 68.

[83] D. E. Khoshtariya, T. D. Dolidze, M. Shushanyan, K. L. Davis, D. H. Waldeck, and R. van Eldik, *Proc. Natl. Acad. Sci. USA* **107** (2010) 2757.

[84] L. S. Wong, F. Khan, and J. Micklefield, *Chem. Rev.* **109** (2009) 4025.

[85] S. Chen and L. M. Smith, *Langmuir* **25** (2009) 12275.

[86] C. Bunte, O. Prucker, T. Koenig, and J. Ruehe, *Langmuir* (2010) ASAP.

[87] C.-C. Wu, D. N. Reinhoudt, C. Otto, A. H Velders, and V. Subramaniam, *ACS Nano* (2010) ASAP.

[88] Z. R. Taylor, K. Patel, T. G. Spain, J. C. Keay, J. D. Jernigen, E. S. Sanchez, B. P. Grady, M. B. Johnson, and D. W. Schmidtke, *Langmuir* **25** (2009) 10932.

[89] L. D. Zusman, *Chem. Phys.* **49** (1980) 295.

[90] D. F. Calef and P. G. Wolynes, *J. Phys. Chem.* **87** (1983) 3387.

[91] J. T. Hynes, *J. Phys. Chem.* **90** (1986) 3701.

[92] H. Sumi, and R. A. Marcus, *J. Chem. Phys.* **84** (1986) 4894.

[93] I. Rips, and J. Jortner, *J. Chem. Phys.* **87** (1987) 2090.

[94] L. D. Zusman, *Z. Phys. Chem.* **186** (1994) 1.

[95] D. N. Beratan, and J. N. Onuchic, *J. Chem. Phys.* **89** (1988) 6195.

[96] R. Kosloff, and M. A. Ratner, *J. Phys. Chem. B* **106** (2002) 8479.

[97] W. Schmickler, and J. Mohr, *J. Chem. Phys.* **117** (2002) 2867.

[98] A. K. Mishra and D. H. Waldeck, *J. Phys. Chem. C*, **113** (2009) 17904.

[99] N. A. Deskins, and N. Dupuis, *Phys. Rev. B* **75** (2007) 195212.

[100] B. M. Hoffman, and M. A. Ratner, *J. Am. Chem. Soc.* **109** (1987) 6237.

[101] V. Davidson, *Acc. Chem. Res.* **33** (2000) 87.

[102] L. J. C. Jeuken, *Biochim. Biohphys. Acta* **1604** (2003) 67.

[103] V. Davidson, *Acc. Chem. Res.* **41** (2008) 730.

[104] A. Kranich, H. K. Ly, P. Hildebrandt, and D. H. Murgida, *J. Am. Chem. Soc.* **130** (2008) 9844.

[105] S. Hammes-Schiffer, and A. V. Soudackov, *J. Phys. Chem. B* **112** (2008) 14108.
[106] D. Beece, L. Eisenstein, H. Frauenfelder, D. Good, M. C. Marden, L. Reinish, A. H. Reinolds, L. B. Sorensen, and K. T. Yue, *Biochemistry* **19** (1980) 5147.
[107] D..E. Khoshtariya, *Biofizika (Sov. Biophysics)* **31** (1986) 391.
[108] N. G. Goguadze, D. E. Khoshtariya, J. M. Hammerstad-Pedersen and J. Ulstrup, *Eur. J. Biochem.* **200** (1991) 423.
[109] B. Gavish, and S. Yedgar, in: *Protein – Solvent Interactions*, R. B. Gregory, ed., Marcel Dekker, New York, 1995, p.343.
[110] H. Frauenfelder, N. A. Alberding, A. Ansari, D. Braunstein, B. Cowen, M. K. Hong, I. E. T. Iben, J. B. Johnson, S. Luck, M. C. Marden, J. R. Mourant, P. Ormos, L. Reinisch, R. Scholl, A. Schulte, E. Shyamsunder, L. B. Soremen, P. J. Steinbach, A. Xie, R. D. Young, and K. T. Yue, *J. Phys. Chem.* **94** (1990) 1024.
[111] T. D. Dolidze, D. E. Khoshtariya, D. H. Waldeck, J. Macyk and R. van Eldik, *J. Phys. Chem. B* **107** (2003) 7172.
[112] L. Qin, N.M. Kostić, *Biochemistry* **33** (1994) 12592.
[113] M. M. Crnogorac, C. Shen, S. Young, Ö. Hansson, and N. M. Kostić, *Biochemistry* **35** (1996) 16465.
[114] M. M. Ivković-Jensen, G. M. Ullmann, S. Young, Ö. Hansson, M. M. Crnogorac, M. Ejdeback, and N. M. Kostić, *Biochemistry* **37** (1998) 9557.
[115] A. Kapturkievicz, and B. Behr, *J. Electroanal. Chem.* **179** (1984) 187.
[116] A. Kapturkievicz, and M. Opallo, *J. Electroanal. Chem.* **185** (1985) 15.
[117] M. Opallo, and A. Kapturkievicz, *Electrochim. Acta* **30** (1985) 1301.
[118] M. Opallo, *J. Chem. Soc. Faraday Trans.* **82** (1986) 339.
[119] W. R. Fawcett, and C. A. Foss, *J. Electroanal. Chem.* **306** (1991) 71.
[120] W. R. Fawcett, and C. A. Foss, *Electrochim. Acta* **36** (1991) 1767.
[121] W. R. Fawcett, and M. Opallo, *J. Electroanal. Chem.* **331** (1992) 815.
[122] M. J. Weaver, and T. Gennet, *Chem. Phys. Lett.* **113** (1985) 213.
[123] T. Gennet, D. F. Milner, and M. J. Weaver, *J. Phys. Chem.* **89** (1985) 2787.
[124] G. E. McManis, M. N. Golovin, and M. J. Weaver, *J. Phys. Chem.* **90** (1986) 6563.
[125] R. M. Nielson, and M. J. Weaver, *J. Electroanal. Chem.* **260** (1989) 15.
[126] R. M. Nielson, G. E. McManis and M. J. Weaver, *J. Phys. Chem.* **93** (1989) 4703.
[127] M. J. Weaver, D. K. Phelps, R. M. Nielson, M. N. Golovin and G. E. McManis, *J. Phys. Chem.* **94** (1990) 2949.
[128] M. J. Weaver and G. E. McManis, *Acc. Chem. Res.* **23** (1990) 294.
[129] M. J. Weaver, *Chem. Rev.* **92** (1992) 463.
[130] W. R. Fawcett, and M. Opallo, *Angew. Chem. Int. Ed. Engl.* **33** (1994) 2131.
[131] X. Zhang, J. Leddy, and A. J. Bard, *J. Am. Chem. Soc.* **107** (1985) 3719.
[132] X. Zhang, H. Yang, and A. J. Bard, *J. Am. Chem. Soc.* **109** (1987) 1916.
[133] R. Pyati, and R. W. Murray, *J. Am. Chem. Soc.* **118** (1996) 1743.

[134] M. E. Williams, J. C. Crooker, R. Pyati, L. J. Lyons, and R. W. Murray, *J. Am. Chem. Soc.* **119** (1997) 10249.
[135] D. E. Khoshtariya, T. D. Dolidze, D. Krulic, N. Fatouros, and D. Devilliers, *J. Phys. Chem. B* **102** (1998) 7800.
[136] D. E. Khoshtariya, T. D. Dolidze, A. Vertova, M. Longhi, and S. Rondinini, *Electrochem. Commun.* **5** (2003) 241.
[137] T. Asano, H. Furuta, and H. Sumi, *J. Am. Chem. Soc.* **116** (1994) 5545.
[138] R. R. Dogonadze, and A. M. Kuznetsov, *Progr. Surf. Sci.* **6** (1975) 1.
[139] N. Sutin, *Acc. Chem. Res.* **15** (1982) 275.
[140] R. J. D. Miller, G. L. McLendon, A. J. Nozik, W. Schmickler, and F. Willig, *Surface Electron Transfer Processes*,VCH, NY, 1995.
[141] M. Galperin, M. A. Ratner, and A. J. Nitzan, *Phys. Cond. Matt.* **19** (2007) 103201.
[142] A. K. Mishra, and S. K. Rangarajan, *J. Mol. Struc. (Theochem)* **361** (1996) 101.
[143] A. K. Mishra, and S. K. Rangarajan, *J. Phys. Chem.* **91** (1987) 3417.
[144] A. K. Mishra, and S. K. Rangarajan, *J. Phys. Chem.* **91** (1987) 3425.
[145] E. P. Vakarin, Y. Duda, and J. P. Badiali, *J. Chem. Phys.* **124** (2006) 144515.
[146] J. Mohr, and W. Schmickler, *Phys. Rev. Lett.* **84** (2000) 1051.
[147] H. A. Kramers, *Physica (Utrecht)* **7** (1940) 284.
[148] P. Debye, *Polar Molecules*, Dover Publishers, New York, 1929.
[149] N. Nandi, K. Bhattacharyya, and B. Bagchi, *Chem. Rev.* **100** (2000) 2013.
[150] Y.-P. Liu and M. D. Newton, *J. Phys. Chem.* **98** (1994) 7162.
[151] Y. I. Kharkats and L. I. Krishtalik, *J. Theor. Biol.* **112** (1985) 221.
[152] I. Muegge, P. X. Qi, A. J. Wand, Z. T. Chu, and A. Warshel, *J. Phys. Chem. B* **101** (1997) 825.
[153] O. Miyashita, and N. Go, *J. Phys. Chem. B* **103** (1999) 562.
[154] M. Cascella, A. Magistrato, I. Tavernelli, P. Carloni, and U. Rothlisberger, *Proc. Natl. Acad. Sci. USA* **103** (2006) 19641.
[155] H. Frauenfelder, P. W. Fenimore, and B. H. McMahon, *Biophys. Chem.* **98** (2002) 35.
[156] P. W. Fenimore, H. Frauenfelder, B. H. McMahon, and F. G. Parak, *Proc. Natl. Acad. Sci. USA*, **99** (2002) 16047.
[157] P. W. Fenimore, H. Frauenfelder, B. H. McMahon, and R. D.Young, *Proc. Natl. Acad. Sci. USA*, **101** (2004) 14408..
[158] A. M. Tsai, D. A. Neumann, and L. N. Bell, *Biophys. J.* **79 (2004)** 2728.
[159] J. J. Hill, E. Y. Shalaev, and G. Zografi, *J. Pharm. Sci.* **94** (2005) 1636.
[160] J. L. Green, J. Fan, and C. A. Angell, *J. Phys. Chem.* **98** (1994) 13780.
[161] C. A. Angell, *Science,* **267** (1995) 1924.
[162] D. A. Cherepanov, L. I. Krishtalik, and A. Y. Mulkidjanian, *Biophys. J.* **80** (2001) 1033.

[163]A. I. Kotelnikov, J. M. Ortega, E. S. Medvedev, B. L. Psikha, D. Garcia, and P. Mathis, *Bioelectrochemistry* **56** (2002) 3.
[164]K. Weber, and S. E. Creager, *Anal. Chem.* **66** (1994) 3164.
[165]L. Tender, M. T. Carter, and R. W. Murrey, *Anal. Chem.* **66** (1994) 3173.
[166]J. C. Imbeaux, and J. M Savèant, *J. Electroanal. Chem.* **44** (1973) 169.
[167]W. Schmickler, *Electrochim. Acta* **20** (1975) 137.
[168]D. E. Khoshtariya, T. D. Dolidze, A. Neubrand, and R. van Eldik, *J. Mol. Liquids* **89** (2000) 127.
[169]D. E. Khoshtariya, T. D. Dolidze, and R. van Eldik, *Phys. Rev. E* **80** (2009) 065101.
[170]N. A. Lewis, Y .S. Obeng, D. V. Taveras, and R. van Eldik, *J. Am. Chem. Soc.* **111** (1989) 924.
[171]N. A. Lewis, R .R. McNeer, and D. V. Taveras, *Inorg. Chim. Acta* **225** (1994) 89.
[172]D.E. Khoshtariya, R. Billing, M. Ackermann, R. van Eldik, *J. Chem. Soc. Faraday Trans.* **91** (1995) 1625.
[173]D. E. Khoshtariya, R. Meusinger, and R. Billing, *J. Phys. Chem.* **99** (1995) 3592.
[174]P. Vath, M. B. Zimmt, D. V. Matyushov and G. A. Voth, *J. Phys. Chem. B* **103** (1999) 9130.
[175]M. B. Zimmt and D. H. Waldeck, *J. Phys. Chem. A* **107** (2003) 3580.
[176]*Inorganic High-Pressure Chemistry. Kinetics and Mechanisms*, R. van Eldik, ed., Elsevier, Amsterdam, 1986.
[177]R. van Eldik, T. Asano, and W. J. Le Noble, *Chem. Rev.* **89** (1989) 549.
[178]A. Drljaca, C. D. Hubbard, R. van Eldik, T. Asano, M. V. Basilevsky, and W. J. Le Noble, *Chem. Rev.* **98** (1998) 2167.
[179]*High Pressure Chemistry*, R. van Eldik, and F.-G. Klärner, eds., Wiley-VCH, Weinheim, 2002.
[180]C. D. Hubbard, and R. van Eldik R., in: *Chemistry at Extreme Conditions,* M. Riad Manaa, ed., Elsevier, Amsterdam, 2005, Chapter 4.
[181]Y. Furukawa, K. Ishimori, and I. Moroshima, *J. Phys. Chem. B* **104** (2000) 1817.
[182]Y. Fu, A. S. Cole, and T. W. Swaddle, *J. Am. Chem. Soc.* **121** (1999) 10410.
[183]T. W. Swaddle, and P. A. Tregloan, *Coord. Chem. Rev.* **187** (1999) 255.
[184]T. W. Swaddle, *Chem. Rev.* 105 (2005) 2573.
[185]T. D. Dolidze, D. E. Khoshtariya, P. Illner, L. Kulisiewicz, A. Delgado, and R. van Eldik, *J. Phys. Chem. B* **112** (2008) 3085.
[186]T. D. Dolidze, D. E. Khoshtariya, P. Illner, and R. van Eldik, *Chem. Commun.* (**2008**) 2112.
[187]D. E. Khoshtariya, T .D. Dolidze, and R. van Eldik, *Chem. Eur. J.* **15** (2009) 5254.
[188]P. W. Bridgman, *The Physics of High Pressure*, Bell and Sons, London, 1931, p. 346.

[189] N. S. Isaacs, *Liquid-Phase High-Pressure Chemistry*, Wiley and Sons, New York, 1981, p.102.
[190] J. V. Sengers, and B. Kamgar-Parsi, *J. Phys. Chem., Ref. Data* **13** (1984) 185.
[191] D. E. Khoshtariya, T. D. Dolidze, P. Lindqvist-Reis, A. Neubrand, and R. van Eldik, *J. Mol. Liquids* **96/97** (2002) 45.
[192] D. E. Khoshtariya, A. Zahl, T. D. Dolidze, A. Neubrand, and R. van Eldik, *Chem. Phys. Chem.* **5** (2004) 1398.
[193] D. E. Khoshtariya, A. Zahl, T. D. Dolidze, A. Neubrand, and R. van Eldik, *J. Phys. Chem. B* **108** (2004) 14796.
[194] N. Agmon, *J. Phys. Chem.* **100** (1996) 1072.
[195] A. Geiger, M. Kleene, D. Paschek and A. Rehtanz, *J. Mol. Liquids* **106** (2003) 131.
[196] D. E. Khoshtariya, E. Hansen, R. Leecharoen, and G. C. Walker, *J. Mol. Liquids* **105** (2003) 13.
[197] J. S. Bashkin, G. McLendon, S. Mukamel, and J. Marohn, *J. Phys. Chem.* **94** (1990) 4757.
[198] D. W. Pierce, and S. G. Boxer, *J. Phys. Chem.* **96** (1992) 5560.
[199] Y. Shibata, H. Takahashi, R. Kaneko, A. Kurita and T. Kushida, *Biochemistry* **38** (1999) 1802.
[200] A. Ansari, C.M. Jones, E.R. Henry, J. Hofrichter, and W.A. Eaton, *Biochemistry* **33** (1994) 5128.
[201] D. Sagnella, J. E. Straub, and D. Thirumalai, *J. Chem. Phys.* **113** (2000) 7702.
[202] K. Heremans, and L. Smeller, *Biochim. Biophys. Acta* **1386** (1998) 353.
[203] K. Heremans, and L. Smeller, in: *Biological Systems under Extreme Conditions*, Y. Taniguchi, H. E. Stanley, and H. Ludwig, eds., 2002, p.53-73.
[204] P. J. Artimiuk, C. C. T. Blake, D. E. P. Grace, S. J. Oatley, D. C. Phillips, and M. J. E. Sternberg, *Nature* **280** (1979) 63.
[205] P. Zavodszky, J. Kardos, A. Svingor, and G. A. Petsko, *Proc. Natl. Acad. Sci. USA* **95** (1998) 7406.
[206] D. Vitkup, D. Ringe, G. A. Petsko, and M. Karplus, *Nat. Struct. Biol.* **7** (2000) 34.
[207] R.D. Lins, C. S. Pereira, and P. H. Hünenberger, *Proteins* **55** (2004) 177.
[208] M. D. Newton, and H. L. Friedman, *J. Chem. Phys.* **83** (1985) 5210.
[209] J. T. Hupp, and M. J. Weaver, *J. Electroanal. Chem.* **152** (1983) 1.
[210] A. Gochev, G. E. McManis, and M. J. Weaver, *J. Chem. Phys.* **91** (1989) 906.
[211] W. R. Fawcett, and C. A. Foss, *J. Electroanal. Chem.* **270** (1989) 103.
[212] A. S. Baranski, K. Winkler, and W. R. Fawcett, *J. Electroanal. Chem.* **313** (1991) 367.
[213] R.C. Newmann, W. Kauzmann, and A. Zipp, *J. Phys. Chem.* **77** (1973) 2687.
[214] Y. Kitamura, and T. Itoh, *J. Solution Chem.* **16** (1987) 715.
[215] R. S. Nicholson, *Anal. Chem.* **37** (1965) 1351.

[216] R. S. Nicholson, and I. Shain, *Anal. Chem.* **36** (1964) 706.

[217] A. J. Bard, and L. R. Faulkner, L. R. *Electrochemical Methods. Fundamentals and Applications*, 2nd ed.; John Wiley, New York, 2001.

[218] R. Greef, R. Peat, L. M. Peter, D. Pletcher, and J. Robinson, *Instrumental Methods in Electrochemistry. Southhampton Electrochemistry Group*, Horwood, New York, 1986.

[219] J.-M. Savéant, *J. Phys. Chem. B* **106 (2002**) 9387.

[220] E. Laviron, *J. Electroanal. Chem.* **101** (1979) 19.

[221] B. D. Fleming. S. Praporski, A. M. Bond, and L. L. Martin, *Langmuir* **24** (2008) 323.

[222] A. M. Bond, K. B. Oldham, and G. A. Snook, *Anal. Chem.* **72** (2000) 3492.

[223] W. Oelssner, F. Berthold, and U. Guth, *Materials and Corrosion* **57** (2006) 455.

[224] T. D. Dolidze, S. Rondinini, A. Vertova, M. Longhi and D. E. Khoshtariya, *Open Phys. Chem. J.* **2** (2008) 17.

[225] P. Krysinski, M. R. Moncelli, and F. Taadini-Buoninesegni, *Electrochim. Acta* **45** (2000) 1885.

[226] J. Calvente, G. Lopez-peroz, P. Ramirez, H. Fernandez, M. Zon, W. H. Mulder, and J. Andreu, *J. Am. Chem. Soc.* **127** (2005) 6476.

[227] S. Rondinini, A. Vertova, and L. Pilan, *Electroanalys* **15** (2003) 1297.

[228] D. A. Brevnov, and H. O. Finklea, *Langmuir* **16** (2000) 5973.

[229] W. R. Fowcett, M. Hromadova, G. A. Tsirlina and R. R. Nazmutdinov, *J. Electroanal. Chem.* **498** (2001), 93.

[230] L. V. Protsailo, and W. R. Fawcett, *Electrochim. Acta* **45** (2000) 3497.

[231] P. Krysinski, and M. Brzostowska-Smolska, *J. Electroanal. Chem.* **424** (1997) 61.

[232] P. Krysinski, M. R. Moncelli, and F. Taadini-Buoninesegni, *Electrochim. Acta* **45** (2000) 1885.

[233] V. Mareček, Z. Samec, and J. Weber, *J. Electroanal. Chem.* **94** (1978) 169.

[234] T. D. Dolidze, and T. R. Agladze, *Bull. Georgian Acad. Sci.* **139** (1990) 317.

[235] R. R. Dogonadze, Y. L. Kharkats, and J. Ulstrup, *J. Electroanal. Chem.* **39** (1972) 47.

[236] R. R. Dogonadze, Y. L. Kharkats, and J. Ulstrup, *J. Electro1anal. Chem.* **39** (1973) 161.

[237] L. D. Zusman, and D. N. Beratan, *J. Chem. Phys.* **105** (1997) 165.

[238] L. D. Zusman, and D. N. Beratan, *J. Chem. Phys.* **110** (1999) 10468.

[239] G. Iversen, E. P. Friis, Y. I. Kharkats, A. M. Kuznetsov, and J. Ulstrup, *J. Biol. Inorg. Chem.* **3** (1998) 229.

[240] L. D. Zusman, *Chem. Phys.* **112** (1987) 53.

[241] G. W. Pettigrew, and G. R. Moore, *Cytochromes C. Biological Aspects*; Springer-Verlag, Berlin, **1987**.

[242] G. R. Moore, and G. W. Pettigrew, *Cytochromes C. Evolutionary, Structural and Physicochemical Aspects*; Springer-Verlag, Berlin, **1990**.

[243] R. A. Scott, in: *Cytochrome C: A Multidisciplinary Approach*, R. A. Scott, and A. G. Mauk, eds., University Science Books, Sausalito, 1996, p.515.
[244] M. Ott, J. D. Robertson, V. Gogvadze, B. Zhivotovsky, and S. Orrenius, *Proc. Natl. Acad. Sci. USA* **99** (2002) 1259.
[245] Z. Hao, G. S. Duncan, C .C. Chang, A. Elia, M. Fang, A. Wakeham, H. Okada, T. Calzascia, Y. J. Jang, A. You-Ten, W.-C. Yeh, P. Ohashi, X. Wang, and T. W. Mak, *Cell* **121** (2005) 579.
[246] E. V. Pletneva, D. B. Fulton, T. Kohzuma, and N. M. Kostic, *J. Am. Chem. Soc.* **122** (2000) 1034.
[247] R. C. Lacey, L. Liu, L. Zang, and M. Y. Ogawa, *Biochemistry* **42** (2003) 3904.
[248] M. Tetreault, S. H. Rongey, G. Feher, and M. Y. Okamura, *Biochemistry* **40** (2001) 8452.
[249] O. Miyashita, M. Y. Okamura, and J. N. Onuchic, *Proc. Natl. Acad. Sci. USA* **102** (2005) 3558.
[250] P. Hildebrandt, and M. Stockburger, *Biochemistry* **28** (1989) 6722.
[251] P. Hildebrandt, *Biochim. Biophys. Acta* **1040** (1990) 175.
[252] H. H. J. de Jongh, J. A. Killian, and B. Kruijff, *Biochemistry* **31 (1992**) 1636.
[253] V. E. Bychkova, A. E. Dujsekina, S. I. Klenin, E. I. Tiktopulo, V .N. Uversky, and O. B. Ptitsyn, *Biochemistry* **35 (1996**) 6058.
[254] E. Sedlak, M. Antalik, J. Bagelova, and M. Fedurco, *Biochim. Biophys. Acta* **1319** (1997) 258.
[255] M. Antalik, J. Bagelova, Z. Gazova, A. Musatov, and D. Fedunova, *Biochim. Biophys. Acta* **1646** (2003) 11.
[256] E. K. J. Tuominen, C. J. A. Wallace, and P. K. J. Kinnunen, *J. Biol. Chem.* **277** (2002) 8822.
[257] F. Daldal, M. Deshmukh, and R. C. Prince, *Photosynth. Res.* **76** (2003) 127.
[258] G. Cheddar, and G. Tollin, *Arch. Biochem. Biophys.* **310** (1994) 392.
[259] E. J. Choi, and E. K. Dimitriadis, *Biophys. J.* **87** (2004) 3234.
[260] P. L. Privalov, and S. L. Gill, *Advan. Protein Chem.* **39** (1988) 191.
[261] P. L. Privalov, *J. Mol. Biol.* **258** (1996) 707.
[262] M. Ohgushi, and A. Wada, *FEBS Lett.* **164** (1983) 21.
[263] Y. Goto, and S. Nishikiori, *J. Mol. Biol.* **222** (1991) 679.
[264] Y. Bai, and S. W. Englander, *Proteins*, **24** (1996**)** 145.
[265] O. B. Ptitsyn, *J. Protein Chem.* **6** (1987) 273.
[266] O. B. Ptitsyn, *Curr. Opinion Struct. Biol.* **5** (1995) 74.
[267] V. N. Uversky, and O. B. Ptitsyn, *J. Mol. Biol.* **255** (1996) 215.
[268] N. Poklar, N. Petrovcic, M. Oblak, and G. Vesnaver, *Protein Science* **8** (1999) 832.
[269] D. E. Khoshtariya, M. Shushanyan, R. Sujashvili, M. Makharadze, E. Tabuashvili, and G. Getashvili, *J. Biol. Phys. Chem.* **3** (2003) 2.

[270] J. N. Onuchic, Z. Luthey-Schulten, and P. G. Wolynes, *Annu. Rev. Phys. Chem.* **48** (1997) 545.
[271] C. L. Brooks, M. Gruebele, J. N. Onuchic, P. G. Wolynes, *Proc. Natl. Acad. Sci. USA* **95** (1998) 11037.
[272] R. Ravindra, and R. Winter, *Chem. Phys Chem.* **5** (2004) 566.
[273] J. R. Winkler, and H .B. Gray, *Chem. Revs.* **92** (1992) 369.
[274] H. B. Gray, and J. R. Winkler, *Annu Rev. Biochem.* **65** (1996) 537.
[275] J. R. Winkler, A. J. Di Bilio, N. A. Farrow, J. H. Richards, and H. B. Gray, *Pure Appl. Chem.,* **71** (1999) 1753.
[276] H. B. Gray, and J. R. Winkler, *Proc. Natl. Acad. Sci. USA* **102** (2005) 3534.
[277] M. Meier, J. Sun, J. F. Wishart, and R. van Eldik, *Inorg. Chem.* **35** (1996) 1564.
[278] J. Sun, C. Su, M. Meier, S. S. Isied, J. F. Wishart, and R. van Eldik, *Inorg. Chem.* **37** (1998) 6129.
[279] M. Fedurco, J. Augustinski, C. Indiani, G. Smulevich, M. Antalik, M. Bano, E. Sedlak, M. C. Glascock, and J .H. Dawson, *J. Am. Chem. Soc.* **127** (2005) 7638.
[280] T. Pineda, J. M. Sevilla, A. J. Roman, and M. Blazquez, M. *Biochim. Biophys. Acta* **1343** (1997) 227.
[281] B. A. Feinberg, X. Liu, M. D. Ryan, A. Schejter, C. Zhang, and E. Margoliash, *Biochemistry* **37** (1998) 13091.
[282] B. S. Russel, R. Melenkivitz, and K. L. Bren, *Proc. Natl. Acad. Sci. USA* **97** (2000) 8312.
[283] Santucci, C. Bongiovanni, G. Mei, T. Ferri, F. Polizio, A. Desideri, *Biochemistry* **39** (2000) 12632.
[284] Y.-S. Kim, L. S. Jones, A. Dong, B. S. Kendrick, B. S. Chang, M. C. Manning, T. W. Randolph, and J. F. Carpenter, *Protein Science* **12** (2003) 1252.
[285] P. Cioni, *Biophys J.* **91** (2006) 3390.
[286] G. Mei, A. Di Venere, F. M. Campeggi, G. Gilardi, N. Rosato, F. De Matteis, and A. Finazzi-Agrò, *Eur. J. Biochem.* **265** (1999) 619.
[287] P. Cioni, and G. B. Strambini, *J. Mol. Biol.* **242** (1994) 291.
[288] A. Priev, A. Almagor, S. Yedgar, and B. Gavish, *Biochemistry* **35** (1996) 2061.
[289] N. Taulier, and T. V. Chalikian, *Biochim. Biophys. Acta* **1595** (2002) 48.
[290] D. N. Dubins, R. Filfil, R. B. Macgregor, and T. V. Chalikian, *Biochemistr* **42** (2003) 8671.
[291] D. H. Murgida, P. Hildebrandt, J. Wei, Y.-F. He, H. Liu, and D. H. Waldeck, *J. Phys. Chem. B* **108** (2004) 2261.
[292] T. G. Spiro. and X. Y. Li, *Resonance Raman Spectra of Heme and Metalloproteins in Biological Applications of Raman Spectroscopy*, vol.3, Wiley, NY, 1988.
[293] S. Hirota, T. Ogura, I. K. Shingawa, S. Yoshikawa, and T. Kitagawa, *J. Phys. Chem.* **100** (1996) 15274.
[294] M. Smith, and G. McLendon, *J. Am. Chem. Soc.* **103** (1981) 4912.
[295] S. R. Yeh, and D. L. Rousseau, *J. Biol Chem.* **274** (1999) 17853.

[296] J. C. Ferrer, J. G. Gullemette, R. Bogumil, S. C. Inglis, M. Smith, and A. G. Mauk, *J. Am. Chem. Soc.* **115** (1993) 7507.
[297] G. Liu, Y. Chen, and W. Tang, *J. Chem. Soc. Dalton Trans.* (**1997**) 795.
[298] C. Fan, B. Gillespie, C. Wang, A. J. Heeger, and K. W. Plaxco, *J. Phys. Cem. B* **106** (2002) 11375.
[299] G. Battistuzzi, M. Borsari, J. A. Cowan, A. Ranieri, and M. Sola, *J. Am. Chem. Soc.* **124** (2002) 5315.
[300] G. Battistuzzi, M. Borsari, A. Ranieri, and M. Sola, *J. Am. Chem. Soc.* **124** (2002) 26.
[301] M. J. Honeychurch, *Langmuir* **15** (1999) 5158.
[302] H. Yamamoto, and D. H. Waldeck, *J. Phys. Chem. B* **106** (2002) 7469.
[303] M. R. Mauk, and A. G. Mauk, *Eur. J. Biochem.* **186** (1989) 473.
[304] J. Xu, and E. F. Bowden, *J. Am. Chem. Soc.* **128** (2006) 6813.
[305] K. L. Davis, and David H. Waldeck, *J. Phys. Chem. B* **112** (2008) 12498.
[306] K. Niki, J. R. Sprinkle, and E. Margoliash, *Bioelectrochemistry* **55** (2002) 37.
[307] G. P. Kreishman, M. J. Simone, R. M. Hines, and E. E. Brooks, *Anal. Biochem.* **138** (1984) 360.
[308] T. Pascher, *Biochemistry* 40 (2001) 5812.
[309] I. Taniguchi, M. Iseki, T. Eto, K. Toyosawa, H. Yamaguchi, and K. Yasukouchi, *Bioelectrochem. Bionerg.* **13** (1984) 373.
[310] T. J. Meade, H.B. Gray, and J. R. Winkler, *J. Am. Chem. Soc.* **111** (1989**)** 4353.
[311] J. S. Zhou, and M. A. J. Rodgers, *J. Am. Chem. Soc.* **113** (1991**)** 7728.
[312] D. R. Casimiro, J. H. Richards, J. R. Winkler, and H. B. Gray, *J. Phys. Chem.* **97** (1993) 13073.
[313] G. A. Mines, M. J. Bjerrum, M. G. Hill, D. R. Casimiro, I.-J. Chang, J. R. Winkler, and H. B. Gray *J. Am. Chem. Soc.* **118** (1996**)** 1961.
[314] J. Luo, K. B. Reddy, A. S. Salameh, J. F. Wishart, and S. S. Isied, *Inorg. Chem.* **39** (2000) 2321.
[315] J. S. Zhou, and M. A. J. Rodgers, *J. Am. Chem. Soc.* **113** (1991)7728.
[316] G. McLendon, *Acc. Chem. Res.* **21** (1998) 160.
[317] H. Mei, K. Wang, N. Peffer, G. Weatherly, D.S. Cohen, M. Miller, G. Pielak, B. Durham, and F. Millett, *Biochemistry* **38** (1999) 6846.
[318] R. E. Overfield, C. A. Wraight, and D. DeVault, *FEBS Lett*, **105** (1979) 137.
[319] M. R. Harris, D. J. Davis, B. Durham, and F. Millett, *Biochim. Biophys. Acta* **1319** (1997) 147.
[320] M. M. Ivković-Jensen, and N. M. Kostić, *Biochemistry* **36** (1997) 8135
[321] N. Legrand, A. Bondon, and G. Simonneaux, *Inorg. Chem.* **35** (1996*)* 1627.
[322] E. Sigfridsson, M. H. M. Olsson, and U. Ryde, *J. Phys. Chem. B* **105** (2001) 5546.
[323] G. Basu, A. Kitao, A. Kuki, and N. Go, *J. Phys. Chem. B* **102** (1998) 2085.

[324] I. Muegge, X. P. Qi, A. J. Wand, Z. T. Chu, and A. Warshel, *J. Phys. Chem. B* **101** (1997) 825.
[325] M. Collinson, E. F. Bowden, and M. J. Tarlow, *Langmuir* **8** (1992) 1247.
[326] A. El Kasmi, J. M. Wallace, E. F. Bowden, S. M. Binet, and R. J. Linderman *J. Am. Chem. Soc.* **120** (1998) 225.
[327] M. C. Leopold, and E. F. Bowden *Langmuir* **18** (2002) 2239.
[328] K. L. Davis, B. J. Drews, H. Yue, D. H. Waldeck, K. Knorr, and R. A. Clark, *J. Phys. Chem. C* **112** (2008) 6571.
[329] H. Yue, D. H. Waldeck, J. Petrovic, and R. A. Clark, *J. Phys. Chem. B* **110** (2006) 5062.
[330] H. Yue, D.H. Waldeck, K. Schrock, D. Kirby, K. Knorr, S. Switzer, J. Rosmus and R.A. Clark, *J. Phys. Chem. C* **112** (2008) 2514.
[331] Y. Z. Du, and S. S. Saavedra, *Langmuir* **19** (2003) 6443.
[332] Q. Zhang, and L. A. Anchor, *Langmuir* **21** (2005) 5405.
[333] L. Wang, and D. H. Waldeck, *J. Phys. Chem. C* **112** (2008) 1351.
[334] L. V. Basova, I. V. Kurnikov, L. Wang, V. B. Ritov, N. A. Belikova, I. I. Vlasova, A. A. Pacheco, D. E. Winnica, J. Peterson, H. Bayir, D. H Waldeck, and V. E. Kagan, *Biochemistry* **46** (2007) 3423.
[335] P. Hildebrandt, F. Vanheck, G. Heibel, and A. G. Mauk, *Biochemistry* **32** (1993) 14158.
[336] I. Bertini, J. G. Huber, C. Luchinat, and M. Piccoli, *J. Magn Reson.* **147** (2000) 1.
[337] P. X. Qi, J. L. Urbauer, E. J. Fuentes, M. F. Leopold, and J. A. Wand, *Struct. Biol.* **1** (1994) 378.
[338] J. Petrović, R. A. Clark, H. Yue, D. H. Waldeck, E. F. Bowden, *Langmuir* **21** (2005) 6308.
[339] S. Gosavi and R. A. Marcus, *J. Phys. Chem. B*, **2000**, *104* (9), pp 2067.
[340] E. Vijgenboom, J. E. Busch, and G. W. Canters, *Microbiology* **143** (1997) 2853.
[341] O. Farver, Y. Blatt, and I. Pecht, Biochemistry **21** (1982) 3556.
[342] O. Farver, and I. Pecht, Proc. Natl. Acad. Sci. USA **86** (1989) 6968.
[343] O. Farver, in: *Protein Electron Transfer*, D.Bendall, ed., BIOS Scientific, Oxford, 1996, p. 161.
[344] O. Farver, Y. Li, M. C. Ang, and I. Pecht, I. (1999) Proc. Natl. Acad. Sci. USA **96** (1999) 899.
[345] K. W. Penfield, A. A. Gewirth, and E. I. Solomon, *J. Am. Chem. Soc.* **107** (1985) 4519.
[346] E. I. Solomon, and M. D. Lowey, *Science* **259** (1993) 1575.
[347] A. G. Skyes, *Adv. Inorg. Chem.* **36** (1990) 377.
[348] T. Adman, *Adv. Protein Chem.* **42** (1991) 145.
[349] A. Messerschmidt, *Struct. Bonding* **90** (1980) 37.
[350] W. E. B. Shepard, B. F. Anderson, D. A. Lewandoski, G. E. Norris, and E. N. Backer, *J. Am. Chem. Soc.* **112** (1990) 7817.

[351] A. J. Di Bilio, M. G. Hill, N. Bonader, B. G. Karlsson, R. M.Villahermosa, B. G. Malmström, J. R. Winkler, and H. B.Gray, *J. Am. Chem. Soc.* **119** (1997) 9921.
[352] H. Nar, A. Messerschmidt, R. Huber, M. van de Kamp, and G. W. Canters, *J. Mol. Biol.* **221** (1991) 765.
[353] H. B. Gray, B. G. Malmström, and R. J. P. Williams, *J. Biol. Inorg. Chem.* **5** (2000) 551.
[354] R. Langen, I. J. Chang, J. P. Germanas, J. H. Richards, J. R. Winkler, and H. B. Gray, *Science* **268** (1995) 1733.
[355] U. Ryde, M. H. M. Olsson, B. O. Roos, J. A. DeKerpel, and K. Pierloot, *J. Biol. Inorg. Chem.* **5** (2000) 565.
[356] B. L. Vallee, and R. J. P. Williams, *Proc. Natl. Acad. Sci. USA* 59 (1968) 498.
[357] B. G. Malmström, *Eur. J. Biochem.* **223** (1994) 207.
[358] K. Pierloot, J. O. A. De Kerpel, U. Ryde, M. H. M. Olsson, and B. O. Roos, *J. Am. Chem. Soc.* **120** (1998) 13156.
[359] U. Ryde, and M. H. M. Olsson, *Int. J. Quantum Chem.* **81** (2001) 335.
[360] M. van Gastel, J. W. A. Coremans, H. Sommerdijk, M. C. Hemert, and E. J. J. Groenen, *J. Am. Chem. Soc.* **124** (2002) 2035.
[361] A. R. Jaszewski, K. Tabaka, J. Jezierska, and J. Kedzierska, *Chem. Phys. Lett.* **367** (2001) 678.
[362] S. Corni, *J. Phys. Chem. B* **109** (2005) 3423.
[363] T. Simonson, *Proc. Natl. Acad. Sci. USA* **99**(2002) 6544.
[364] J. W. A. Coremans, O. G. Poluektov, E. J. J. Groenen, G. W. Canters, H. Nar, and A. Messerschmidt, *J. Am. Chem. Soc.* **119** (1997) 4726.
[365] J. Winkler, P. Wittung-Stafshede, J. Leckner, B. G. Malmström, and H. B. Gray, *Proc. Natl. Acad. Sci. USA* **94** (1997) 4246.
[366] J. Leckner, P. Wittung-Stafshede, N. Bonander, B. G. Karlsson, and B. G. Malmström, *J. Biol. Inorg. Chem.* **2** (1997) 368.
[367] P. Wittung-Stafshede, M. G. Hill, E. Gomez, A. J. Di Bilio, B. G. Karlsson, J. Leckner, J. R. Winkler, H. B. Gray, and B. G. Malmström, *J. Biol. Inorg. Chem.* **3** (1998) 367.
[368] M. H. M. Olsson, G. Hong, and A. Warshel, *J. Am. Chem. Soc.* **125** (2003) 5025.
[369] S. S. Skourtis, I. A. Balabin, T. Kawatsu, and D. N. Beratan, *Proc. Natl. Acad. Sci. USA* **102** (2005) 3552.
[370] B. M. Hoffman, L. M. Celis, D. A. Cull, A. D. Patel, J. L. Seifert, K. E. Wheeler, J. Wang, J. Yao, L. K. Kurnikov, and J. M. Nocek, *Proc. Natl. Acad. Sci. USA* **102** (2005) 3564.
[371] D. N. LeBard, and D. V. Matyushov, *J. Phys. Chem. B* **112** (2008) 5218.
[372] J. Zhang, Q. Chi, A. M. Kuznetsov, A. G. Hansen, H. Wackerbarth, H.. E. M. Christensen, J. E. T. Andersen, and J. Ulstrup, *J. Phys. Chem. B* **106** (2002) 1131.
[373] R. Rinaldi, and R. Cingolani, *Physica E* **21** (2004) 45.
[374] N. M. Marshall, D. K. Garner, T. D. Wilson, Y.-G. Gao, H. Robinson, M. Nilges, and Y. Lu, *Nature* **462** (2009) 113.

[375] Q. Chi, J. Zhang, E. P. Friis, J. E. T. Andersen, and J. Ulstrup, *Electrochem. Commun.* **1** (1999) 91.

[376] Q. Chi, J. Zhang, J. U. Nielsen, E. P. Friis, I. Chorkendorff, G. W. Canters, J. E. T. Andersen, and J. Ulstrup, *J. Am. Chem. Soc.* **122** (2000) 4047.

David H. Waldeck

David H. Waldeck was born in Cincinnati, OH on September 5, 1956. He obtained a BS in chemistry from the University of Cincinnati (1978) and a PhD in chemistry from the University of Chicago (1983). He was an IBM postdoctoral fellow at the University of California, Berkeley from 1983 to 1985. In 1985 he moved to the University of Pittsburgh, where he is now a Professor and Chair of the Chemistry Department. His research interests include condensed phase reaction dynamics (homogeneous and heterogeneous), solute/solvent interactions (structural and dynamical characteristics), relaxation process in solids, electron transport in supramolecular and nanometer structures, and nanophotonics.

Dimitri E. Khoshtariya

Dimitri E. Khoshtariya, MS, PhD, DSci, was born in Tbilisi, Georgia in 1953. He obtained his Masters in Biophysics from the Tbilisi State University in 1975 and PhD in Physical Chemistry from the A.N. Frumkin Institute of Electrochemistry in Moscow, Russia in 1980, under the direction of L.I. Krishtalik. After a number of postdoctoral and senior positions in Georgia and abroad, in 1996 he was appointed as a Head of Research Laboratory (since 2006 – Department of Molecular Biophysics) at the Institute of Molecular Biology and Biophysics, Tbilisi, Georgia. Since 2009 he combines this position with that of a Research Director of the Institute for Biophysics and Bionanosciences at the Department of Physics, Tbilisi State University. D.E. Khoshtariya notably contributed to the mechanistic understanding of chemical and biological electron transfer in electrochemical nanodevices, the outer-sphere optical versus thermal electron transfer, and the interplay of dynamic and stochastic reactive modes in protein function. Research interests of D.E. Khoshtariya also cover electron transfer in room-temperature ionic liquids, protein folding, stability, and conformational flexibility (in the context of their function), hydrogen-bonding in liquid and protein-bound water. D.E. Khoshtariya has been awarded several prestigious fellowships and research grants from the Alexander von Humboldt Foundation (Germany), Ministère de l'Education Nationale et de la Recherche (France), Swedish Natural Sciences Research Council, National Research Council (USA), NATO, Landau Network – Centro Volta (Italy), and the Volkswagen Foundation (Germany).

3

Microbial Fuel Cells – Scalability and their Use in Robotics

John Greenman[1,2], Ioannis Ieropoulos[1,2] and Chris Melhuish[2]

[1]*School of Life Sciences, Faculty of Health & Life Sciences, Frenchay Campus, University of the West of England, Coldharbour Lane, Bristol, BS16 1QY, UK*
[2]*Bristol Robotics Laboratory, University of the West of England, Bristol Business Park, Coldharbour Lane, Bristol, BS16 1QD, UK*

I. INTRODUCTION

> "…and because there are bred certain minute creatures which cannot be seen by the eyes, which float in the air and enter the body through the mouth and nose and there they cause serious diseases…"
>
> Cato and Varro: On Agriculture 1, XII Loeb.

This is probably the first written and relatively accurate description of microorganisms, made by the Roman scholar Marcus Terentius Varro (116-27 B.C.). It was then Anthony van Leeuwenhoek (ca. 1677) who, for the first time, observed *little animals* under his version of the microscope (a collection of powerful magnifying glasses), and who in a letter to the Royal Society

N. Eliaz (ed.), *Applications of Electrochemistry and Nanotechnology in Biology and Medicine I*, Modern Aspects of Electrochemistry 52,
DOI 10.1007/978-1-4614-0347-0_3, © Springer Science+Business Media, LLC 2011

(1683) gave for the first time a description of dental plaque as the first example of what we would now term a *microbial biofilm*.

This chapter describes some of the basic concepts of biofilms and explains what biofilm electrodes are and where they can be useful. It then moves on to explain the microbial fuel cell technology, which comprises biofilm electrodes and finishes with a description of the applications in which both biofilm electrodes (B-E) and microbial fuel cells (MFCs) can and have been used.

Biofilms have been the centre of attention of many scientists over the last century, and have gained such reputation that allowed them – from quite early on – to be employed in real (field) applications, including vinegar production and wastewater treatment (trickling filters). At present, the interest in biofilms is continuously increasing, and as systems they have proved to be most useful in numerous applications. One of these is MFCs, which is the thrust of this chapter.

II. BIOFILMS

Two *modes of existence* are generally recognised with regard to microorganisms – biofilm mode or planktonic mode. The biofilm mode is where cells stick to surfaces and grow as adhesive layers or films. The surfaces that they stick to may include almost all types of (wetted) materials (e.g., metals, plastics, minerals, organic debris, carbon-veil electrodes) and this can also include other microbes (via co-agglutination, or co-aggregation) where they may form *flocks* or accrete to a biofilm that is already present. Although some biofilms are fairly compact (such as dental plaque that forms around tooth surfaces in animals), others may be loose, stringy or slimy in nature. It is thought that biofilms are the predominant mode of life for the vast majority of microbial species on the planet, but it is clear that some species are also well adapted for existence in planktonic mode, where they are freely suspended in liquid medium (sea-water, lakes, rivers, shaken-broth cultures) and where all the cells are subject to the forces of hydrodynamic flow. For example, many species possess specialized features (flagella), allowing them to be actively motile. Others appear to be well adapted to both modes; up-regulating flagella production in broth culture and down-regulating (and *dropping* off of flagella) when growing in biofilms.

Biofilms are composed of cells and exocellular polymeric substances (EPS) that may be produced by the microbes themselves (e.g., capsular polysaccharides) or can be present in the biofilm by adsorption from the input fluid (for example, the adsorption of salivary glycoproteins found as a component of dental plaque).[1] Although laboratory models may consist of a single pure mono-species, the vast majority of microbial biofilms are mixed communities and, depending on the physicochemical conditions of the specific examples, may be highly diverse, consisting of hundreds of different species. The term *biodiversity class* has been used to describe the biodiversity of biofilms.[2] Class-I: single species (e.g., *Streptococcus* species growing on heart valve in endocarditis); class-II: two to seven species (e.g., *Acetobacter* biofilms in vinegar production; mixed *Enterobacteriaceae* growing on a urinary catheter); class-III: seven to 30 or so (e.g., Bioremediation consortia); class-IV: thirty to 500 species (e.g., dental plaque) and class-V: 500–1000 or more (e.g., anaerobic sludge or soil communities). The role of biofilms in health and disease and the complex interactions that occur within biofilms (competition, synergy, quorum sensing, genetic interactions) is a vast area of research and the reader is referred to other reviews on these subjects.[3,4,5]
Biofilm experiments are usually accomplished by setting up an *in vitro* model. There are many such models described in the literature (for reviews see Refs. 2 and 6) and these are generally designed to study different aspects regarding biofilms. For example, batch culture models can be used for studying initial adhesion or early events in biofilm formation, but (in common with all closed systems) it cannot be used for studying the main metabolic processes that occur in established (mature) biofilms since it runs out of nutrients (or builds up toxic products) and rapidly goes into decay phase. In nature, the vast majority of biofilms are formed in conditions of continuous or periodic nutrient replacement and can be regarded as open systems, and a number of laboratory models are designed along the same lines. Mathematical models that regard biofilms as homogeneous steady-state films containing a single species have been proposed.[7] This model has evolved to cover dynamic multi-substrate/multispecies biofilm computer mod-

els.[8,9,10] An approach using discrete cellular automata (to simulate the rules that govern the lives or *properties* of microbial cells) has also been employed to model biofilms.[11,12] These allow the simulated biofilm structure to evolve as a self-organization process emulating how real bacterial cells organize themselves into biofilms. These models produce realistic, structurally heterogeneous biofilms. Mathematical models of diffusion/reaction in biofilms (incorporating Fick's laws of diffusion) have also been described[13,14] as have models to predict the effects of antimicrobial activity within biofilms.[15,16] MFC and biofilm electrodes (of the types described in this chapter) can be grouped under the general description of continuous flow matrix perfusion systems. In this model, following transfer of inoculum, cells attach themselves to the substratum (the matrix) and sterile perfusate supplies the cells with nutrients and buffering (set or controlled by the operator). The matrix population begins to grow until a matrix yield limit (a population saturation point) is reached. The biofilm continues to grow, but growth accumulation appears to be matched by the loss of cells from the system; cells that are shed into the perfusate output. By judicious choice of perfusate composition and flow rate, the biofilm can be maintained in quasi-steady state for many weeks, months or years. It appears that the behaviour of a matrix perfusion biofilm has certain similarities to a chemostat planktonic model in that *concentration* of the growth-limiting substrate dictates the yield of cells held in the biofilm whilst the perfusate flow rate (*supply rate* of limiting nutrient) dictates the growth rate of the biofilm. Microscopic observation of colonized and non-colonized matrices using environmental scanning electron microscope (ESEM, Fig. 1) suggests that none of the biofilm cells are more than 20 or 30 μm from a capillary perfusion channel and with sufficient hydrodynamic flow of medium, the matrix can maintain biofilms where solute and product transfer are not significantly impinged-upon by diffusion. The system is analogous to a vascular bed, supplying all internal parts of the body equally well. The nearest description from a chemical engineering view is that the system behaves as a plug-flow reactor[17] with biotransformations occurring by whole cell catalysis attached to a stationary matrix within the system. Catalytic activity is thus renewed by continuous endogenous production. The system is nonaccumulative and can maintain steady state conditions for as long as it is kept perfused.

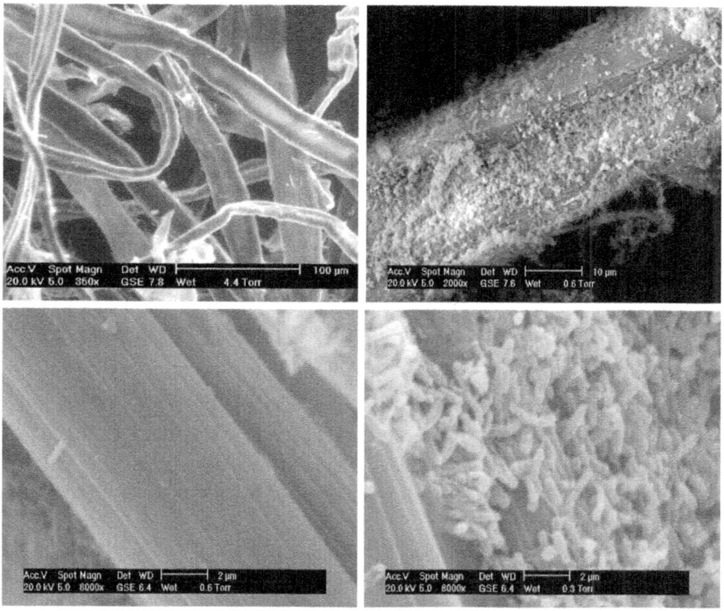

Figure 1. Top left: Sorbarod cellulose matrix (sterile control); top right: Sorbarod colonised with *Streptococcus* species; Bottom left: carbon fibre veil electrode (sterile control); bottom right: *Geobacter sulfurreducens* biofilm formed on electrode inside a microbial fuel cell.

III. PERFUSION BIOFILMS

Continuous culture of microbes in biofilm mode has been previously described by many workers and groups around the world.[18,19] The nearest equivalent to the biofilm model described herewith is a system described as a matrix perfusion model (e.g., the so called *Sorbarod perfusion model* first described by Hodgson et al.[20] Monitoring of the dynamic state was restricted to sampling of perfusate and destructive sampling of replicate biofilms at various times using viable counting methods (plate counts) to quantify cells. This has obvious disadvantages in that due to viable counts requiring incubation, data from experimental intervention are not available until 24–48 hours following the sample. Nevertheless, reports have shown that biofilms of some species can reach quasi-steady-state after 24–48 hours and can be maintained in quasi-steady state for a number of days.[21,22] The idea of continuously monitoring perfused matrix biofilms during their operation has also been described, using

(a) pH electrodes[23]
(b) sulphide sensitive electrodes,[24] and
(c) detection of bioluminescence (from a genetically engineered lux-luminescent target species) measured using a photon camera.[25,26]

Most work described to date has centred on pathogens (*Pseudomonas aeruginosa*, *Staphylococcus aureus*) or mixed cultures of dental plaque. Finally, the anodic side of a microbial fuel cell (of the types used by the authors) is another example of a perfusion biofilm system (see later).

IV. MISCONCEPTIONS REGARDING BIOFILMS

There are many workers who believe that biofilms form without shedding daughter cells; that new cells are produced by *growth from within* but that they remain attached until the biofilm becomes too extended for outer cells to remain in attachment against the liquid shear forces. Outer cells are then eroded in clumps and re-entrained into the perfusate flow. However, data produced from Sorbarod experiments suggest that this is not the case and that cells are continuously released throughout the inoculation and initial colonisation phases and that growth and release of daughter cells is a smooth rather than a stochastic process.[21,24,25]

V. DYNAMIC STEADY STATE

A true dynamic steady state with regard to any microbial cultures is of course only a theoretical construction whereby cells are imagined to grow at constant rate in response to a constant supply of nutrients and all other physicochemical factors are constant. The nearest equivalent in planktonic culture is the empirical steady state achieved by using a chemostat. Monod (1950) proposed the model as a useful research tool to produce balanced growth, whilst Novick and Szilard (1950) recognised evolution (by generation and selection of mutants) and used the model as a way of studying this latter phenomenon.[27,28]

Different models have been proposed to describe the growth kinetics of cells growing in steady state when limited by a single growth-limiting nutrient. The Monod model is favoured since the parameters involved (μ, μ_{max}, s and K_s) have real-world counterparts whose meaning and empirical measurement are experimentally accessible. In the chemostat, $\mu = D$, and $\mu = \mu_{max}\ s/K_s + s$, where D = dilution rate (h^{-1}), μ = specific growth rate (h^{-1}), and μ_{max} = the maximum specific growth rate at saturating substrate concentrations (s). K_s is the substrate affinity constant and 's' is the concentration of substrate that allows cells to grow at $^1/_2\mu_{max}$. An extended model[29,30] considers the energy required by the cells in order to remain viable even when the growth rate is zero (also known as the maintenance energy) and expresses this portion of the total energy in terms of the growth limiting substrate concentration; i.e., the minimal substrate concentration required for growth (s_{min}):

$$\mu = \mu_{max}\ s - s_{min}/K_s + s - s_{min} \tag{1}$$

This model provides a good fit for *Escherichia coli* grown in chemostats with glucose as the limiting nutrient at different temperatures and different dilution rates.[31] For a well perfused biofilm in steady state (i.e., non accumulative), the specific biofilm growth rate (μ) is given by:[6]

$$\mu = \frac{\text{Rate of elution of daughter cells}\ (h^{-1})}{\text{Total biofilm population}} \tag{2}$$

There is little reason to dispute that, at least as a first approximation, the relationships between growth rate, growth yield and concentration of growth limiting substrate (as previously proposed by

Monod) do not apply equally well to a perfused biofilm, providing it is not limited by diffusion.

VI. BIOFILM-ELECTRODE STEADY STATE

Empirical demonstration of the metabolic steady state from a *Geobacter* perfusion biofilm-electrode has been previously reported.[32] Mixed culture (sludge-derived inoculum) MFCs are also capable of long-term steady states, following a slow maturity over a number of weeks. Thereafter they produce steady-state outputs for steady-state physicochemical environmental conditions (see Fig. 2). In this example, small scale (1 mL) MFCs running in continuous flow were inoculated with mixed sludge cultures of bacteria and their output was monitored for long periods (weeks). The stability of the output from these MFC units was comparable with single (pure) culture anodophilic species such as *Geobacter*.

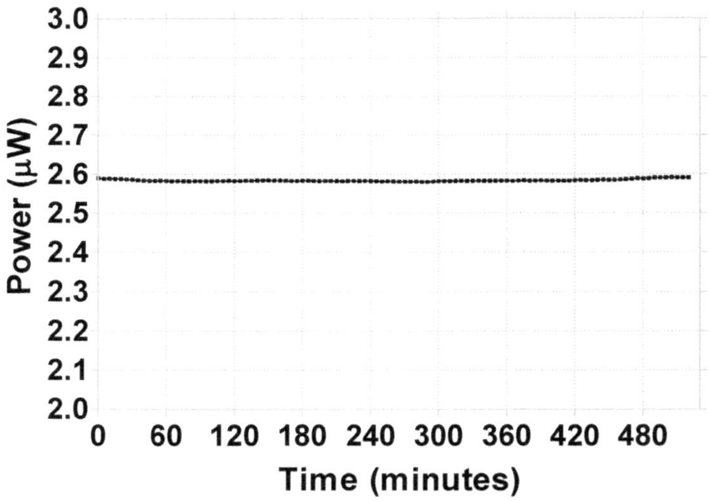

Figure 2. Measured metabolic steady state under defined conditions.

VII. BIOFILM STEADY STATES (QUASI STEADY STATE)

The descriptive *quasi* is used because of the comparison of a biofilm model with the planktonic equivalent (the chemostat). In a chemostat, a steady state occurs when all components (cells, substrate molecules, product molecules) are produced or utilised at the same rate. An analysis of the cell residence times shows a Gaussian distribution, with points clustered tightly around the mean residence time.[17] This dynamic state is maintained for as long as the physicochemical environment is maintained (assuming no internal changes such as genetic shifts). For a biofilm, there are two distinct populations, the planktonic phase (detached cells) and attached cells. An analysis of the cell residence time gives a much more complicated picture of distributions; a short interval for unattached cells, but a very long interval for the inner layers of a well embedded biofilm. The system is still homogeneous over a large scale (e.g., a 1 cm^3 cube of matrix) but heterogeneous at a smaller scale. It is a homogeneous mix of heterogeneous compartments; a mosaic of micro-environments. For these reasons the steady state is termed *quasi* steady state. For a Sorbarod or a carbon veil electrode matrix, there is a fairly even distribution of cells, matrix substratum and fluidic channels or spaces. It is assumed that the system is not particularly diffusion-limiting to nutrients or products and that with appropriate buffering large gradients of pH will not occur.

VIII. CONTROL OF THE PHYSICOCHEMICAL ENVIRONMENT

In a model flow system it is easy to accurately control the physicochemical environment. Stock solutions of perfusate can be made from pure chemicals and all macronutrients, mesonutrients and micronutrients can be defined. For example, buffers can be used to control pH, which in turn can be verified by (pH) electrode measurements. Other parameters such as temperature can be controlled by a thermostat/heater and osmolarity may be adjusted by the addition of salt or (none metabolically active) sugars or glycols. During an experiment, inhibitors, inducers and repressors can simply be weighed out and added as pure compounds and the electrode potential (V) can also be controlled (online) by the external resistive load and measured, and finally, flow rate can be set and measured by appropriately calibrated perfusion pumps. In many cases a fully defined medium is not required as long as the same medium is employed for all experiments. Types of responses that may be expected to be seen are shown diagrammatically in Fig. 3. It is still unknown what responses are likely to be produced when two or more parameters are changed simultaneously.

IX. ENERGY SPILLING MODE

It has been demonstrated that excess substrate can cause uncoupling between anabolism and catabolism, which leads to energy spilling.[33] When (some) microbial species are grown under carbon-energy excess conditions (i.e., N-limited conditions), the metabolic rate may remain high even though the growth rate of cells might be relatively low. Under these conditions, cells cannot re-oxidise NADH (or NADPH) and some species resort to energy spilling mechanisms. It is likely that anodophiles or electricigens on electrodes would simply get rid of excess reducing power by producing higher electrode output.

X. BIOFILM ELECTRODES

Electrodes used generally in electrochemical systems can be of many forms. These materials, depending on the application, may be solid 3-dimensional shapes, which offer high conductivity (low resistivity), but also low surface area to volume ratio. On the other hand, electrodes of high surface area to volume ratio usually are inherently resistive (low conductivity). For applications such as MFCs a *golden mean* is required whereby the electrode material should be of sufficiently high surface area to volume ratio to allow colonisation, high matrix yield and access to nutrients when perfused, but also high conductivity. One such material which appears to fulfil these criteria (and is thus widely used by the MFC community) is the carbon veil electrode.

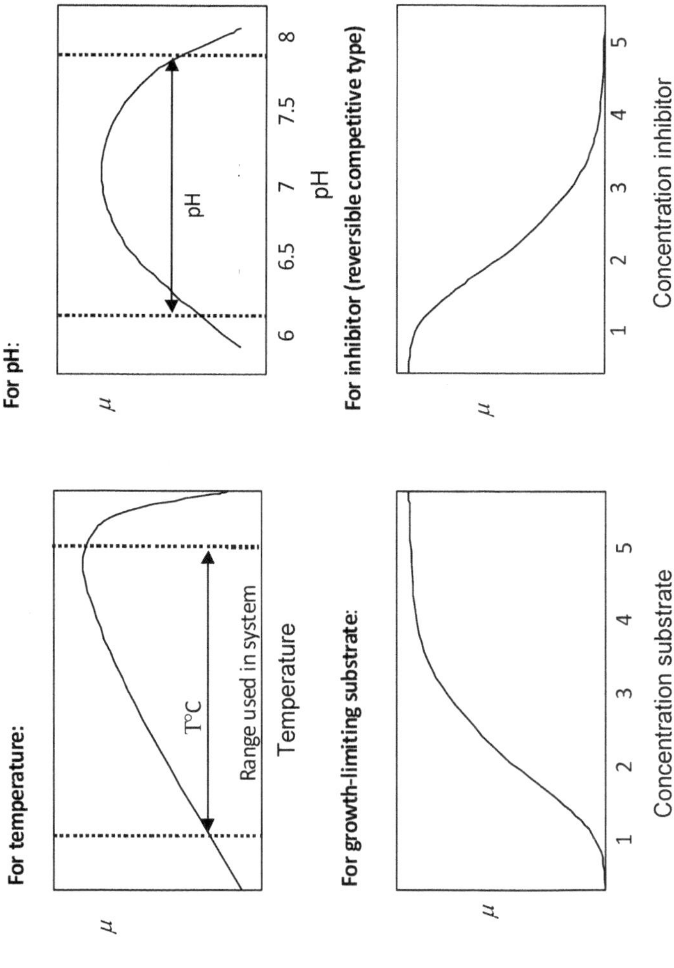

Figure 3. Expected biofilm responses to some exemplar physicochemical conditions.

XI. ALTERNATIVE ENERGY

1. Energy from Natural Substrates

Energy from the environment can come from natural substrates such as fruits or vegetables, food waste or even dead insect biomass. These substrates can be utilised to produce useful energy in a number of ways. For example, herbivorous animals metabolise raw vegetables into biochemical energy that is used for their maintenance. The metabolic cycle of biological systems is highly complex and, moreover, extremely difficult to replicate. The energy used to drive animals is derived from electron-rich carbohydrates, which are ultimately oxidised by a chain of metabolic steps. The collective process, in aerobes or facultative anaerobes, is called respiration with oxygen (O_2), being the end terminal electron acceptor. The end result of respiration is chemical energy in the form of NADH and ATP, carbon dioxide (CO_2) and water.[34] There are, of course, several organisms that are capable of breaking down carbohydrates in the absence of oxygen and donating electrons to organic electron acceptors (fermentation) or inorganic electron acceptors (anaerobic respiration).[34] The major inorganic alternative end terminal electron acceptors are sulphate, nitrate, CO_2 and also metals. For mammalian cells, the types of substrates that can be utilised is limited to sugars, amino acids and organic acids (acetate and butyrate).[34] However, whole animals can utilise a wider range of substrates by means of digestive transformation such as hydrolysis of some polymeric substrates into utilisable monomers. Furthermore, some animal species, although lacking the enzymes necessary for the break-down of many polymeric substrates, are able to utilise these through symbiotic associations with microbial cells. Thus, ruminant animals are able to feed on cellulose from hay due to the symbiosis that exists between the anaerobic bacteria in the rumen of the animal and the animal itself. In nature, the symbiosis between animals and microorganisms is an existing successful partnership and therefore a good example to exploit in practical engineering. There are a number of ways that the bacterial metabolism can be utilised to produce useful energy. For example, hydrogen can be produced by microorganisms that can reduce free protons into molecular hydrogen (H_2). Methane can also be produced by methanogenic microorganisms of the Archaea kingdom, for example *Methanobacterium thermautotrophicum*.[35,36,37] This is mainly the result of the fermentation of simple organic carbon compounds (acetate) or the oxidation of hydrogen. Both hydrogen and methane can then be used as fuels in chemical fuel cells, with conversion efficiencies of the order of 80 and 65%, respectively.

It is also possible to convert the microbial biochemical energy directly into electricity. The available technology for performing this is based on the microbial fuel cell (MFC). In this system, as described in the following sections, electrons are extracted directly from the microbial metabolic cycle and diverted towards an electrical load via electrodes, thus establishing current flow. Direct electricity generation from microbes is an interesting option, worthy of study, with some significant advantages over the bio-hydrogen or bio-methane production.

XII. CONVENTIONAL CHEMICAL FUEL CELLS

Although fuel cells appear to be a somewhat recent technological development, their history dates back to the mid 1800's. In 1838, the Swiss Professor Christian Friedrich Schoenbien (1799-1868) discovered the fuel cell effect and in 1839 his close friend Sir William Grove, a Welsh lawyer who later became a scientist, physically demonstrated the apparatus.[38] Grove developed two important systems (known as the Grove-cells). The first, which was used to power the early telegraph due to its high current output, employed a platinum electrode immersed in a nitric acid solution and a zinc electrode immersed in zinc sulphate. The output from this system was 12 A at 1.8 V. In 1800, two British scientists, William Nicholson and Anthony Carlisle, demonstrated how electricity can decompose water to its two elements. Grove's second fuel cell applied the reverse to this principle of electrolysis: i.e., separate oxygen and hydrogen were combined to produce electricity and water. This is the basic description of a simple hydrogen fuel cell.

A fuel cell, in general, is a high efficiency electrochemical device used as an alternative to conventional methods of generating electricity.[39] It converts chemical energy to electrical energy from a fuel, without direct combustion. Electricity is generated by a chemical catalytic reaction, which decomposes the fuel molecule, extracting the electrons, and diverts these to an external electrical circuit to produce current (work). Unlike batteries, which are limited by the internal capacity of their constituent materials, fuel cells have the advantage of generating electricity from an external fuel that can be continuously supplied.

Fuel cells are an alternative to conventional energy plants or sources. Depending on the electrolytes, electrodes, catalysts and fuels used they can offer a clean and environmentally friendly way of producing energy, for almost any kind of application.[40] Hydrogen (H_2) is the most commonly used fuel extracted from various hydrocarbons (gasoline, methane, diesel) and water for the Regenerative Fuel Cell, but any fuel cells utilising hydrogen or methane that originate from fossil fuels, will still contribute to global warming. An extensive overview of fuel cell fundamentals and application is given by Carrette et al.[40]

There are numerous examples from over the world where workers have been using chemical fuel cells as a replacement for batteries. Hydrogen, for example, has been reported to be extracted from biomass derived carbohydrates such as glucose.[41] The process is reported to be greenhouse gas neutral; however, it still requires high temperatures, of the order of 227°C (500 K) for the reaction to occur.

One of the common characteristics of normal fuel cells is that the operating temperature and pressure have to be at high levels for the reactions to take place. Temperature, for example, can range between 80–1000°C, depending on the type of electrolyte used, and although the lower end can be considered as low/moderate within the fuel cell community, it is still relatively high for mobile applications such as robotics. Similarly, higher than normal levels of pressure are commonly required for the chemical reactions to occur, which in some cases result in lower fuel cell performance and fuel crossover via the membrane.[42]

1. Fuel Cell Principle of Operation

The anode half-cell is the compartment of the fuel cell in which the electrons are generated and defines the direction in which they flow. In contrast with conventional electrolytic and battery cells and accumulators, the anode is the source of electrons and consequently the negative terminal of the system. The conductive material in the anode compartment for the transfer of electrons from a fuel to the external load is termed the anode electrode. This can either be of similar or dissimilar material to the cathode electrode.

The cathode half-cell defines the final destination of the electrons and the direction in which they flow. This is the compartment of the fuel cell that acts as the electron sink and keeps the system balanced. It is the positive terminal of the system. A conductive material for the transfer of electrons from the external load to the catholyte is also found in this half-cell and is known as the cathode electrode.

A polymer electrolyte, which in its most commonly used form is a membrane material, exists between these two half-cells, providing a physical barrier, but more important – allowing electrically charged particles such as protons to pass from one electrode (anode) to the other (cathode).

The fuel cell principle of operation is based on the extraction of electrons as a result of the chemical reaction of separating a fuel, using a catalyst, in the anode half-cell. The electron flow is the result of the electrophillic attraction from the cathode electrode and the charged motion of cations (e.g., H^+) towards the anionic groups of the polymer electrolyte (proton exchange membrane). Faraday assumed that electron flow was due to the movement of charged ions; anions are those moving towards the anode and cations are those moving towards the cathode. Hence, the electrons, which are diverted out of the anode, flow through the external circuit in order to reach and reduce the cathode to complete the circuit, by recombining with the incoming cations. The quantity of electrons flowing through the external circuit is the electricity being produced, or current. When there is no load connected to the fuel cell, electrons cannot flow out of the system, and therefore the open-circuit voltage can be measured. This is a direct indication of the potential difference between the two half-cells and it is greatly affected by the type of electrolytes, electrodes and catalysts being used.

Catholyte and anolyte are the liquid electrolytes (for some fuel cells) in the two cathode and anode half-cells, respectively. Electrolysis defines the passage of electric current through a molten chemical or an aqueous solution. Hence, the anolyte is the medium that allows the transfer of electrons from the catalyst to the electrode, and the catholyte allows the taking up by the cathode electrode of incoming electrons and cations. In some fuel cells, where there is enough oxygen supply to the cathode, electrons recombine with cations (protons) and oxygen to form water.

2. Types of Fuel Cells

There are various types of fuel cells, distinguished by the types of electrolyte or catalyst employed, the kind of fuel consumed, or the nature of their operation, and these can be categorised as follows:

(a) Alkali Fuel Cell, Molten Carbonate Fuel Cell (MCFC), Phosphoric Acid Fuel Cell (PAFC), Solid Oxide Fuel Cell (SOFC) and Proton Exchange Membrane Fuel Cell (PEM);
(b) Direct Methanol Liquid Feed Fuel Cell (DMLFFC), Methane Fuel Cell, Carbon Fuel Cell, Microbial Fuel Cell (MFC) and Regenerative Fuel Cell (RFC).

Fuel cells, depending on their type, can consume various types of energy source – fuel – in various forms. Hydrogen (H_2) is the most commonly used fuel extracted from various hydrocarbons (gasoline, methane, propane, diesel, ethanol, methanol) and water via electrolysis for the Regenerative Fuel Cell.

The Regenerative Fuel Cell is in a way different to other fuel cells in the sense that it can be used as an energy storage unit, i.e., a battery. As a fuel cell it operates in a similar way to any other fuel cell. Hydrogen is used as the fuel at the anode and oxygen is used as the oxidant at the cathode, with platinum catalysts and the main by-products are heat and water. The unique property of this system is that it can be combined with a conventional electrical power source (mainly hydroelectric, solar or nuclear) as a hybrid, which is used to dissociate (electrolyse) the produced water to hydrogen and oxygen. These can be fed back to the *fuel cell* stage of the system for the generation of power.

A hydrocarbon is a compound consisting of hydrogen and carbon molecules only. In order for hydrogen to be extracted from such compounds, it is necessary to go through a reforming process, prior to usage since hydrogen is impure and difficult to store. The reforming process is the major drawback in the hydrogen technology. It is a hazardous and time consuming process involving complicated multi-stage systems, high energy consumption and high financial costs that have a negative effect on the high conversion efficiency (80%) of the hydrogen fuel cell. The steam reforming process is the one most commonly used and it involves the extraction of hydrogen from natural gas (methane, CH_4). The initial phase involves the disintegration of methane into H_2, carbon dioxide (CO_2) and carbon monoxide (CO), after exposing CH_4 to high temperature and pressure steam. Reforming processes, in general, always operate under extremely high temperatures (600–800°C) and pressure.[43,44]

XIII. METHANE AND CARBON FUEL CELLS

Direct methane consumption is also actively investigated, as an alternative to hydrogen. This kind of fuel cell is more promising and desirable since no reforming process is necessary; however, it is less efficient than the hydrogen fuel cell. Methane fuel cells are known to be prone to CO_2 accumulation[45] from the reaction of CH_4 and O_2, occurring at temperatures of the order of 700-800°C, even though lower temperatures (500°C) have also been reported.[46]

There are several environmental issues involved with chemical fuel cells, especially in the case of methane. Methane is a greenhouse gas that can be produced by both natural and anthropogenic processes and forms part of the Earth's equilibrium system. Despite the fact that its concentration has more than doubled over the last 200 years, methane (along with CO_2) is vital to keep the Earth's temperature from reaching freezing levels. Researchers from around the world (ICP-GISS, NASA, EIA-DOE, EPA) argue that the concentration increase in greenhouse gases such as CO_2, CH_4 and nitrous oxide (N_2O) is the result of anthropogenic processes. Carbon dioxide, for example, has increased by 30% over the last 100 years.[47] Utilisation of methane as a fuel will probably contribute to the already increasing CO_2 emissions. Hence, any CO_2 production from fossil fuels is undesirable and should be unacceptable.

XIV. MICROBIAL FUEL CELLS

A microbial fuel cell is a *bio*-electrochemical transducer that converts *bio*-chemical energy to electrical energy. The basic principle of operation, i.e., the extraction of electrons from an aqueous fuel source and their transfer onto electrode surfaces, has certain similarities to a conventional fuel cell. Microbial cells are used for the catalytic reactions in the anode chamber, hence the term biocatalyst. Of the microbes commonly employed in MFCs, the majority are mesophiles, hence the operating temperatures are close to ambient room conditions. A study by Lee et al. indicated the presence of a truly wide range of genera and species within activated sludge as inoculum.[48] As shown in Fig. 4, a wide range of organic substrates or nutrients of the types found in agriculture or food wastes can be utilised as fuel, depending on the type(s) of microorganism used.[49]

Substrates can be soluble small molecules (e.g., acetate) or soluble or insoluble macromolecules, depending on the types of species present in the anodic biofilm. Since microbes cannot directly transport substrates with a large molecular weight across

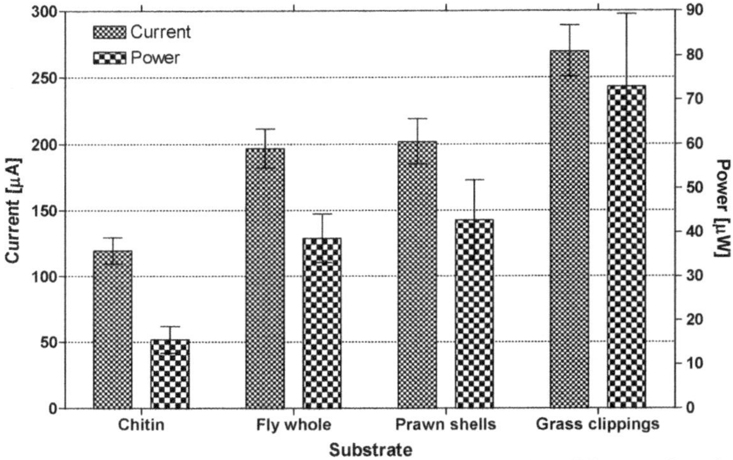

Figure 4. Comparison of mean current and power output when different polymeric natural and unrefined energy sources were utilised by otherwise identical MFCs. Error bars show ± SD.

their cytoplasmic membrane, it is assumed that many of the species present can produce hydrolytic enzymes such as proteases (to digest proteins), amylases (to break down starch), cellulases (to break down cellulose), pectinases, chitinases and many other depolymerisation enzymes, which enable them to utilise these types of substrate. It is highly unlikely that any single species would produce the full range of hydrolytic enzymes, but if a community is sufficiently diverse – there will be species that can benefit from the different types of carbon-energy sources available (with just the right set of enzymes) to be enriched within the community and hydrolyse the feedstock for the benefit of the other species (e.g., electricigens), which may not have a wide spectrum of enzymes.

MFCs fall under the proton exchange membrane (PEM) fuel cell category, since that is the solid electrolyte used in the system. The PEM is used to provide a pathway to cations such as hydrogen ions to reach the cathode and also to physically separate the anode and cathode. An MFC in its analytical form is shown below in Fig. 5.

1. MFC Principle of Operation

The principle of operation is based on bacterial activity and metabolism, the extraction of a portion of the bio-chemical energy generated by microbes and its conversion to electricity. MFCs often consist of two compartments, the anode and the cathode separated by a PEM. In the anode chamber, bacteria anaerobically oxidise organic substrate (fuel), generating electrons and releasing protons. As in the case of conventional fuel cells, electrons travel to the cathodic compartment via the external circuit and protons migrate through the PEM. Protons and electrons combine at the cathode, reducing oxygen (electron acceptor) to water. Oxygen is potentially the most effective electron acceptor for a MFC due to its high redox potential, availability, low cost and zero chemical waste products (the only by-product being water). Provided that the microbes are continuously fed and there is constant oxygen supply at the cathode, a MFC will continuously produce power. However, there are a number of limiting factors that influence the energy generation and performance of a MFC.[50] These include the

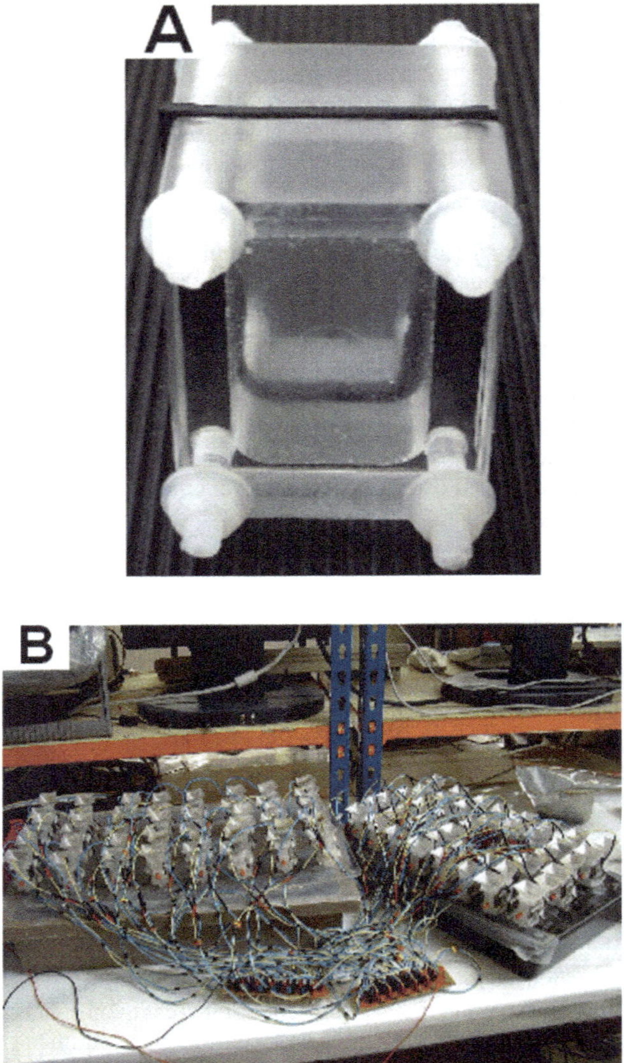

Figure 5. (A) Microbial fuel cell (analytical form); (B) stacks of rapidly prototyped small-scale MFCs.

rates of fuel oxidation and electron transfer to the anodic electrode by the microbes, the migration of protons to the cathode through the membrane, and the oxygen supply and reduction reaction at the cathode. In general, these processes taking place inside the anode and cathode half-cells are known as **redox** reactions and are shown in Fig. 6.

2. The Cathode System

Two main systems are known to be used in the cathode half-cell of MFCs; the first, commonly used, works with ferricyanide ($K_3Fe[CN]_6$), and the other with oxygen (O_2). Ferricyanide is a good laboratory standard used in analytical studies that is highly efficient, to start with, but degrades with operation time. This is mainly due to the continuous reduction of the oxidising agent (in the cathode) by the incoming electrons and also the taking up of the hydrogen ions (protons or H^+), which result in the need for replenishment. Oxygen, for the gas (O_2)-diffusion cathode, on the other hand, is available from free air and offers a stable and, in some cases, improved performance over operation time. It does not suffer from H^+ ion accumulation and concomitant decrease in pH, which usually has an adverse effect on the overall MFC performance. Electrons combine with H^+ ions and O_2 molecules to form water. The only condition is that the electrode must be kept moist in order for the H^+ ions to flow. This initially may be done manually, in the case of a completely open-to-air system, but henceforward water generation keeps the half-cell self-sustaining.

Although the origins of MFCs go back to Galvani and his demonstration of animal electricity,[51] the earliest report regarding MFCs[52] was the first description of the ability of microorganisms to transform organic substrates (chemical energy) into electricity. In the early stages of MFC development, artificial mediators were widely used in the anodic chamber.[53,54,55,56] These are becoming obsolete since the discovery that some species of microbes can produce electro-active metabolites (e.g., sulphide), which act as natural mediators.[57] In conjunction with this, complex mixed species consortia were reported to outperform any monoculture-based MFCs in terms of power output[50,57,58,59,60,61] and at the same time widen the range of utilizable organic substrates.[61] Moreover, the

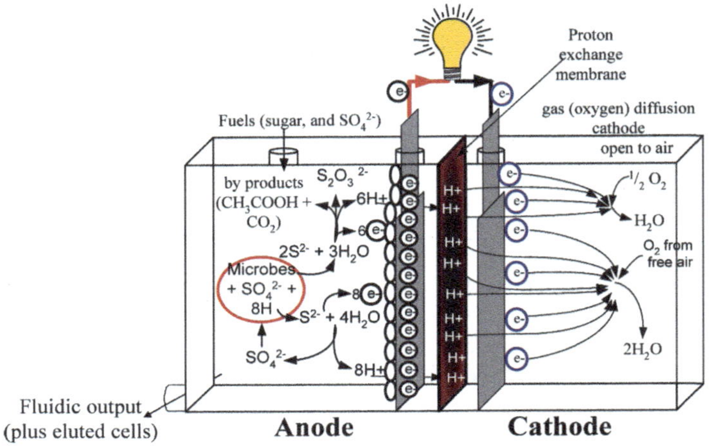

Figure 6. MFC reactions. The oxidation reaction rate is dictated by bacterial species and population, concentration/type of substrates and their supply rate. The reduction reactions are governed by the electrode material, its O_2-diffusion and electrical/ionic properties. Both reactions may be limited by external resistance (load).

discovery of anodophiles (or electrophiles) such as *Geobacter* or *Rhodoferrax* species[62,63] that conduct electrons directly to the anode through outer cell membrane cytochromes and nanowires, has further marginalized the use of artificial mediators. The discovery of nanowires is of enormous interest to the MFC community and a number of workgroups are focusing in this exciting area.[64,65] The majority of MFC experiments reported in the literature are batch culture systems with the disadvantage of discontinuity; the relatively recent development of continuous flow MFCs[50,66,67] has enabled longer-term performances to be realized. Developments have also been reported for the cathodic systems, moving from those based on chemical electrolytes (e.g., ferricyanide) as the terminal electron acceptor[56,68] to ones based on oxygen electrodes, which offer the advantage of self-sustainability, provided they are kept hydrated in oxygenated environments.[61,69] MFCs possess high substrate to electricity conversion efficiencies (up to 96%)[62] but one disadvantage is the low energy transformation rates, which limits this technology to only low-power applications. The typical sustainable voltage output from a 25-mL-sized MFC with an oxygen-diffusion cathode and plain carbon veil electrodes is of the order of 0.5 V (open circuit). Higher open-circuit values of 1 V have also been reported from individual small-scale MFCs under special conditions,[70,71] which are closest to the theoretical maximum of 1.14 V (derived from the main lowest anodic intracellular redox couple NADH↔NAD$^+$ and the highest cathodic oxidizing redox couple $O_2 + 4e^- + 4H^+ \leftrightarrow 2H_2O$ [+0.82 - {-0.32} = 1.14V]). Thus, in order to produce sufficient voltage (1.5 V) and/or power – to reside within the operating range of silicon-based circuitry – it is necessary to either scale-up one single unit (but this cannot increase voltage) or connect multiple small units together. For conventional (chemical) fuel cells and batteries it is well known that a large number of units connected together will produce more power than a single unit.[72] The same applies to MFCs, as shown by EcoBots-I and -II, which employed stacks of 8 MFCs connected together[49] and also more recently demonstrated in a continuous flow stack of 6 units.[73]

Up until recently it was still unknown whether increasing the volume of a single MFC unit, for example 20-fold, has any advantage over joining 20 single small units together (with equivalent total anodic volume). It has now been reported that stacks of small-scale MFCs are more efficient and give higher (*normalised*) power outputs than individual units of the same size.[74,75] Furthermore, issues are encountered when connecting multiple units together under continuous flow, as this can

(a) be done in a number of different ways such as series, parallel and series/parallel combined,
(b) result in losses via the conducting fluidic connections of interconnected units, and
(c) affect the performance depending on whether the units are serially fed in cascade or individually fed from the source.

MFCs are still in their infancy, but with a great potential to bring about innovation and become true alternatives to fossil fuel energy generation. It is an up and coming technological area and the interest from the scientific community is rapidly increasing. This is clearly because fossil fuel combustion is costly and cutting back on carbon emissions is even more expensive; renewables, on the other hand, are freely available. The MFC technology has additional merits for energy generation. The electrons in the anode half-cell are generated by the break-down of the given fuel. In cases where the fuel is waste or wastewater, then its molecules are broken down, i.e., degraded or, using water industry terms – treated by the microbes. It is therefore only natural that this technology can be used for the utilisation of waste. In their present state of development, MFCs may not be of equal calibre to high voltage/power renewable energy sources (e.g., solar panels, wind turbines); however, they do offer the advantages of *simultaneous waste clean-up and electricity production.*

Fuel cells that generate energy from the breakdown of renewable substrates operate within the current carbon cycle. The CO_2 produced as waste is equivalent to that fixed in the first place, which would have been released to the biosphere with or without the intervention of these devices. MFCs are one example of such ecologically friendly systems.

3. MFC Stacks and Scalability

MFCs are substantially less powerful when compared to more mature technologies (batteries or photovoltaics). Current research in the field is showing that individual units are thermodynamically limited, producing relatively low energy output levels, emphasising the need for scale-up. It has previously been demonstrated[74,75] that more efficient energy harvesting takes place in small-scale MFC units and, thus, there is a natural drive for miniaturisation and multiple-unit stack development. The latter carries its own challenges, some of which are associated with internal resistance, cell polarity reversal and electron leakage across both structural materials and fluidic links. It therefore becomes vital how such stacks may be configured electrically and fluidically. One example, which has been described[74] as a better way of connecting a stack for high performance and stability, is shown below in Fig. 7.

In essence, MFC units need to be in isolation from each other, to prevent any electron leakage through the various futile paths. The schematic in Fig. 7 is illustrating multiple MFC units connected in a series/parallel mode and fed from individual feedstock bottles. The latter is simply a representation of the isolation element that is required and symbolically shows how MFC units can be efficiently maintained. One common feedstock bottle may also be used, provided that the manifold output is of sufficient length to introduce a fluid resistance that is higher than the circuit resistance and thus stop electrons from *leaking* through the feedstock tubes.

4. Internal Resistance

In any electrochemical, power or flow system, internal resistance is a centrally important parameter that governs the performance. It is defined as the inherent ability of that system to resist the flow of electrons (in the cases of electrochemical and power systems) or fluid (in the case of a flow system). Generally speaking, the lower the internal resistance, the higher the electrical output, which is also true of MFCs. This parameter can be measured by a variety of methods e.g., electrochemical impedance spectroscopy (EIS), current interrupt method and polarisation technique. The latter is applying Kirchoff's law for impedance matching in conjunction with Jacobi's law for maximum power transfer (MPT). Data from two of these methods are shown in Fig. 8.

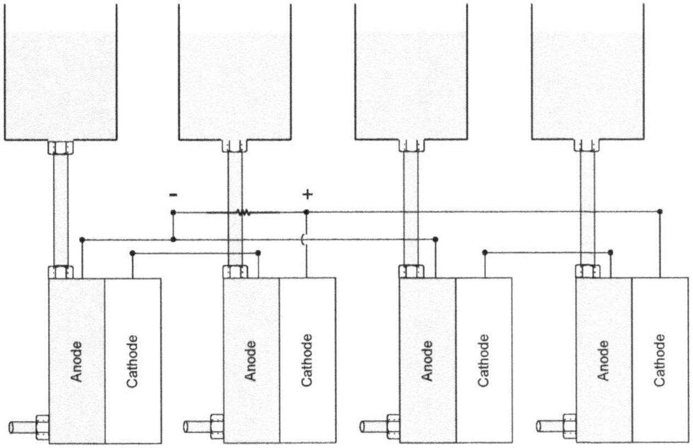

Figure 7. Optimised configuration for MFC stacks, which comprises a combination of series (to step up voltage) and parallel (to step up amperage) electrical connections as well as fluidically isolated units (adapted from Ieropoulos et al. 2008).[74]

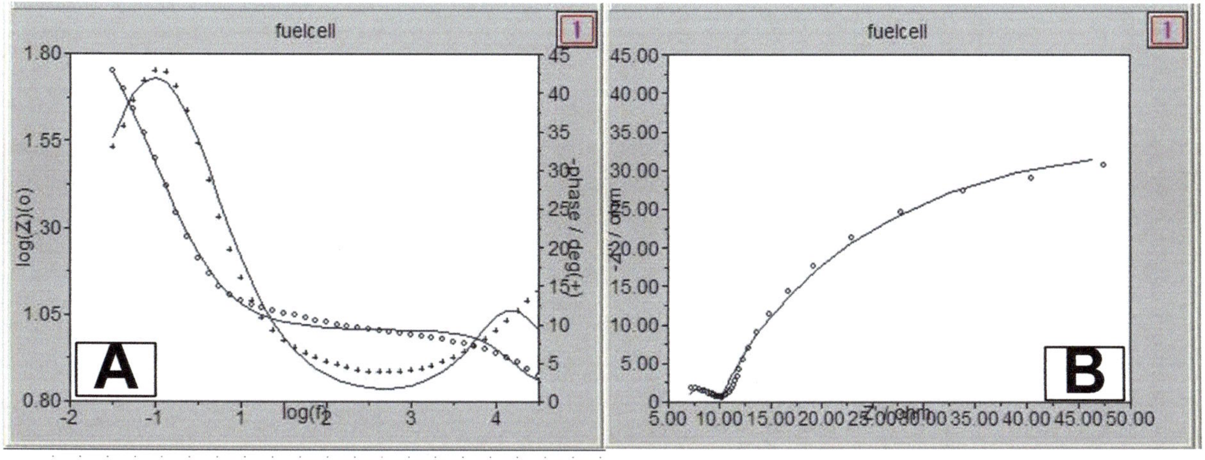

Figure 8. A and B show the electrochemical impedance spectroscopy experiment carried out on an individual working MFC; C (raw data) and D (processed data) show results from a polarisation experiment on an individual working MFC. Dotted lines (D) are second order polynomial non-linear regression fits.

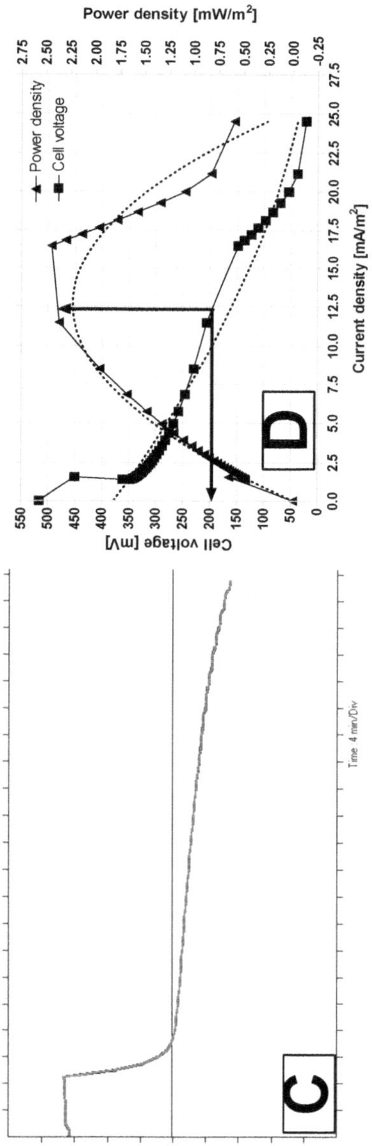

Figure 8. Continuation.

It is interesting to note that the two methods can produce very different values of internal resistance for the same system. The Z-plot (B) shows that the internal resistance from EIS is approx. 12 Ω, which is in contrast to the value calculated (1.3 kΩ) after completing a polarisation sweep (C & D). The arrows on D, show the point of peak power (maximum power transfer), which is produced at half the open circuit voltage ($\frac{1}{2}V_{O/C}$) that is in turn produced when the external resistive load is equal to the internal resistance (impedance matching). For conventional batteries, electrochemical cells and chemical fuel cells, this is almost always the case. As can be seen from D above, peak power can occur at a voltage slightly lower than $\frac{1}{2}V_{O/C}$ and this may be related to a dynamic open circuit voltage.[76] This is part of on-going investigations. EIS is a well-established method for calculating the ohmic part of the internal resistance, which is only one of three parts of the total internal losses – the other two being activation (at low current densities) and mass transfer/kinetic (at high current densities). Polarisation sweeps on the other hand allow the holistic quantification of the total internal losses, without selectively identifying which part is contributing how much to the total internal losses.

5. Cell Reversal in Stacks: An Explanation Why it is Happening

In physics, polarity reversal is a phenomenon often observed in networks (stacks) of multiple electrical devices – such as batteries – connected together. The reversal takes place when the units are connected in series, because this is when the voltage differential pressure is at its highest and, thus, is more sensitive to any changes in the internal state of each of the interconnected units. Electrical flow obeys the same laws, and is therefore analogous to, fluid flow dynamics. A good way of understanding cell polarity reversal is by analogy with the fluid dynamics perspective.

Let us consider a system whereby 3 identical water tanks are connected together (Fig. 9), so that water pressure due to gravity drives water out of the first tank into the second, and out of the second into the third. The capacity of each water tank is (hypothetically) 100 L, and due to pipe sizes and surface tension, water can only flow out from each tank at a rate of 90 L/h. This means that the overall network volume is 300 L, but the flow rate throughout

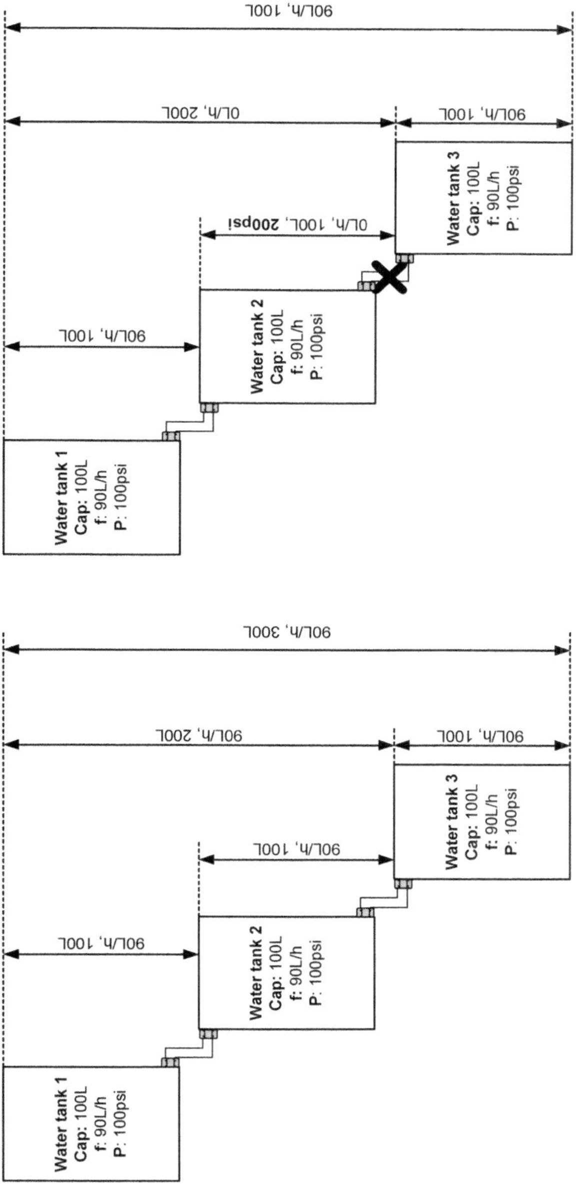

Figure 9. Water circuit analogy with electric circuit: schematic diagram of a water flow system under blockage and *polarity/flow* reversal.

the system is constant at 90 L/h. If water tank #2 develops a blockage at the outlet pipe, then water is no longer flowing out of this tank. The outflow from water tank #1 continues at 90 L/h, which means that in 6 minutes and 42 seconds water tank #2 will have reached its maximum capacity. If the outflow of tank #2 continues to be blocked, and assuming that the thickness of the outer casing of each water tank is sufficient to withstand a pressure equivalent to 100,000 L (9,678.4 atm), then the only way for the water to go is back out from the inlet, i.e., the flow becomes reversed. The volume of each of the tanks remains at 100 L, and it will take another 6 minutes and 42 seconds before the flow of water tank #1 reverses, and 1 hour before water tank #3 empties. It should also be noted that the pressure at the outlet of tank #2 increases.

This analogy is what happens in a network of batteries or MFCs. Whether due to

(a) bacterial cell starvation as a result of a blocked feed flow into a particular unit,
(b) heavy loading as a result of shorting,
(c) membrane fouling,
(d) biofilm overgrowth, or
(e) faulty cathode

the internal resistance will change disproportionately to the other units, and the voltage pressure in a series connected network, will *force* a reversal in polarity.

6. Applications

(*i*) *Wastewater Treatment*

The ability of MFCs to utilise waste organic material may potentially be used to *treat* this type of waste. Anaerobic sludge, as well as containing a rich diverse flora suitable for inoculating MFC, can also be used (as a fine suspension) to *feed* MFCs as a utilisable carbon-energy source (i.e., fuel) in its own right.[76] MFCs have also been shown to utilise algal-waste (what remains of algal cells after biofuel has been extracted from them). The ability of MFCs to *treat* waste from environments that are grossly polluted with toxic organic and metal ions (e.g., as occurring in landfill leachate) has also been demonstrated,[77,78] although not yet implemented on a large enough scale to show the economic advantages of this approach compared to conventional waste treatment methods.

(*ii*) *Denitrification*

Nitrate is at present very difficult and expensive to remove from wastewater. It has been reported that certain microbes e.g., *Thiobacillus denitrificans*, *Micrococcus denitrificans*, *Paraococcus denitrificans* and *Pseudomonas* can reduce nitrates or nitrites to nitrogen-containing gases. Several workers have reported that biotic MFC cathodes inoculated with a mixed culture (possibly containing some if not all of the above bacteria) can remove nitrate from wastewater streams.[79,80,81]

(*iii*) *Hydrogen*

It has also been reported that under certain conditions, pure hydrogen can be produced at the MFC cathode.[82] This is for an abiotic cathode poised (voltage *held*) at +250 mV by an external power supply and depleted completely of oxygen. The poising of the system in the absence of oxygen allows for the incoming H^+ and e^- to react with water (found as an electrolyte) to produce hydrogen gas. This is again another useful advantage that the MFC technology has to offer and extensive investigations are underway to turn this into a viable commercial application.

(*iv*) *Power Packs*

It is difficult to compare MFCs with conventional chemical batteries (of the sort used typically in everyday low-power applications) since one uses endogenous fuel (the battery) whilst the other (like a conventional chemical fuel cell) can be constantly fed from exogenous fuel and will continue to generate electricity long after the battery runs out. The data in Table 1 are from a single MFC running continuously for one year compared with a single alkaline battery cell.

Table 1
Comparison between a Microbial Fuel Cell (Bristol Robotics Laboratory, UWE) and a Standard AA Battery.

Type	$V_{O/C}$ (V)	Capacity (Ah)	Energy (Wh)	Weight (g)	Energy density (J/g)	Cost (£)	Durability
Alkaline cell (AA)	1.5	2.8	4.2	25	604	0.30	Will run out
Microbial fuel cell	0.8	1.05 (~120 μA per yr)	0.1752 (~20 μW per yr)	30	21	3.00	Will keep on going

7. Sensing Using B-E

Microbial cells are replete with receptors for many types of nutrient compounds within their environment. When cells grow in continuous culture, it is assumed that they receive all the essential classes of nutrients (all the required macro-, meso- and micronutrients: carbon-energy, nitrogen, phosphate, K^+, Mg^{2+}, essential vitamins and amino acids). The growth rate of cells in continuous culture is limited by whichever class of essential compound is in greatest *need* by the cells. All that is required to detect and measure a particular growth-limiting nutrient is to feed the culture on defined (or semi-defined) medium which contains all nutrients required for growth with the exception of the growth-limiting nutrient [the target analyte]. As well as being essential, the target analyte must be within the range at which it is growth rate limiting. Using this principle[83,84] determined that both the growth rate and final yield were related to the concentration of limiting substrate. Microbes in planktonic batch culture mode have been used to assay particular vitamins and amino acids for many years (> 50 years) including *Lactobacillus viridescens* (to assay for thiamine), *Lactobacillus casei* (riboflavin), *Saccharomyces uvarum* (vitamin B6) and *Lactobacillus leichmannii* (for Vit B12).[85,86,87,88] In conventional vitamin or essential amino acid microbiological assays, sterile samples containing the target analyte (along with a series of standards and control samples) are added to the appropriate microbial cultures and the amount of growth (both growth rate and final yield as measured by optical density) is proportional to the concentration of analyte in the sample.

It has been shown that MFCs can be used to measure biological oxygen demand (BOD)[89,90,91] as well as the presence of toxic pollutants.[92] There are now commercial products based on this technology (Korbi Inc. [S. Korea] which supplies two products – one for measuring BOD and one for detecting water pollution in the form of heavy metals or toxic organic compounds). The sensitivity of the cathode (liquid flow type) to the concentrations of dissolved oxygen has also been shown.[71]

8. Processing Using B-E

In a stack of MFC it is conceivable that single units can be arranged in a connectionist network whereby some units modulate the behaviour of the other units around it. Ideas for using biofilm electrodes as conventional logic gates have been proposed.[32] MFC units can be configured to do conventional binary logic gate such as AND, OR, NOT, Buffer, NAND, NOR, XOR, XNOR and the logic can be implemented via chemical or electrical inputs. The units have neuronal-like properties (being modulated by chemicals), and transistor-like properties (being modulated by voltage). The physicochemical parameters of the biofilm electrode include temperature, pH, nutrients, minerals, flow rate, redox, overpotential and inhibitory compounds. The outputs include electrode voltage and amperage and may be used to operate conventional silicon binary controllers and switch or control actuators.

A theoretical arrangement of Biofilm-Electrodes has been proposed in order for a robot to achieve a Pavlovian-type reflex[22,32] but these have not been built in practice. It thus remains to be seen if such intelligent systems can be built in reality to perform real logic functions, e.g., logic gates (AND, OR, NOT, XOR), decoders (Turn ON 1 of 8 units using only 3 inputs), counters, memory, toggle switches, clocks, and learning circuits (connectionist matrix), and utilised in practical information processing (known as Unconventional Computing).

9. EcoBots, Symbots, AI and AL

Employment of MFC technology on a robot would result in a kind of *artificial symbiosis* and represent a novel area for research and development.

10. Artificial Symbiosis

Association between two species of living systems may take many forms including symbiosis, which describes an obligate association between two species that benefit from, but also depend upon, each other. There are many other forms of interaction between two species and the whole spectrum of association may be divided into two groups: one, in which neither of the two parties is harmed, and one in which either or both species are harmed. The latter category describes different forms of antagonism and associations could be of the competition form in which both species are harmed, amensalism in which one of the species is harmed with no effect on the other, and parasitism in which one party benefits from damaging the other. Predation also belongs to this latter category, in which eventually one of the parties is killed by the other. In the former category of associations, apart from symbiosis and protocooperation, there is also commensalism, which describes a beneficial interaction for the one party with no effect on the other. Neutralism also belongs to this category, in which the close spatial association between two species has no effect on each other.

The envisaged system in this work is one that potentially mimics symbiosis and since one part of this association is artificial, the term *artificial symbiosis* is more appropriate. This would be the result of the integration between the robot and the microbes in the MFCs. The microorganisms, as part of their normal activities, would be supplying energy to the robot which, in turn, would be performing various tasks, one of which would be to collect more material from which the microbes may extract energy. This seemed to be the most interesting approach in the pursuit for building an autonomous robot.

11. Robotic Autonomy

The notion of autonomy has emerged in the general context of artificial systems and is now widely used in the areas of artificial intelligence, artificial life, biologically inspired robotics, and practical engineering. In these areas, autonomy has a different meaning. Generally, it is used to describe robots which are able to carry out tasks assigned by the human operator, but with minimum (or non-continuous) human intervention or supervision. Although robots may be designed, built and programmed by the human operator, the fact that they can carry out tasks with minimum or at least non-continuous human intervention or guidance, has allowed the attribution of such a fundamentally anthropocentric concept to such an emphatically artificial process.

Robots that can execute pre-assigned tasks or even improve their performance through feedback from the environment (*learn*) during that process, independent of the energy supply, have also been termed autonomous, although it has previously been stated that if such systems are based on classical symbol processing operations, then they are merely automatic.[93] These robots have what is known as computational autonomy, or automaticity.[93] Computational autonomy, in biological systems, defines the ability of an agent to carry out actions, which may or may not be related to the collection of energy, such as looking out for and avoiding predators or mating. Even when animals hunt for food, which is energy related, they somehow estimate their body's energy reserves and weigh them out against the energy requirements of their predatory actions. This indicates a form of homeostatic computation.

Computational autonomy in robots therefore implies that agents are able to perform computations, e.g., process information and carry out tasks, which may or may not be related to energy management. However, as with animals, these agents need energy to continue operating. This may suggest that overall robotic autonomy could be limited depending on the availability of energy. Solar-powered robots operating in an environment with access to daylight can have the computational ability to perform tasks and in addition collect their energy. These robots are, however, dependent on solar radiation and their autonomy and overall operation may be compromised in the dark. Hence, there is a strong link between computational ability and energy collection.[94,95,96,97]

In animals, autonomous behaviour is the result of the combination of computational autonomy and energetic autonomy.[94,95,96,97] Energetic autonomy refers to the ability of an agent to maintain itself in a viable state for prolonged periods.[94,95,98,99,100] For whole animals, energetic autonomy (or self-sustainability) as is the case with all kinds of behaviour, requires computation, which is mainly a feature of a nervous system and a result of homeostasis, different to that exhibited by single cells, tissues and organs. It is, however, the only behaviour that *powers* the computation module. The two co-govern the agent's overall behaviour, as shown in Fig. 10. This could be considered as a possible flow chart representation of the behavioural model of, e.g., carnivorous mammals. The thin brighter lines represent the energy feed into each behavioural module while the thicker darker lines represent the information and control *bus* for each of the modules.

Until recently, in the field of autonomous robots, more emphasis has been given to computational autonomy alone, rather than energetic autonomy. From the oversimplified design shown above, it is suggested that although different behaviours are the result of homeostatic computation, they are all dependent on the energy currency of the system, including the computation module. This indicates the potential importance of energy management in robotics. Some of the computations that may be incorporated by agents towards self-sustainability may involve parameter calculations such as distance to the next energy source and economisation of the on-board energy supply to avoid reaching *lethal* limits.[95,98,99]

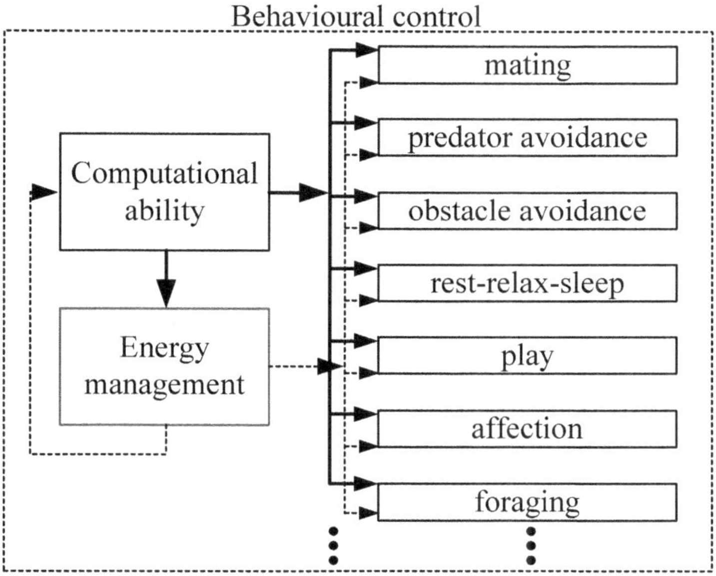

Figure 10. Flow chart of the different possible behavioural modules.

For the reasons mentioned above, it was decided to address the problem of robotic autonomy from the perspective of energy management as this encompasses computational ability as well. Autonomy for robots could simply be considered as a higher degree of freedom than the ability to compute towards a resulting behaviour.

A very basic illustration of how such systems may work is given in Fig. 11. The flow chart on the left (Fig. 11a) shows that the robot's sole objective may be to keep the energy level above a certain threshold based on explicit or implicit computational information. Alternatively, the robot may be required to execute a task, which is independent of the collection of energy and in addition to sustaining itself (Fig.11b). Both models result in behaviours that are informed by computation.

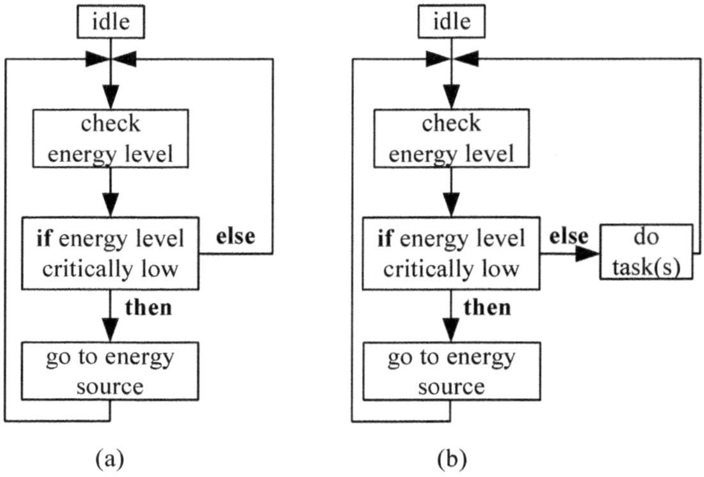

Figure 11. Two behaviour models resulting from computational information that are (a) self-sustained, and (b) self-sustained with a token task.

Robots may be designed to operate in a range of habitats where they could exploit various forms of energy sources. This design may be based on a pre-defined environment criterion. Furthermore, robots may be categorised by their energy supply mechanisms. For example, one category of robots could be battery-powered (Ro_B), a second could be solar-powered (Ro_S) and a third and final could be robots that can directly extract energy from natural sources in their environment (real food substrates) (Ro_F). Robots may also be categorised by the level (or type) of autonomous behaviour that they exhibit. They can either be exclusively computationally autonomous (Au_C) or energetically autonomous (Au_E) which also implies some degree of computational ability. Natural environments offer access to real food and may be crudely classified into three categories. The indoors habitat (Ha_I), for example a room offering access to the national grid through the mains, access to solar radiation through window(s) and access to real food from the fruits or vegetables left on a table. The outdoors habitat (Ha_O), which may be characterised as the environment offering access to solar radiation during daylight hours, possible access to the national grid through outdoors mains sockets and access to real food substrates such as fruits, vegetables or insects. The third and final type of habitat may be represented by a remote area (Ha_R), for example the ocean floor or tropical forest, which can be characterised as the environments offering access to raw substrates. In the case of the ocean floor these may be plankton or marine vegetation (seaweed) and in the case of tropical forests these may be vegetation, sludge and insects. The three identified scenarios hereinabove are merely representations of some possible habitats and therefore cannot cover the entire range of natural environments.

Robots in the real world may be considered fully autonomous if they are able to operate consistently in a changing environment. As with biological systems in nature, this implies the ability to gain information about the environment, execute set tasks and navigate if mobile, extract energy from the environment and avoid harmful (to people, property or itself) situations, and even repair themselves without human intervention.

For the three different habitat scenarios, autonomy depends on the availability of food, light or electricity from the mains. Assuming that these are equally available, then all three robots could be potentially energetically autonomous. Naturally occurring substrates are less dependent on humans and therefore robots Ro_F may be at an advantage especially in the cases of Ha_O and Ha_R (the method for utilising raw substrates will be explained later).

The distinction between robots made here was based on the state-of-the-art technology in robots. Truly autonomous robots, however, would be those agents that have diversity in the method of collecting their energy and will almost certainly outperform robots that are reliant on a single source of power. Robots Ro_F may have a greater potential towards autonomy if they are designed to utilise a wide variety of naturally occurring substrates without crossing the limits of the human society.

Robots in environments such as those described above will be faced with similar problems to those faced by living organisms and, like their animal counterparts, will need to select appropriate actions from a behavioural repertoire. In other words, they must solve their own version of the action selection problem. Distance, for example, and type of terrain to a food-source will be converted to energy expenditure. The onboard energy reserve together with the energy that can be produced from the onboard food supply will be compared to this energy consumption. Gain from the goal food-source may also form a calculated parameter. If, for example, the gain from a goal that is further away is higher than that from another closer to the robot, but there is not enough energy to reach, then the robot will go for the latter. This will be in order to collect enough energy to reach the higher gain goal. In a sense, this will be mimicking *opportunistic* behaviour found in animals like raccoons and rats. The habitat chosen should be one with abundance of the energy source that the robot system requires so that internal parameters can remain at equilibrium, which is somewhat akin to the homeostatic mechanisms found in living organisms. For example, the agent may be designed with built-in modules for attracting its food (e.g., fly trap for flies).

This may serve as a paradigm shift to the methods currently employed by roboticists in designing and developing mobile autonomous robots. It will possibly give rise to novel complications and challenges not considered before, since energy collection and management may become vital in the development of such robots. The onboard energy supply, for instance, may not be enough for continuous operation but instead only suffice for periodic actuation. This potentially opens a *waiting* window, of variable time length, which could be exploited in various ways. It could be used to incorporate other behavioural aspects that may not be very energy expensive. This *waiting* window will probably result in a *pulsed* mode behaviour, which could be offering the option of doing nothing in order to remain operational. Human beings and animals in nature seem to have a similar behavioural pattern, and in addition exploit onboard biochemical energy reserves when food is not available. Hence, a reserve energy supply may be considered onboard such a robot, so as to avoid falling below the *lethal* limits in the case of lack of energy sources in the environment. This does not necessarily mean that the reserve is going to be of electrical nature. It may be possible to have bio-chemical energy reserves that can be readily converted to electricity with minimal losses during storage.

The example of robots, habitats and autonomy, more than anything, raises the issue of energy extraction from the environment, which may potentially prove to be a key factor in the design and development of autonomous robots. For instance, there may be a need for a cleaning system to be designed for an area of the ocean where there has unfortunately been an oil spill. A robot could be designed with a system capable of collecting and cleaning the polluted waters by extracting and possibly utilising the hydrocarbons present in the polluted samples. Such a mechanism may be based on electromechanical parts, or it may even be based on hydrocarbon-utilising microorganisms (*Vibrio*, *Pseudomonas*, *Micrococcus*, *Nocardia* and *Acinetobacter*, *Methanococcus*).[101,102]

12. Symbot

Symbot is short for Symbiotic Robot and it refers to the beneficial integration between a live part and an artificial mechatronic part, on a robotic platform. Thus, a Symbot can potentially mimic symbiosis, which is obligatory and mutually beneficial. Since one part of this association would be an artifact, it was decided to attribute the term artificial symbiosis to the system. This would come as the result of the integration between the robot and the microbes in the MFCs. The microorganisms, as part of their normal activities, supply energy to the robot which, in turn, performs various tasks, one of which would be to collect more material from which the microbes may extract energy. This seemed to be the most interesting approach in the pursuit for building an autonomous robot called **EcoBot** (short for Ecological Robot) referring to a class of energetically autonomous robots that can remain self-sustainable by collecting their energy from (mostly) waste in the environment. The MFCs aboard the robot degrade the waste and produce electricity at the same time. The by-products from this process are CO_2 and water. The amounts of these compounds add up to what would have been produced from the natural biodegradation of the waste in the first place had it not been intercepted by the robot. Thus, the technology is part of the immediate carbon cycle of our planet and is not reliant on fossil fuels.

The mechatronic part of EcoBot (pumps, motors, chassis plus silicon circuitry) has to provide the robot with its means to function. This includes the main systems for getting fuel (feeding) and for maintaining the microbes within their MFC units. Some form of intelligent processing is required to control the extraction, storage and distribution of the electrical energy. Typically, the robot works in pulsed behaviour mode so that it can utilise stored energy (sufficient for its function) in pulses of high energy and then must use the interval time between actuator firings to re-charge its capacitor bank (store). In the future it may be possible to utilize the microorganisms on board for many different purposes in addition to making electrical energy. These include their incorporation into specialized units (biofilm-electrodes) that can sense the physicochemical environment. When used in sensing mode it is also possible for one group of units to sense the state or condition of the others. Units have transistor-like properties (being controlled by electrode overpotential) or neuron-like properties (being controlled by chemicals in the input stream). Within a connectionist system, these properties may be exploited to the full so that in the long term the EcoBots may operate intelligently using biological computing rather than with conventional discrete binary silicon-based systems with attendant algorithms.

The EcoBot series of robots were built to show the potential application of MFC systems to give robots energy autonomy and to demonstrate the feasibility of building Symbots; a type of robot that is made from around the symbiotic relationship between microbes and mechatronic units. Table 2 lists the development of

EcoBots through the current series. In all cases, microbes in the MFCs produced all the electricity for the mechatronics and were the sole energy sources aboard the robot. Since all robots were capable of moving themselves by wheels, it implies an ability to lift and shift the full weight of the MFC and fluids on board along with the weight of the chassis and other components.

XV. DISCUSSION

Although at first sight it might seem that MFCs can be described (or explained) using classical ideas from electrode chemistry alone, it soon becomes clear that for many species of anodophiles (those that directly conduct electrons onto the obliging anode) it is the reaction of the microorganism with its physicochemical environment rather than the reactions at the electrode surface that dictates the biotransformation properties of the whole system. In sensing terms, it is the microbes in the biofilm electrode that act as both, specific receptor and first transducer (turning the environmental/condition change into a change in growth rate) and the electrode is simply a second transducer, converting the main output (growth rate) into an electrical signal that can be measured as MFC output. Electrode chemistry cannot explain how the system can (for example) operate on acetate feedstock in the first instance and subsequently run on an alternative substrate, nor can it explain how the system as a whole can operate at different pH values or temperatures and be inhibited by different biocides.

Biofilm electrodes with highly diverse flora can be used to break down and mineralise a wide range of organic compounds (including dilute organic materials generally regarded as waste) and turn these into electricity. The system works within the immediate carbon-cycle, not producing additional carbon dioxide than (a) that fixed by photosynthesis in the first place, and (b) what would have changed into carbon dioxide anyway by alternative oxidation (e.g., by natural microbial degradation or burning by humans).

Table 2
EcoBots-I, -II and III

EcoBot 1
• 8 MFCs containing *E. coli* • 1 behavioural task i.e., photo-taxis • World's second robot to employ MFCs • World's second robot to utilise a refined organic substrate • World's first robot NOT to include conventional power supply
EcoBot-II
• 8 MFCs containing sewage sludge microbes • World's 1st to exhibit 4 kinds of behaviour: • Sensing • Processing • Actuation (photo-taxis) • Communication • World's first robot to utilise raw unrefined substrate e.g., flies or rotten fruit • World's first robot to employ the O_2 cathode
EcoBot-III (BREADbot) (Fig. 12)
• World's first to demonstrate true cycles of full energy autonomy: Part of the robot's behaviour is to ensure that the microbes are fed and maintained. Can feed on flies but also liquid feedstock which it collects from its arena along with H_2O to hydrate its cathodes. • 48 MFCs onboard • Power output on average 1.5 mW • Energy per firing 2 J • Ran through a period of nine days (320 actuations) before requiring any maintenance • Internal sensing (fluid level, voltage of MFCs, position in arena, type of actuation) • Powered 4 pumps, 3 motors, UV light module, wireless TX, and a complete microprocessor based controller governing energy harvesting • Pulse mode behavior for main actuators but continuous for electronic controllers

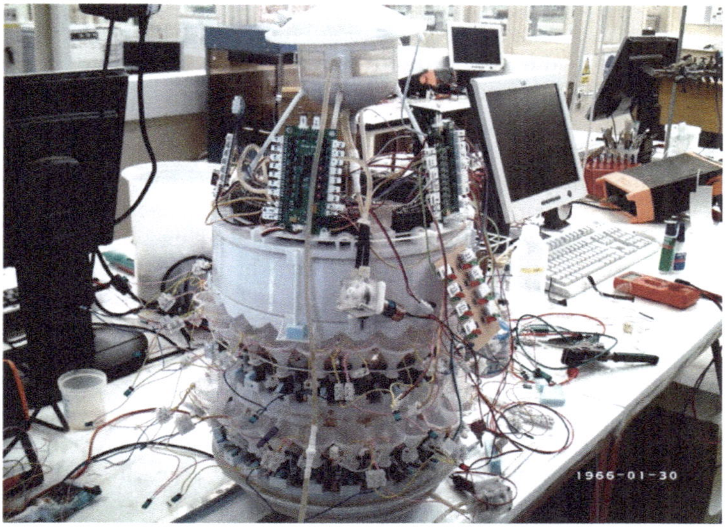

Figure 12. Early prototype of EcoBot-III (BREADBot). The whole chassis is made from Acrylonitrile Butadiene Styrene (ABS) rapid prototype material. The stack of Microbial Fuel Cells powering this robot can be seen (small funnels) from the mid-point down.

MFCs become more energy dense as they become smaller and a stack of small cells is better than a single large chamber. The reason for this is the nature of the electrode which has to interact with a high microbial population requiring a large surface area. A large surface area cannot be created with a smooth block of solid electrode, thus the electrode has to be convoluted (e.g., carbon felt or carbon veil). However, in this form it possesses high electrical resistivity which increases as the MFC and its electrodes get bigger. Furthermore, it is a matter of internal resistance balance; an individual MFC is subject to its internal resistance, which is the result of the combination of all elements inside a MFC. Internal resistance (or losses) is made up of three main components: activation losses (low current densities), ohmic losses (medium current densities – linear part of curve) and mass transfer/kinetic losses (high current densities). A polarisation sweep method of measuring internal resistance accounts for all three major types of internal losses, whilst this is not true of other methods including EIS and current interrupt. The internal resistance is a combination of the nature and composition of electrodes, electrolytes, (bio)-catalysts and membranes. When connected as part of a network (or stack) these internal resistances balance-out depending primarily on the electrical configuration but also on the fluidic linkage. The internal resistance balance is, of course, independent of size; however, the internal resistance of individual MFCs is heavily governed by the MFC volume, electrode surface area and the ratio of the two. It therefore appears that small may be best. A plurality of MFCs is required to step up voltage by a series configuration of connections. If the MFCs are connected in series, the fluidic junctions can also conduct electrons that oppose any useful increases in voltages, so special arrangements are required to fluidically isolate the units. For producing practical amounts of electrical power (e.g., to run small-scale devices such as digital clocks, calculators, robots and electronic gadgets), a large number of small MFC are required in stacks. Stacks of MFCs give rise to their own types of difficulty such as the phenomena of polarity reversal.

MFCs have many potential applications including sensing, utilisation or biotransformation of waste products into more useful (e.g., electricity) or less damaging products (e.g., de-nitrification) and their use may include a new way of energising robots. A new type of robot (Symbot; symbiotic association between microbial life-forms and mechatronic systems, as exemplified by the EcoBot series of robots) can be made to achieve full energy autonomy. In the future, the symbiotic association between microbes and machines may be extended by having specialised microbes for use in sensing units (to sense the robots' environment), or can be used in a stack of MFC biofilm electrodes (configured by a connectionist system) that can help process information on board the robot (non-silicon-based circuitry).

REFERENCES

1. P. D. Marsh, J. R. Hunter, G. H. Bowden et al., *J. Gen. Microbiol.* **129** (1983) 755.
2. C. H. Sissons, L. Wong, and Y. H. An. in *Handbook of bacterial adhesion, principles, methods, and applications*. Ed. by Y. H An and R. J. Friedman, Humana Press: Totowa, NJ, 1999, p. 133.
3. D. G. Allison, *Community structure and co-operation in biofilms*, Cambridge University Press, Cambridge, 2000.
4. P. S. Stewart and J. W. Costerton, *Lancet*, **358** (2001) 135.
5. C. M. Waters, B. L. Bassler, *Annu. Rev. Cell. Dev. Biol.* **21** (2005) 319.
6. J. Greenman, P. Spencer, C. McKenzie et al., *Oral Diseases*, **11** (2005) 14.
7. B. E. Rittmann, P. L. McCarty, *Biotech. Bioeng.* **22** (1980) 2343.
8. O. Wanner, and W. Gujer, *Biotech. Bioeng.*, **28** (1986) 314.
9. B. E. Rittmann, and J. A. Manem, *Biotech. Bioeng.* **39** (1992) 914.
10. O. Wanner and P. Reichert, *Biotech. Bioeng.* **49** (1996) 172.
11. J. W. T. Wimpenny and R. Colasanti, *FEMS Microbiol. Ecol.* **22** (1997) 1.
12. C. M. Picioreanu, C. M. van Loosdrecht and J. J. Heijnen, *Biotech. Bioeng.* **58** (1998) 101.
13. G. H. Dibdin, *Arch. Oral Biol.* **26** (1981) 515.
14. G. H. Dibdin, *Adv Dent Res* **II** (1997) 127.
15. G. H. Dibdin, S. J. Assinder, W. W. Nichols and P. A. Lambert, *J. Antimicrob. Chemother.* **38** (1996) 757.
16. P. S. Stewart, *Antimicrob. Agents Chemother.* **40** (1996) 2517.
17. S. J. Pirt, *Principles of microbe and cell cultivation*, John Wiley & Sons, New York, 1975.
18. D. G. Allison, T. Maira-Litra´n and P. Gilbert, *Methods Enzymol* **310** (1999) 232.
19. A. Gjaltema, L. Tijhuis, M. C. M. van Loosdrecht and J. J. Heijnen, *Biotech. Bioeng.* **46** (1995) 258.
20. A. E. Hodgson, S. M. Nelson, M. R. W. Brown and P. Gilbert, *J. Appl. Bacteriol.* **79 (**1995) 87.
21. P. Spencer, J. Greenman, C. McKenzie et al. *Appl. Microbiol.* **103** (2007) 985.
22. J. Greenman, I. Ieropoulos and C. Melhuish, *Int. J. Unconv. Comp.* **4** (2008) 23.
23. C. McKenzie, Thesis, University of the West of England, 2007.
24. J. Greenman, C. McKenzie, S. Saad et al. *J. Breath Res.* **2** (2008) 046005 (7pp).
25. R. M. S. Thorn and J. Greenman, *Appl. Microbiol.* **107** (2009) 2070.
26. C. N. H. Marques, V. C. Salisbury, J. Greenman, K. E. Bowker and S. M. Nelson, *Antimicrob Chemother* **56** (2005) 665.
27. J. Monod, *Ann. Inst. Pasteur*, **79** (1950) 390.
28. A. Novick, L. Szilard, *Science* **112** (1950) 715.
29. S. J. Pirt, *Proc. Roy. Soc. B* **163** (1965) 224.
30. S. J. Pirt, *Arch. Microbiol.* **133** (1982) 300.
31. K. Kovárová, Thesis, Swiss Federal Institute of Technology, 1996.
32. J. Greenman, I. Ieropoulos, C. McKenzie, and C. Melhuish, *Int. J. Unconv. Comp.* **2** (2006) 249.
33. J. B. Russell, *J. Mol Microbiol Biotech*, **13** (2007) 1.
34. L., M. Prescott, J., P. Harley and D., A. Klein, *Microbiology 3rd Edition*, Wm. C. Brown Publishers, Dubuque, IA, 1995.

35. R. Boopathy, *Appl Environ Microbiol*, 62 (1996) 3483.
36. B., P. Lomans, R. Maas, R. Luderer, H., J., M., Pol, A. Op den Camp, C. van der Drift and G. D. Vogels, *Appl. Environ. Microbiol.* **65** (1999) 3641.
37. K. Watanabe, Y. Kodama, N. Hamamura, and N. Kaku, *Appl. Environ. Microbiol.* **68** (2002) 3899.
38. U. Bossel, *The birth of the fuel cell 1835-1845,* European Fuel Cell Forum, Oberrohrdorf, Switzerland, 2000.
39. K. Kordesch and G. Simader, *Fuel Cells and their Applications*, John Wiley & Sons/VCH, 1996.
40. L. Carrette, K. A. Friedrich and U. Stimming, *Fuel Cells (Weinheim, Ger)*, **1** (2001) 5.
41. R. D. Cortright, R. R. Davda and J. A. Dumesic, *Nature*, **418** (2002) 964.
42. A. Heinzel and V. M. Barragán, *Power Sources*, **84** (1999) 70.
43. G. Barbieri and F. P. Di Maio, *Ind. Eng. Chem. Res.* **36** (1997) 2121.
44. D. Wang, S. Czernik and E. Chornet, *Energ Fuel*, **12** (1998) 19.
45. S. Park, J. M. Vohs and R. J. Gorte, *Lett Nature*, **404** (2000) 265.
46. T. Hibino, A. Hashimoto, T. Inoue et al. *Science*, **288** (2000) 2031.
47. A. B. Robinson, N. E. Robinson and W. Soon, *JPandS*, **12** (2007) 79.
48. J. Lee, N. T. Phung, I. S. Chang et al. *Microbiol. Lett.* **223** (2003) 185.
49. C. Melhuish, I. Ieropoulos, J. Greenman and I. Horsfield, *Auton Robots*, **21** (2006) 187.
50. G. C. Gil, I. S. Chang, B. H. Kim et al. *Biosens. Bioelec.* **18** (2003) 327.
51. M. Piccolino, *Brain Res. Bull.* **46** (1998) 381.
52. M. C. Potter, *Proc. Royal Soc.*, **84** (Part B) (1912) 260.
53. J. L. Stirling, H. P. Bennetto, G. M. Delaney, J. R. Mason, S. D. Roller, K. Tanaka et al. *J Biochem Soc Trans*, **11** (1983) 451.
54. H. P. Bennetto, G. M. Delaney, J. R. Mason et al. in *Alternative Energy Sources VII,* 4, *Bioconversion/Hydrogen,* Hemisphere Publishing Corporation, New York, 1984, p. 143.
55. H. P. Bennetto, G. M. Delaney, J. R. Mason et al. *Biotechnol. Lett.* **7** (1985) 699.
56. G. T. R. Palmore and G. M. Whitesides, in *Enzymatic Conversion of Biomass for Fuels Production*, Ed. by M. E. Himmel, J. O. Baker and R. P. Overend, *Am. Chem. Soc. Symp.* Series No. 566, Oxford University Press, 1994, p. 271.
57. W. Habermann and E.-H. Pommer, *Appl. Microbiol. Biotechnol.* **35** (1991) 128.
58. D. H. Park and J. G. Zeikus, *Biotechnol. Bioeng.* **81** (2003) 348.
59. K. Rabaey, G. Lissens, S. D. Siciliano and W. Verstraete, *Biotechnol. Lett.* **25** (2003) 1531.
60. H. Liu, R. Ramnarayanan and B. E. Logan, *Environ. Sci. Technol.* **38** (2004) 2281.
61. B. E. Logan, M. Cassandro, K. Scott et al. *Water Res.* **39** (2005) 942.
62. D. R. Bond and D. R. Lovley, *Appl. Environ. Microbiol.* **69** (2003) 1548.
63. S. K. Chaudhuri and D. R. Lovley, *Nat. Biotechnol.* **21** (2003) 1229.
64. Y. A. Gorby, S. Yanina, J. S. McLean et al. *PNAS*, **103** (2006) 11358.
65. G. Reguera, K. P. Nevin, J. S. Nicoll et al. *Appl. Environ. Microbiol.* **72** (2006) 7345.
66. J. K. Jang, T. H. Pham, I. S. Chang et al. *Process Biochem.* **39** (2003) 1007.
67. S. Cheng, H. Liu and B. E. Logan, *Environ. Sci. Technol.* **40** (2006) 2426.
68. G. M. Delaney, H. P. Bennetto, J. R. Mason et al. *J. Chem. Tech. Biotechnol.*, **34B** (1984) 13.
69. K. H. Kang, J. K. Jang, T. H. Pham et al. *Biotechnol. Lett.* **25** (2003) 1357.

70. B. C. Jong, B. H. Kim, I. S. Chang et al. *Environ. Sci. Tech.* **40** (2006) 6449.
71. I. Ieropoulos, C. Melhuish and J. Greenman, *Bioinsp. Biomim.*, **2** (2007) S83.
72. J. Larminie and A. Dicks, *Fuel Cell Systems Explained.* John Wiley & Sons Ltd, Chichester, 2003.
73. P. Aelterman, K. Rabaey, T. H. Pham et al. *Env. Sci. Tech.* **40** (2006) 3388.
74. I. Ieropoulos, J. Greenman and C. Melhuish, *Int. J. Energy Res.* **32** (2008) 1228.
75. I. Ieropoulos, J. Greenman and C. Melhuish, *Bioelectrochem.* **78** (2010) 44.
76. I. Ieropoulos, J. Winfield and J. Greenman, *Biores. Tech.* **101** (2010) 3520.
77. J. Greenman, A. Gálvez, L. Giusti and I. Ieropoulos, *Enz. Microb. Technol.* **44** (2009) 112.
78. A. Gálvez, J. Greenman and I. Ieropoulos, *Biores. Tech.* **100** (2009) 5085.
79. Z. He and L. T. Angenent, *Electroanalysis*, **18** (2006) 2009.
80. P. Clauwaert, K. Rabaey, P. Aelterman et al. *Env. Sci. Tech.* **41** (2007) 3354.
81. Y. H. Jia, H. T. Tran, D. H. Kim et al. *Bioprocess Biosyst. Eng.* **31** (2008) 315.
82. H. Liu, S. Grot and B. E. Logan, *Env. Sci. Tech.* **39** (2005) 4317.
83. J. Monod, *Annu. Rev. Microbiol.* **3** (1949) 371.
84. T. Bauchop and S. R. Elsden, *J. Gen. Microbiol.* **23** (1960) 457.
85. E. E. Snelle, *Physiol. Rev* **28** (1948) 255.
86. L. M. Henderson and E. E. Snelle, *J. Biol. Chem.* **172** (1947) 15.
87. G. H. Spray, *Clin. Sci.* **14** (1955) 661.
88. B. P. Kelleher and S. D. Broin, *J. Clin. Pathol.* **44** (1991) 592.
89. I. S. Chang, J. K. Jang, G. C. Gil et al. *Biosen. Bioelectron.* **19** (2004) 607.
90. H. Moon, I. S. Chang, K. H. Kang et al. *Biotechnol. Lett.* **26** (2004) 1717.
91. A. Kumlanghan, J. Liu, P. Thavarungkul et al. *Biosens. Bioelec.* **22** (2007) 2939.
92. M. Kim, S. M. Youn, S. H. Shin, et al. *J. Environ. Mon.* **5** (2003) 640.
93. L. Steels, *Robot. Auton. Syst.* **15** (1995) 3.
94. E. Spier and D. McFarland, in *From Animals to Animats 4, SAB'96*, MIT Press, 1996, p. 255.
95. D. McFarland and E. Spier, *Robot. Auton. Syst.* **20** (1997) 179.
96. J. Greenman, O. Holland, I. Kelly et al. *Mechatron.* **13** (2003) 195.
97. I. Kelly and C. Melhuish, in *6th European Conference on Artificial Life (ECAL'01)*, Springer-Verlag, 2001, 10 pgs.
98. D. McFarland, in *From Animals to Animats 1, SAB'90*, MIT Press, 1990, p. 22.
99. E. Spier and D. McFarland, *J. Theor. Biol.* **189** (1997) 317.
100. S. Wilkinson, *Ind. Robot*, **28** (2001) 213.
101. K. Watanabe, Y. Kodama, N. Hamamura et al. *Appl. Environ. Microbiol.* **68** (2002) 3899.
102. J. Ikävalko, Arctic Operational Platform (ARCOP), FIMR, 2005 project report available online at: http://www.arcop.fi/reports/D4232.pdf.

John Greenman

Professor John Greenman is Professor of microbiology in the School of Life Sciences, University of the West of England (UWE) Bristol. His first degree and PhD were in Microbiology at the University of Liverpool. His post-doctoral work was at Leeds University and involved microbiology of cutaneous surfaces. He then joined Bristol Polytechnic (now UWE Bristol) in 1979 and was appointed Research Professor in 2001. His research interests include constantly monitored in vitro biofilm models, microbial fuel cells (MFCs) both used as sensors and power units to energise robots. His most recent work has applied biofilm models to the study of human diseases (caries, periodontal disease, wound infection) as well as to water treatment. His work (with Dr. Ieropoulos) on MFCs has led to a breakthrough in MFC scalability and energy management.

Ioannis A. Ieropoulos

Dr. Ioannis A. Ieropoulos is a Research Fellow at the Bristol Robotics Laboratory and has produced EcoBots I and II for his PhD. His first degree was in Electronics Engineering and his MSc was in Communication Systems and Signal Processing which was obtained from the University of Bristol. For the last seven years he has been working on autonomous robotics and further improving the MFC technology as part of his post-doctoral Research Fellowship appointment. His work (with Prof. Greenman) on MFCs has led to a breakthrough in MFC scalability and energy management.

Chris Melhuish

Professor Chris Melhuish is Director of the Bristol Robotics Laboratory (BRL), and holds Professorial chairs at both the University of Bristol and the University of the West of England. He has over 20 years experience in designing advanced robotic equipment. His last 12 years have been spent working on autonomous robot systems and has for the last 6 years led the BRL team in self-sustaining robotics research. He is an adviser to the EU in ICT and Future Emerging Technologies.

4

Electrochemical Coating of Medical Implants

Regina Guslitzer-Okner and Daniel Mandler*

Institute of Chemistry, The Hebrew University of Jerusalem, Jerusalem 91904, Israel

I. INTRODUCTION

Electrochemistry deals with charge transfer across interfaces. It is one of the oldest fields of chemistry dated back to Faraday, yet, its importance to basic research as well as industrial applications cannot be overestimated. One of the first applications of electrochemistry was for coating conducting surfaces with either metals, e.g., electroplating, or by inorganic or organic substances, such as oxides and polymers. The fact that electron transfer is limited to a few nanometers from the interface has long ago been recognized as a means of coating conducting surfaces possessing complex geometry. Coatings ranging from a monoatomic layers, e.g., via under potential deposition, to microns thick have been carried out, which affected the physical and chemical properties of the coated surface. Hence, electrochemically deposited films have been used to inhibit corrosion, accommodate enzymes and biosensors, con-

*To whom correspondence should be addressed.

N. Eliaz (ed.), *Applications of Electrochemistry and Nanotechnology in Biology and Medicine I*, Modern Aspects of Electrochemistry 52,
DOI 10.1007/978-1-4614-0347-0_4,

trol optical properties and in numerous others applications. Among the advantages of electrocoating are the fine control of the thickness, the ability to co-deposit simultaneously more than one substance and the fact that deposition occurs exclusively on conducting surfaces, thus enabling to coat patterned structures, such as printed and integrated circuits.

During the years different electrochemical coating techniques have been developed including conventional electrodeposition (ED), electrophoretic deposition (EPD), microarc deposition (MAD), electropolymerization (EP) and electrografting (EG). These will be described briefly later. The range of substances and materials that have been electrochemically deposited spans from metals, oxides, and ceramics to inorganic and organic polymers. The latter were either conducting polymers or insulators including also biopolymers, such as chitosan.

Electrochemically deposited films have also been used in medicine, such as for coating needles with hard materials; however, the approaches were similar to those applied in other fields. We are aware of only a short review dealing with electrochemical deposition of medical devices, which is a chapter in Paunovic and Schlesinger well known book on ED.[1] This chapter reviews the various studies of electroplating of medical devices, such as implants. There are obviously numerous patents in this field, which we do not intend to cover in this review.

Reviewing the different studies that are related to medical devices can follow different categorization paths (Fig. 1). The studies can be divided according to the electrochemical approach used, e.g., electroplating, electrophoretic deposition, etc. A different division can follow the nature of the substrate to be coated, namely, titanium, stainless steel, and other alloys or the nature of the medical device, such as stent. Finally, the description of previous works can be divided based on the nature of the deposit itself that varies between metals to organic polymers and ceramics.

To avoid overlapping and confusion, we will first review briefly the different methods of electrochemical deposition and then briefly mention the various substrates electrochemically coated. The major part of this review will focus on the coating itself, that is on the nature and mechanism of the electrochemical deposit.

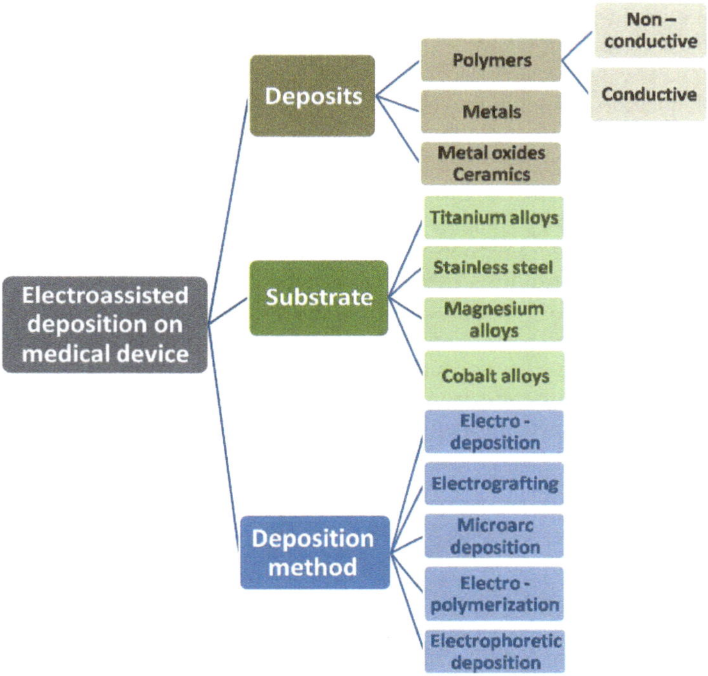

Figure 1. Categorization of the various studies involving electrochemical coating of medical devices and implants.

II. THE DEPOSITION METHOD

1. Electrodeposition (ED)

ED is also termed electroplating and is probably the oldest method, which was also used by Faraday and many of the pioneers in electrochemistry. The method involves the formation of a coating on a conducting substrate as a result of passing current and driving an oxidation-reduction process. The process can be carried out in electrolyte solutions as well as in molten salts and ionic liquids. Electroless deposition, whereby instead of passing current from an external device the process is driven by another oxidation-reduction process in the solution, is also well-known and widely used[2-5] for example in the microelectronic industry.

The simplest approach for ED comprises the reduction and deposition of metal ions (Eq. 1) that can be carried in aqueous solution for the more noble metals, e.g., gold, silver and copper, or in non-aqueous electrolytes and molten salts for more reactive metals, such as zinc, aluminum and magnesium. ED of metals has been extensively studied and proceeds through a complicated mechanism of reduction, nucleation and nuclei growth.[6-12] The nature of the resulting deposit, e.g., smoothness and adhesion, depends on many parameters such as the kinetics of the process, the interaction between the metallic film and the substrate, and therefore hundreds of procedures have been developed for each metal and differ in the electrolyte, temperature, additives added to the electrolytic bath and the applied electrochemical perturbation that causes deposition,

$$M^{n+}_{(aq)} + ne^- \rightarrow M_{(s)} \tag{1}$$

ED can also proceed through the application of anodic currents that usually require positive potentials. Such an approach relies on altering the solubility of soluble species upon oxidation. Metal hydroxides of higher oxidation states are regularly less soluble, which causes their precipitation according to the following scheme (Eqs. 2 and 3):

$$M^{n+}_{(aq)} \rightarrow M^{n+1}_{(aq)} + e^- \tag{2}$$

$$M^{n+1}_{(aq)} + (n+1)OH^-_{(aq)} \rightarrow M(OH)_{n+1(s)} \tag{3}$$

For example, $Fe(OH)_3$ readily precipitates upon electrochemically oxidizing soluble Fe^{2+} under appropriate conditions[13-14].

ED of metal hydroxides can be accomplished via a similar approach where instead of oxidizing the metal ions at positive potentials, cathodic currents cause the reduction of water and the local elevation of the pH in aqueous solutions (Eq. 4). This method has widely been used[15-19] for the deposition a variety of metal oxides and hydroxides, e.g., $Ni(OH)_2$, TiO_2, because it has many advantages such as its applicability with also reactive electrodes, such as stainless steel and Al without driving their oxidation and dissolution:

$$H_2O_{(l)} + e^- \rightarrow OH^-_{(aq)} + \frac{1}{2}H_{2(g)} \tag{4}$$

2. Electrophoretic Deposition (EPD)

EPD is a widely used industrial process that includes also sometimes the following terms electrocoating, e-coating, cathodic electrodeposition and electrophoretic coating or painting. The essence of this process, EPD, is the migration and deposition of charged species, either dissolved or suspended under an electrical field[20-25]. Therefore, EPD is considered to be a two-step process, whereby in the first step the charged species are forced to move towards the electrode (electrophoresis), while in the second step, the substance is deposited and forms a coherent layer. EPD yields in most cases a powder compact layer that is required to follow a densification step such as sintering or curing in order to obtain a fully dense material.

The velocity of the species is determined by the net charge enclosed in the liquid sphere which travels along with the species, rather than by the net charge on the surface. The potential at the surface of shear is termed the zeta-potential or electrokinetic potential.

The kinetics of EPD has been thoroughly studied and formulated by Hamaker (Eq. 5):

$$\frac{dY}{dt} = f \cdot \mu \cdot E \cdot S \cdot c \tag{5}$$

where Y is the yield of deposition (Kg), t, is time (s), μ is the electrophoretic mobility ($m^2\ V^{-1}\ s^{-1}$), S the surface area (m^2), c is the species concentration ($Kg\cdot m^{-3}$) and f (≤ 1) is the factor of the species that travel to the electrode and eventually deposited. This equation is nowadays accepted as describing the basis for the kinetics of electrophoretic deposition.

EPD has been applied for coating a wide variety of conducting surfaces with charged polymers (polyelectrolytes), dyes and pigments, ceramics and nanoparticles.[22-23,26-32] EPD is particularly attractive for coating *electroinactive* materials, which cannot be electrochemically deposited upon oxidation or reduction. EPD is by far the most common commercially used electrochemical deposition method in particularly in aqueous solutions although non-aqueous EPD processes have also been developed and reported[24]. Typically high voltage is applied (>100 V) for EPD, which causes the evolution of hydrogen (in the cathode) and oxygen (in the anode). The advantage of non-aqueous processes is the elimination of gas evolution. Other advantages of EPD are the high uniformity of the coating and the excellent control of its thickness, the ability to coat complex geometries, its simplicity, speed and generic nature that allows coating a wide range of substances with different materials and low cost. In addition, the process carried out in aqueous solution is environmental friendly and safe.

EPD has been applied in particular for making ceramic coatings. It is claimed that EPD can be applied to any solid that is available as a fine powder (e.g., <30 μm particle size) or colloidal suspension, including metals, polymers, ceramics and glasses. It has been widely used to coat automobile bodies and parts and many other industrial products. EPD has been used to produce customized micro and nanostructures such as functional gradients and laminates, through suspension control during processing. During the last two decades EPD has been increasingly employed to produce advanced ceramic coatings on solid substrates in order to tailor the substrate properties. EPD has been utilized to deposit materials with improved wear and oxidation resistance, to deposit bioactive coatings for biomedical implants and to produce functional coatings for electronic, magnetic and related applications. Finally, EPD has been employed in combination with electroplating for producing metal/ceramic composites.

3. Microarc Deposition (MAD)

MAD, also termed microarc (or micro-arc) oxidation (MAO) or plasma electrolytic oxidation, was primarily developed is a Russian in the late 1980's[33-35] although previous reports in which *anodic spark deposition* were reported in the 1970's.[36-37] MAD is used for producing hard and corrosion resistant coatings on Al, Mg, Ti and other light metals. The method is based on the formation of a microarc plasma discharge on the surface of an oxidizable metal, which is used as the anode, in an aqueous alkaline solution (2–3%). High voltage (150–200 V) is usually applied to form the microarc plasma, which causes the oxidation of the surface and the formation of different micro-structure and phase composition coating.

The localized high temperature, high pressure and high current density that are formed in the course of the microarc oxidation results in the formation of anodic films with special properties. Moreover, the films can incorporate different substances, which were added to the solution. The method has been used for treating the light and passivating class metals, such as Al, Ti and Zr and their alloys. Furthermore, microarc coating of these metals with different oxides, e.g., barium titanate.[38] Since the beginning of this millennium MAD has been widely applied for coating of Ti and its alloys by hydroxyapatite.[39-53]

MAD is particularly appealing for producing ceramic coatings with higher performance, namely, superior microhardness, abrasion resistance and corrosion resistance. The advantages of the MAD process are the ability to grow very thick coatings (up to a few hundreds of microns), the applicability of this method to high Si and Cu Al alloys, environmentally friendly coating bath, the formation of colorful coatings, e.g., Ti, and the fact that pre and post-treatments are usually not required.

4. Electropolymerization (EP)

EP is an electrochemical method of polymerizing electroactive organic monomers followed by theirs deposition on conducting surfaces. Once the insoluble oligomer is formed in the vicinity of the electrode surface it irreversibly deposits on it. Electropolymerization can lead to electronically conducting as well as non-conducting polymers. While the thickness of conducting polymers can be electrochemically grown to form relatively thick (tens of microns) layers, non-conducing polymers often block electron transfer and cease growing after a few nanometers. The field of conducting polymers was stimulated by the discovery of the high conductivity of doped polyacetylene some 30 years ago. Scientists were challenged by the synthesis and studied new conjugated polymers.[54-58] Polypyrrole, polyaniline, and polythiophene are typical representatives of this class of materials.

The polymerization commences by electrochemical oxidation of the monomers on the electrode surface. Electrooxidation leads to the formation of radical cations, which usually dimerize and form oligomers followed by deposition of a film on the electrode surface. Polymerization can proceed in aqueous or organic solutions. The films typically have good adhesion and electrical contact to the electrode surface. When supported by an electrode surface, the films can be electrochemically cycled between the oxidized, conducting state, which normally increases the conductivity of the polymer films by a few orders of magnitude and bring the conductivity of the layer in the range of semiconductors or even metals (10^3–10^5 $S \cdot cm^{-1}$), and the reduced form into an insulating state. Thicker films can be produced in the oxidized, conducting state and can be peeled off from the electrode surface to yield free-standing, electronically conducting films.

The polymerization and in particular the conductivity mechanism of organically conducting polymers has been extensively studied.[54,59-60] All conducting polymers have in common a conjugated π-system, which is responsible for the electronic conductivity once it is chemically or electrochemically oxidized. The other physical and chemical properties are derived from the organic skeleton and can be modified and tailored, which coined also conducting polymers as synthetic metals. Electropolymerization has been used for designing a variety of electroactive devices, such as biosensors, batteries, ion exchange membranes and corrosion inhibition surfaces. The application of conducting polymers for coating medical implants will be described in detail below.

5. Electrografting (EG)

The EG method involves the electrochemical formation of covalent bonding between polymers and conducting surfaces. In contrary to EP, electrografting generates electroactive species that reacts with the electrode surface and forms a covalent bond. Sometimes this reaction can initiate a polymerization reaction, however, in contrary to electropolymerization, polymerization self-propagates. In most cases, EG proceeds through a cathodic (reduction) process that allows grafting of a wide variety of conducting surfaces and prevents oxidation of the electrode. The resulting organic films are generally insulating, highly adherent, and optically transparent.

Lecayon et al. showed in the early 1980's that strongly adhesive films (from monolayer to 50 nm thick) can be formed on any conductive surface as a result of applying a negative potential in anhydrous solutions of vinylic monomers such as acrylonitrile and methyl methacrylate.[61]

EG was reviewed by Palacin et al.[62] and later by Gabriel.[63] Several electrografting reactions have been developed and studied and include:

- Electro-initiated polymerization by reductive electrolysis of electron-deficient alkenes, e.g., methacrylates, in an anhydrous medium (Fig. 2). This reaction is initiated via one-electron reduction of the vinyl monomer, which is followed by bonding of the radical-anion species to the metal. Propagation proceeds through the repeated addition of the monomer to chemisorbed anionic species.
- Radical cleavage of primary amines, aryl acetates, and alcohols (in anodic processes). The mechanism of this reaction (Fig. 3) is complex and usually involves as the preliminary step the formation of a radical cation, e.g., $R\text{-}NH_2^{+\cdot}$.[64] The latter loses a proton to form a radical, which attacks and binds to the electrode surface. The formation of by products such as polymerization often occurs.
- The recent most popular approach involves the reduction of diazonium salts (in a cathodic process, Fig. 4). This reaction was primarily studied by Savéant et al.[65] and more recently attracted the attention of many laboratories[66-70] due to the fact that almost any metal or carbonaceous material can be grafted using aryl diazonium, i.e., phenyl-N_2^+.

Moreover, the process is driven by applying negative potentials and can be carried out in aqueous and non-aqueous solvents. The reduction of the aryl-diazonium is instantaneously followed by cleavage of the nitrogen-carbon bond to form molecular nitrogen and highly-reactive phenyl radicals. Common metals, such as Fe, Ni, and Cu, were successfully coated by cathodic EG, after previously pretreated for removing the native oxide on the surface.

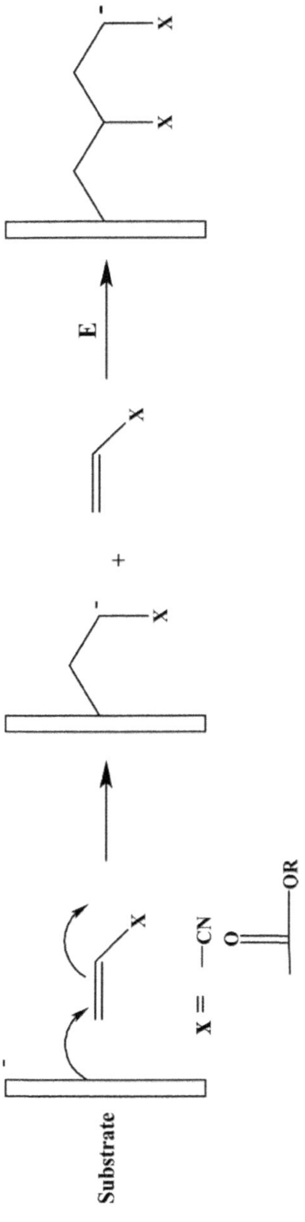

Figure 2. Schematic mechanism for electrografting of acrylic monomers.

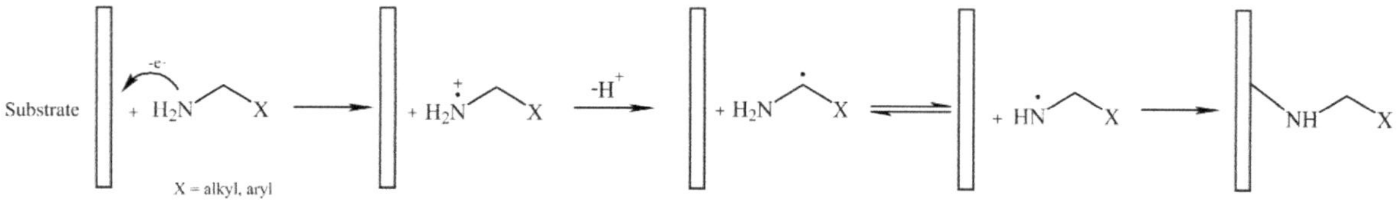

Figure 3. Schematic mechanism of electrically induced radical cleavage and grafting of primary amines.

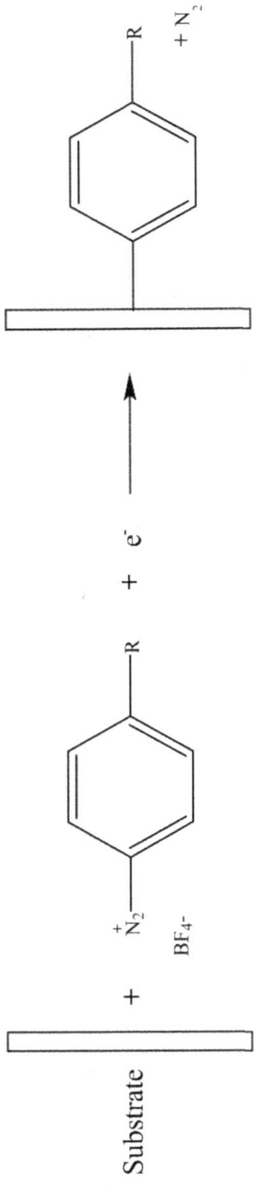

Figure 4. Schematic mechanism of electrically induced radical cleavage and grafting of diazonium salts.

On the other hand, EG onto Zn and Al is quite difficult because the parent superficial oxides resist electrochemical reduction.

Electrografted coatings are applicable in different fields, such as protective coatings on corrosive metals, primers for further surface chemical modification, ion-exchange membranes or lubricants.

III. THE SUBSTRATE

Medical implants are man-made devices that are meant to replace or support missing or damaged biological structures. They are made of different materials[71-74] according to their functionality and can be shaped as pins, rods, screws, hips or as a complete device such as an artificial pacemaker. Medical implant can be made of a variety of materials ranging from organic polymers to metals. A wide range of considerations, such as biocompatibility, biodegradability, mechanical strength and flexibility, stability and durability and obviously ease and price of manufacturing, are taken into account when determining the material which the medical implant is manufactured from. Typically, medical implant and devices will be made of metals, e.g., screws and pins, if superior mechanical strength is required. Furthermore, in case that the medical implant has complex geometry, such as a cardiovascular stent, metals are also preferred.

ED and coating is limited to conducting surfaces and therefore, we will briefly review the materials used for manufacturing only *conducting* medical implants and devices. It should be noted that it is possible to coat non-conducting surfaces by electroless deposition,[4] yet, the latter is almost only limited to the deposition of metals. Although conducting medical implants and devices do not have to be made of metals but also of carbonaceous materials, e.g., carbon nanotubes, we are not aware of non-metallic conducting implants. The metals of which medical implants and devices are made of can be divided into four categories: stainless steel, titanium and its alloys, magnesium alloys and cobalt alloys.

1. Stainless Steel

Stainless steel is the traditional name coined for iron alloys with a minimum of 10.5% chromium. Additional elements are added to enhance the structure of stainless steel and its properties such as formability, strength and cryogenic toughness. The metals, which are added, are Ni, Mo, Ti and Cu while the non-metals are C and N.

The main characteristic of stainless steel is its corrosion resistance for a particular application or environment. Hence, the selection of a specific *type* and *grade* of stainless steel is a function of its corrosion resistance properties. The latter is a result of the formation of chromium oxide passive layer on the surface

Stainless steel is widely used in biomedical applications. This is due to favorable combination of mechanical properties, corrosion resistance and cost effectiveness when compared with other metallic implant materials. The biocompatibility of implants of high quality stainless steel has been proven by successful human implantation for decades. Stainless steel applications as a implantable device is mostly distributed in orthopedic[75] and cardiology.[76] 316L Stainless steel, an austenitic material, is the only stainless steel that is used in biomedical applications. It contains a considerable amount of Ni (10–14%), which might cause Ni allergy;[77] therefore, the research and development of Ni-free stainless steels is in progress. One example is Bio Dur 108 stainless steel that contains up to 0.05% of nickel and higher content of nitrogen. The latter also contributes to its high levels of strength, ductility and corrosion resistance.

2. Titanium and its Alloys

Titanium was discovered close to 500 years ago and has high strength, toughness, durability, low density, corrosion resistance and biological compatibility, making it useful in a variety of applications. It weighs forty percent less than carbon steels and can be strengthened by alloying it with elements such as aluminum and vanadium. Moreover, it is nonmagnetic, nontoxic and has good heat transfer characteristics. Ti becomes corrosion resistance towards acids upon passivation by anodization. These properties make titanium and its alloys indispensable in a wide range of applications such as petrochemistry and marine and biomedicine.

Because of its corrosion resistance, titanium and its alloys are used extensively in many medical implants and devices, such as prosthetic devices, artificial heart pumps, pacemakers, heart-valves and bone or hip-joint replacements or bone splints. The most widely used titanium alloy, Ti-6Al-4V, has approximately 6% of aluminum and 4% of vanadium and is biocompatible, corrosion resistance and most importantly possesses the capacity for joining with bone and other tissue that is termed *osseointegration*. This causes Ti alloy implants to last longer than those made of other materials. Furthermore, the mechanical and physical properties of titanium alloys combine to provide implants which are highly damage tolerant. Reconstructive titanium plates and mesh that support broken bones are also commonly used today. In dentistry, titanium 'root' is introduced into the jaw bone with time subsequently allowed for osseointegration. Titanium has a wide range of applications in cardiology, like defibrillators, as the carrier structure for replacement heart valves, and for intra-vascular stents.

3. Nitinol

NiTi, is a nickel and titanium alloy where both metals are presented in roughly equal amounts. Nitinol alloys combine two closely related and unique properties: shape memory and superelasticity. Shape memory refers to the ability of Nitinol to undergo deformation at one temperature and then recover its original, undeformed shape upon heating above its *transformation temperature*. Superelasticity occurs at a narrow temperature range just above its transformation temperature; in this case, no heating is necessary to cause the undeformed shape to recover. It can be ten times more elastic than the best stainless steel alloy. Because of its extraordinary ability to accommodate large strains, coupled with physiological and chemical compatibility NiTi alloys became the materials of choice for medical device engineering and design.

4. Magnesium

Magnesium as an alkali earth-metal possesses significant advantages, such as high electrical and thermal conductivity, relatively high natural abundance and is easily recycled.[78] Probably, the most important characteristic of Mg is its high strength to weight ratio, which makes it a valuable asset in transportation and aviation industry.[79-80] In spite of the dramatic increase of Mg consumption since the beginning of the 21st century, its utilization has not reached full capacity because of its high chemical reactivity and tendency to corrode.[81-84] The very negative standard potential of Mg ($E^{0'}$ = –2.37 V vs. NHE) makes it unstable with respect to water (eq. 6) which results in the formation of a protecting oxide layer in ambient. Yet, the native oxide layer does not prevent its pitting corrosion,[83-85] facilitated mostly by halides, which results in destructive effects:

$$Mg_{(s)} + 2H_2O_{(l)} \rightarrow Mg(OH)_{2(s)} + H_{2(g)} \qquad (6)$$

Mg alloys such as AZ91 and AZ31 now being studied for implantable devices.[86-91] The medical application of these alloys is different from conventional metallic devices, due to its ability to corrode rapidly and biodegrade. This, of course, goes against the trend to use ever more corrosion-resistant materials for long-term implantable devices. A magnesium-based medical device can adjust a degree of corrosion within a wide range of period such that the device can maintain a sufficient strength only during a desired period and disappear within a desired period. Biodegradable magnesium-alloy stent is a new generation of medical implant metal materials with good biosafety and is expected to replace the traditional medical titanium alloys and stainless steel which are non-degradable metals in the field of orthopedics and cardiovascular stents.[86, 89, 92]

5. Cobalt Alloys

Cobalt is a hard, lustrous, gray metal, which occurs in various metallic-lustered ores, for example cobaltite (CoAsS), but is mainly produced as a by-product of copper and nickel mining. Cobalt is used in the preparation of magnetic, wear-resistant, and high-strength alloys. Cobalt-based alloys are also corrosion and wear-resistant.[93] Special cobalt-chromium-molybdenum alloys are used for prosthetic parts such as hip and knee replacements,[94] for dental prosthetics, where they are useful to avoid allergies to nickel.[95] Some special steels also use cobalt to increase heat and wear-resistance.

Co alloys are advantageous for the fabrication of the parts of medical devices subjected to wear, such as the heads of artificial hip joints. These alloys when annealed have

low strength and hardness, and considerable toughness and good corrosion resistance.[96] Co alloys for biomedical applications are grouped into two categories—cast alloys and wrought alloys—that contain, in general, a considerable amount of Ni to improve the castability and workability of the material.

Cobalt-chromium alloys are important biomaterials because of their corrosion resistance (an order of magnitude greater than stainless steel) and high mechanical properties. Cobalt-based alloys F75 (Co–Cr–Mo),[97] F90 (Co–Cr–W–Ni),[98] and F562 (Co–Ni-Cr–Mo) which conform to ASTM standards have been widely used for orthopedic, cardiovascular (ultra-thin stent struts),[99,76] and dentistry applications.[100] These alloys have excellent mechanical properties, high corrosion resistance, wear resistance, and fatigue strength.[100]

Now, it is time to survey the literature according to the type of deposit (Fig. 1).

IV. NATURE OF DEPOSIT

The nature of the deposit is a function of primarily the required properties and functionalities of the medical implant or device. Evidently, a stent that is placed in an artery and needs to release drug will be coated by a biocompatible matrix that can accommodate the drug. On the other hand, a coated implanted screw is required to maintain its strength, which dictates that its coating will be made of much stronger materials, e.g., ceramics or metals. A general requirement is biocompatibility although this term is not well defined and depends on the biological environment. The immune system is sophisticated that it is often not adequate to describe the biocompatibility of a single material in relation to a single cell type or tissue. Hence, there was an attempt to introduce sub-section for the definition of biocompatibility such as biocompatibility of long-term implanted devices, biocompatibility of short-term implantable devices and biocompatibility of tissue-engineering products.[73] These sub-definitions are related to devices rather than to materials.

As described above, electrochemistry can be used to deposit a wide variety of materials ranging from organic polymers to ceramics and metals. The only requirement is that the substrate, namely, the bare implant or medical device is electrically conductive. Surveying the literature reveals that electrochemistry has been mostly used for coating implants with organic polymers (conducting and non-conducting), biomaterials and in particularly hydroxyapatite and other phosphates, sol-gel materials (organic-inorganic polymers) and ceramics (inorganic materials). There are relatively few studies dealing with implants electrochemically coated with metals. Therefore, the coatings are divided into the categories described next starting with organics through inorganics and ending with metals.

1. Polymers (Organic Conducting and Non-Conducting)

(*i*) *Conducting Polymer Deposits*

Electropolymerization (EP) of conducting polymers on metallic surfaces is a well known method. Among different applications, such as batteries, electronic devices and biosensors, implant coating is a new field of research. There were some critical factors concerning bare metal implantation, which has driven the development of coatings for implants. The major disadvantages of bare metallic implant are dissolution of the metal and mechanical stress, which can cause serious damage to the hosting tissues. Thus, surface modification is an attractive method to tailor the surface properties of the implant. Among the plethora of conducting polymers being exploited and employed currently for biomedical applications, polypyrrole holds a key position since it offers several advantages including good specific conductivity, chemical stability, polymerizability and compatibility with mammalian cells.[101-102] The advantage of EP of conducting polymers is the ability to coat tiny and complex geometrics with thin, continuous and adherable films. Polypyrrole derivatives can be used as a protective as well as active coating for metallic devices, depending on the polymer backbone substituent functionalities. There are a number of publications dealing with polypyrrole coatings. Applications are found in three major categories: orthopedic joints, electrode implants and cardiovascular implants, while the substrates are usually stainless steel or titanium alloys.

De Giglio at al. electrodeposited polypyrrole on platinum and titanium alloys and proposed its surface grafting with cysteine as a precursor for bioactive interfaces.[103] Moreover they suggested modification of polypyrrole with biologically active agents to stimulate the positive interaction with the bone tissue.[104] A polypyrrole derivative, such as pyrrole-3-acetic acid was successfully electropolymerized on titanium and grafted with amino acid residues through the carboxylic functionality on the polypyrrole surface.[105]

In cardiology, drug eluting stents (DES) are already dominating the market.[106-108] DES are based on coating a stent (usually made of metal alloy) with a few micrometer thick coating, which accommodates the drug and at the same time increases the biocompatibility of the stent. There are numerous methods and materials used for coating cardiovascular stents.[76,109] Weiss el al. suggested that polypyrrole and its N-derivatives could be potential coating materials for stainless steel stents.[110] Specifically, aliphatic esters of *N*-(2-carboxyethyl)pyrrole (PPA) with different chain length were synthesized, characterized and electropolymerized on stainless steel surface. The electrochemical parameters of these derivatives were strongly depended on the N-substituent size. Unsurprisingly, the kinetics of long chain substituents was sluggish. The resulted polymers exhibited facile kinetics undergoing the electrochemical transformation between the oxidized (conducting) to and from the reduced (non-conducting) state in pure electrolyte medium. The anodic peak currents, $i_{p.a}$, showed linearity versus scan rate, suggesting a diffusion controlled doping/ undoping process. The ED of these derivatives was studied in detail and explained in terms of nucleation and growth mechanism[111]. The results were fit to the well-established metal electrocrystallization theory.[10,112-114] In general, the hydrophilic monomer such as carboxylic acid (PPA) undergoes progressive nucleation (rate of nucleation is much faster than the rate of nuclei growth), suggesting good interaction of moderately hydrophilic stainless steel surface with the hydrophilic deposit. Conversely, the EP of hydrophobic ester derivative proceeds through instantaneous nucleation, whereby nuclei growth dominates over nucleation. The nature of the deposits could be varied by combination of different N-pyrrole derivatives. Taking into account the chemical and physical behavior of the ED precursors, allows controlling and tuning the structure of the electrodeposited coating and adjusting it for different purposes and specifically for medical implants. Electrochemical co-deposition of polypyrrole derivatives on a stent surface was shown by Okner et al.[115] (Fig. 5).

The stent was coated with N-pyrrole derivatives, as a protective and paclitaxel releasing coating. The adhesion, hydrophobicity, drug (specifically, paclitaxel) loading and release and biocompatibility of the stent coating were studied. The optimal combination of hydrophilic and hydrophobic N-pyrrole derivatives were found to result in adherable (hydrophilic interaction with the stent surface) and stable stent coating with maximal paclitaxel load of 1 $\mu g/mm^2$. In addition, histopathological evaluation did not reveal any evidence of an active inflammation, morphological or histopathological changes. Khan et al. reported on BSA derivatized polypyrrole deposition on stainless steel surfaces. The coating was formed by two-step modification: EP of N-succinimidyl ester pyrrole and further surface modification with BSA.[116] This coating exhibited improved biocompatibility in terms of thrombus formation, platelet adhesion and hemolysis as compared with bare and polypyrrole coated stainless steel.

Since the commercial DES suffer from the lack of adhesion to the stent (metal) surface, the certain derivatives combination was suggested as a primer layer to facilitate better interaction between drug releasing matrix and stent surface.[117]

There are a number of publications dealing with polypyrrole electrocoating on metal electrode implants neurons stimulation in nervous systems. The purpose of introducing polypyrrole was to increase the electrical conductivity ($\sim 10^2$ $S \cdot cm^{-2}$) and morphology of the neuron-electrode interface. Building the neuroactive conducting scaffolds by polypyrrole coating as a functional conductive and stable platform opens a wide range of application in this field.[118,119] Cui at al. carried out *in vivo* studies with electrode implant coated with co-deposited polypyrrole-peptide composite.[120] It was shown that the coatings established strong connections with the neuronal structure up to three weeks.

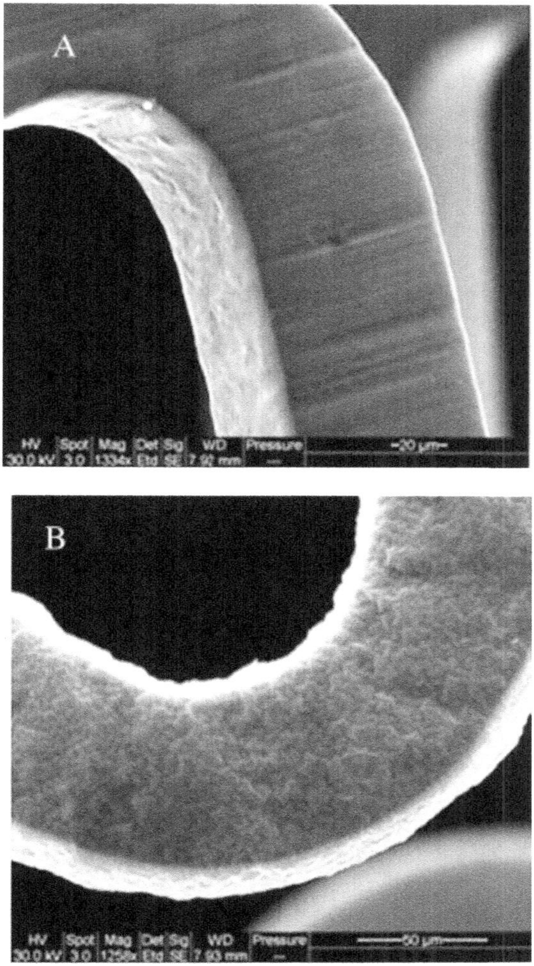

Figure 5. Scanning electron microscopy images of 316L stainless steel stent surface before and after electrodeposition of pyrrole derivative using cyclic voltammetry. Reprinted from Ref. 115, Copyright (2009), with permission from John Wiley and Sons.

Over-oxidized polypyrrole-coated electrodes were tested for *in vivo* dopamine sensing.[121] Coated electrodes implanted in a rat brain generated an increased electronic signal for several hours in comparison with uncoated electrodes. These promising results for polypyrrole coated electrodes still requires better understanding and more research to control and optimize film characteristics for long-term performance.[122] Biocompatibility issues of the implanted electrodes made of different metallic and metal coated polymeric electrodes triggered the development of drug release electrode coatings. The drug was incorporated in polypyrrole via EP of pyrrole and was released in PBS using cyclic voltammetry. The polypyrrole provided not only the drug delivery platform but also a biocompatible barrier between the implant and surroundings tissue.[123-124]

Hence, we can conclude that conducting polymers have still not been widely employed for coating medical devices and in particular implants. It is very likely that the wide range of commercially available monomers ranging from pyrrole to thiophene derivatives that can be electropolymerized at moderate potentials without harming the underlying implant will attract more attention in the future.

(*ii*) Non-Conducting Polymer Deposits

Electrografting (EG) is a method where a conductive surface is modified by charging it in the presence of an organic (non conductive) substance. For acrylic derivatives it is an electronically generated radical anion that propagates the polymerization.

Acrylates, methacrylates, thiols, amines[64,125] and triple bond[126] have been successfully electrografted on various conducting surfaces. In this method, the expected thickness of the organic film ranges between a monolayer to hundred of nanometers if the organic molecule can be polymerized under cathodic conditions[62] and the film does not block electron transfer. The formation of strong bonding between carbonaceous electrode materials and deposits opened intensive research work towards covalent coupling between organic molecules and carbon as well as metallic surfaces in general.[65,67-70,127-128] The aim has been to form M-C (where M stands for metal) bonding instead of binding through metal oxide layer to form M-O-C. The appropriate pretreatment of metal surface by cathodic current, reduces the oxide layer and reveals the pure metal surface for direct bonding through M-C. Recently, this method was suggested for permanent modification of metal implants. Poly(methyl methacrylate) (PMMA) and polyacrylic acid (PAA) electrografted on titanium-based orthopedic implants acted as both an effective bioactive surface and an anti-corrosion barrier.[129-130] PAA-based coatings can be functionalized with bioactive molecules and promote positive reactions with the biological system interfacing the implant while considerably reducing ion release into surrounding tissues.

The need of covering metallic implant surfaces with biodegradable polymers was accomplished by priming the surface with electrografted organic monomers, oligomers, or macromonomers through the covalent attachment of various precursors employing different polymerization mechanisms. Ring opening polymerization (ROP) of D,L-lactide and caprolactone showed promising results by coupling to acrylic ester primer attached to stainless steel and tantalum surfaces.[131-132] These biocompatible and biodegradable polyesters were suggested as potential coatings for drug eluting stents. Electrografted biocompatible thin coatings can provide a primer between the metal implant surface and the biodegradable drug eluting matrix for better surface compatibility and adhesion. Recently Langios et al. electrografted poly(lactic acid), with a molar mass of 600 $gr \cdot mol^{-1}$ having an acryrlic functionality on the metallic stent.[133]

Ignatova et al. investigated the decrease in the fibrinogen adsorption by modifying the stainless steel surface with first EG acrylic derivatives followed by polymerization via atom transfer radical polymerization (ATRP) different polymer families.[134] The same method was proposed for creating a hyperbranched polymer, which contained quaternized bromides or amines that are known for antibacterial properties.[135-136]

Non polymerizable electrografted molecules also showed potential biomedical applications. Diazonium salts derivatives are electroactive molecules that can be cleaved to aryl radicals as a result of electron transfer at the cathode. The highly reactive radicals that are formed bind to the electrode surface. Typically, these molecules create a monolayer, but their coupling occurs spontaneously as a consequence of a transfer reaction of the radical species formed to the grafted layer originally deposited.[136] A novel approach for significantly increasing the adhesion of drug eluting polymer coatings onto stents involved priming the surface with electrochemically deposited diazonium salts with the characteristic structure of $R\text{-}N_2^+X^-$, where R is an organic residue and X^- is an anion. Specifically $C_{12}H_{25}$-phenyldiazonium was investigated and found to increase polymeric coating stability and durability in physiological medium. The diazonium primer was tested and showed inert properties.[137]

Protein antifouling surfaces[138-139] usually contain derivatives of polyethylene glycol, known as excellent anticoagulant agents. The ability to directly modify the implant surface (specifically those in contact with blood) with PEG derivatives is of outmost importance and challenging. Figure 6 shows schematically the EG of acrylate end-capped PEG that was carried out on glassy carbon and stainless steel.[140] The one-step EG of PEG macromonomers created comb-like architecture where the PEG chains oriented in parallel to the surface, because the acrylate backbone propagated perpendicularly. The resulting structure was superior for antifouling.

Tanaka et al. suggested covalent bonding of amino terminated PEG to titanium oxide surface through N-HO-Ti bond. This was formed by electrochemically induced migration of amino-terminated PEG to the cathode. The architecture of the coating could be varied by PEG mono or bi-amine functionalization to cause binding of the polymer from both ends to the surface or form brush monolayers.[141-142]

Very recently, a novel approach, whereby pH responsive organic nanoparticles coagulate on a conducting surface as a result of applying positive potential, has been demonstrated by us (Fig. 7).[143] Specifically, latex nanoparticles stabilized by sodium oleate in aqueous solutions were deposited by applying a positive potential that oxidized the water and caused the decrease of pH on various conducting surfaces. It was found that the applied potential, its duration and the concentration of the dispersed nanoparticles governed the deposition characteristics of the coating. This generic approach allowed coating objects, e.g., coronary stents, with complex geometries with thickness ranging from nanometers to microns and therefore can be utilized for coating medical and other devices as well as for controlling drug release (Fig. 8).

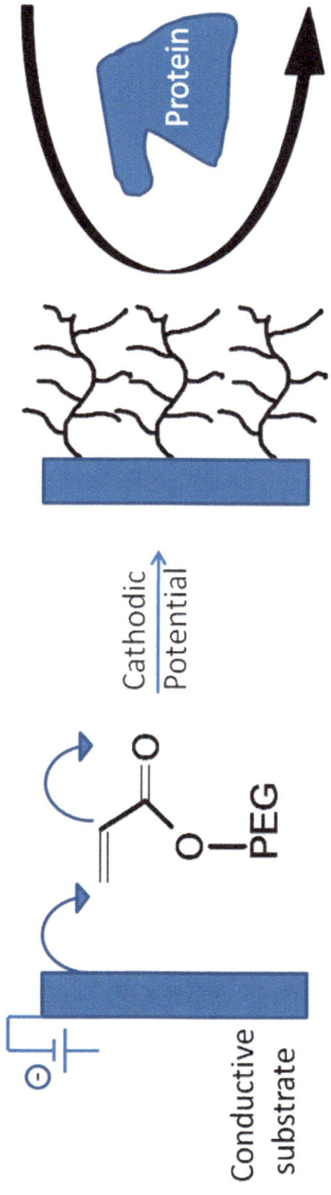

Figure 6. The approach for electrografting an acrylate end-capped PEG. Reproduced with permission from the Angewanted Chemie International Edition[140].

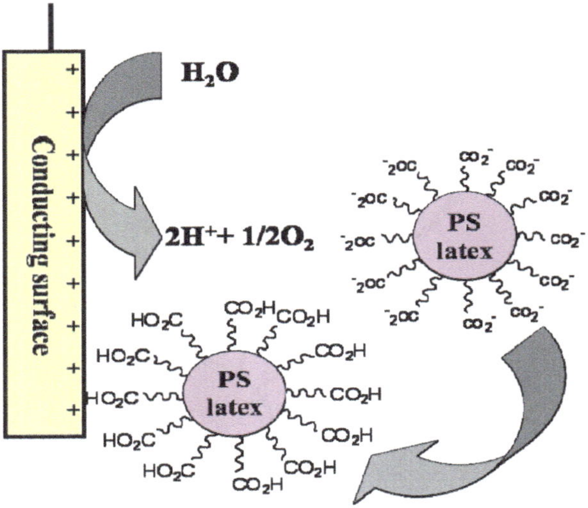

Figure 7. Electrochemical induced deposition of latex nanoparticles by anodic deposition. Reprinted from Ref. 143, Copyright (2010) with permission from the Elsevier.

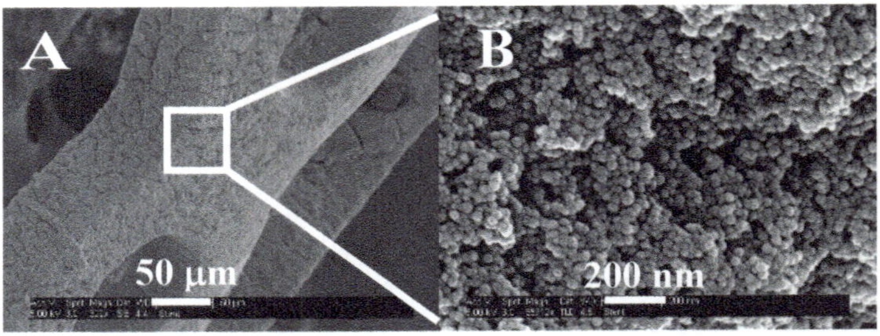

Figure 8. HR-SEM images of a latex film electrochemically deposited on a stainless steel stent under a constant potential of 1.3 V for 10 min: (A) The stent after deposition (B) higher magnification of the deposited film. Reprinted from Ref. 143 with permission from the Elsevier.

2. Sol-Gel (Inorganic-Organic)

Sol-gel deposition is a chemical solution process, typically comprising metal alkoxides.[144-147] This wet process starts from a protic solution consisting of organo-metallic precursors to form an integrated network (gel) of either discrete particles or polymers. Typical precursors are metal alkoxides or metal hilides, which undergo various forms of hydrolysis followed by polycondensation reactions. Thus, the sol evolves towards the formation of a gel-like biphasic system containing both a liquid and a solid phase whose morphologies range from discrete particles to continuous polymer networks.

An interesting and commercially important class of polymers can be synthesized by sol-gel method are alkoxysilane derivatives. These silicon-analogs of alcohols are also unstable, and in most cases quickly polymerize to produce a sol-gel film.[148] This process is associated with *two-step* sol-gel technology.[144,148-149]. The reactions take place in a two-step procedure (hydrolysis and condensation) and are quite complex, but can be represented as follows:

$$RSi(OR')_{3(l)} + H_2O_{(l)} \rightarrow RSi(OH)_{3(l)} + R'OH_{(l)} \text{ (unbalanced)} \quad (7)$$

$$RSi(OH)_{3(l)} \rightarrow R_nSi_nO_pH_q + nH_2O_{(l)} \text{ (unbalanced)} \quad (8)$$

In general, the hydrolysis of metal alkoxides is catalyzed better at low pH followed by polycondensation of hydrolyzed species at higher pH (Fig. 9).

Sol-gel based composites have found applications in numerous fields such as chemo- and biosensors,[150,151] batteries,[152] solid state catalysts,[153] chromatography,[154] optics[155] and recently in medicine.[156,157] The sol-gel materials can be prepared from tetrahydrolyzed silane precursors $Si(OR)_4$ or functional trialkoxysilanes $R_nSi(OH)_m$ (n = 1–3, m = 3–1). The substituent has a significant effect on the chemical, physical and 3D structure of the sol-gel material. The sol-gel can also be modified with a number of dopants to produce unique properties, such as porous matrix, thermal stability and flexibility.

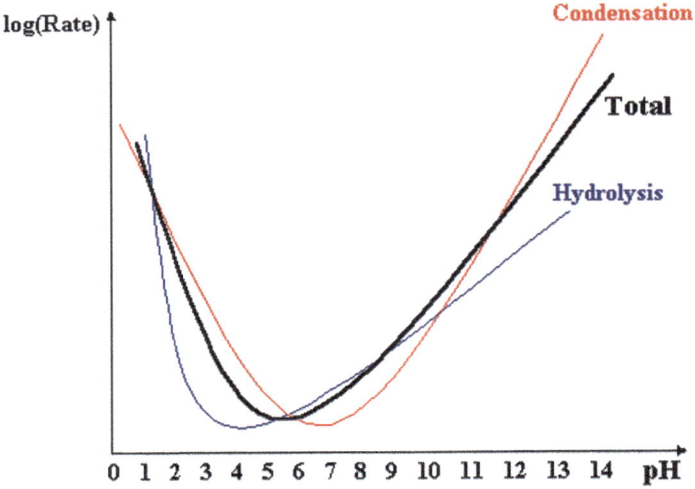

Figure 9. pH dependence of hydrolysis and condensation rates of sol-gel precursors.

One of the important applications of the sol-gel technology is thin film formation.[149] However, the conventional methods for depositing sol–gel films on metals, i.e., dip-coating and spin-coating are applicable to flat surfaces only, producing relatively thick coating (dip-coating) and cannot cope with the complex structure of different metallic devices, such as stents.

Recently, a novel approach for the preparation of sol-gel films has been developed by us[158] based on ED. The electrochemical generation of the condensation catalyst (OH^-) induced by applying a constant negative potential on the electrode surface, resulting in local pH increase around the electrode surface. This allowed the deposition of the sol–gel film on the conducting parts only and controlling the thickness of the deposited film by potential and time variation. Figure 10 represents schematically the electrochemically assisted deposition of hydrolyzed silanes.

Originally, the electrodeposition of alkoxy metals, such as silicon, titanium[159] and zirconium[160] was demonstrated by Shaham and coworkers. The films were electrodeposited on different met-

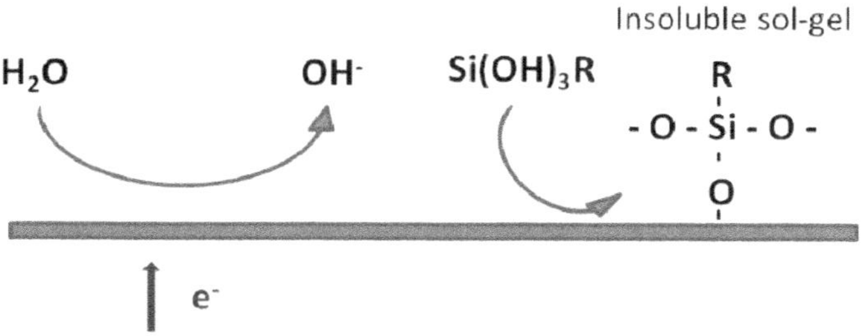

Figure 10. Schematics of the electrochemically induced deposition of sol-gel approach.

als and metal oxide substrates. Methyltrimethoxysilane was electrochemically deposited on indium-tin oxide (ITO) and gold electrodes, by applying positive and negative potentials. Lately, the electrochemical deposition of sol-gel materials was successfully applied on aluminum.[161-162] Collinson and co-workers suggested electroencapsulation of redox molecules and organic dyes. The entrapped species were homogeneously distributed inside the silica matrix.[163] The thickness of the resulted films was strongly depended on the applied potential. In addition, the relationship between thickness of the sol-gel film and electrode nature, time of electrodeposition was studied.[158] The electrodeposition is restricted to the conducting surfaces only; therefore this method is applicable to complex structures, different patterns, consisting of conducting and non-conducting surfaces.

The deposited film was insoluble, uniform and strongly adhered to the substrate. In addition to the physical interaction (physisorption) of the metal oxide surface with the deposit, evidence was provided by FTIR spectroscopy to the existence of Si-O-Fe bonding.[164,165] Taking into account that sol–gel films can bear different organic functionalities (using monomers with non-hydrolysable moieties) makes it a very appealing approach for designing films, free of mobile dopants. Such films, made of hydrophobic organosilanes, can provide protection against corrosion of metals by creating a physical and chemical barrier between the metal and its environment.[166] Electrodeposition of silica with biologically active materials was reported by Luo et al. Specifically, they introduced a new method to design amperometric glucose oxidase based biosensor.[167] Walcarius and co-workers suggested physical entrapment of hemoglobin or glucose oxidase in the sol-gel material in the course of the electrodeposition process.[168] Moreover, different analytical applications of electrically generated sol-gel films were introduced.[169]

Deposition of organosol-gel thin films on metal medical devices is a fast developing method. There are number of reports employing mainly dip-coating of stents with sol-gel films.[170-171] Yet, the application of ED for coating medical devices and in particularly implants is still quite rare. Recently ED of PEG through silane modified chain-ends was demonstrated by us.[172] Stainless steel was the deposited platform and the system is schematically depicted in Fig. 11. This method was limited to short modified polymer chains (less then Mw 1000) due to the 3D structure of this type of polymers in the deposition solution. Electrodeposition in combination with thermal treatment was compared to the conventional dipping method and showed to be superior for the formation of excellent anti-corrosion coatings. Stability of the films in corrosive medium was also improved by this combination. The electrodeposited films were strongly attached to the metal through M-O-Si bond, while the thermal treatment completed the condensation. Once coated on stainless steel, PEG-silane polymer was subjected to fresh blood samples in order to examine its platelets repelling ability. The results obtained by a method mimicking the physiological blood flow conditions showed a significant reduction of platelets adhesion and thus activation of PEGylated surface.

This study was followed by the electrochemically assisted co-deposition of sol-gel thin film on stainless steel[173]. Specifically, electrodeposition of films based on 3-aminopropyltrimethoxysilane (APTS), and its co-deposition with propyltrimethoxysilane (PrTMOS) and phenyltrimethoxysilane (PhTMOS) has been accomplished by applying negative potentials. The films were characterized by various surface techniques, such as AFM and SEM, which disclosed the structural changes induced by altering the deposition solution composition and the applied potential. Codeposited APTS:PhTMOS did not show any structural differences from their electrodeposited homopolymers, while Nano Scratch Test clearly revealed the changes in the elastic and adhesion properties, suggesting the formation of an APTS:PhTMOS composite.

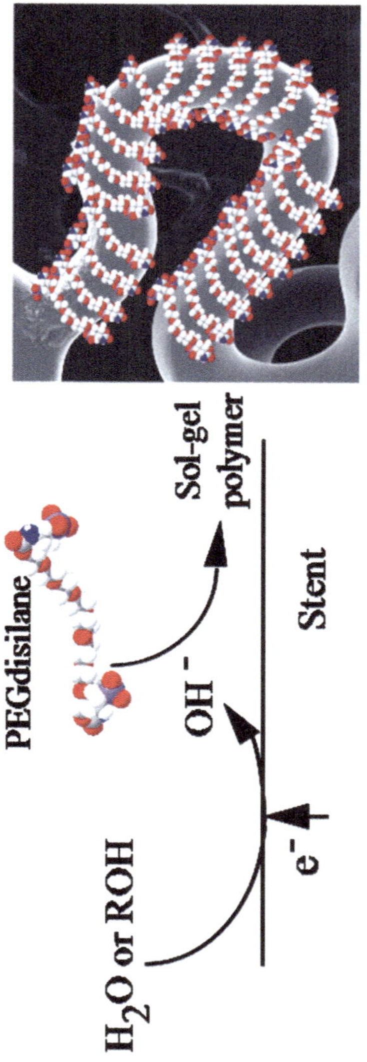

Figure 11. Electrodeposition of PEGylated silane on stainless steel surface. Reproduced with permission from the New Journal of Chemistry.

Electrochemical impedance spectroscopy of the films showed good resistance towards penetration of hydrophilic species, such as hexacyanoferrate. Clearly, the combination of different precursors, which makes it possible tailoring the chemical and physical properties of the film, has the potential of coating medical devices, in particular, coronary stents.

3. Oxides and Ceramics (Inorganic)

Ceramics are inorganic, non-metallic solids that are made by heating and subsequent cooling. Usually the term ceramics refers to crystalline materials as opposed to non-crystalline, i.e., amorphous, glasses. Ceramics are categorized into oxides, such as titania and silica, non-oxides including carbides, borides, nitrides and silicides and composites, which are made of combinations of oxides and non-oxides. Ceramic materials are usually brittle, hard and strong in compression, whereas weak in shearing and tension. They are quite resistive towards acids and bases and withstand very high temperatures (>1000 °C). Hence, ceramics are applied when good mechanical and/or thermal properties are sought.

ED of ceramic materials has been widely reported.[18,23-24,30,38,174-176] There are a few excellent reviews by Zhitomirsky,[24,26,177-178] Kamath[18] and others. ED can be performed by cathodic or anodic methods. The ED of ceramics can proceed through the generation of base, typically OH^-, as was discussed above (Section II). This is achieved by passing cathodic currents and reducing either a protic solvent or the electrolyte, e.g., NO_3^-, ClO_4^-. Accordingly, the electrochemical synthesis of precursors to polycrystalline powders of ZrO_2, TiO_2 and $ZrTiO_4$ and has been reported by different groups.[179-186] Similarly, the deposition of calcium phosphate and hydroxyapatite (HA) has been accomplished (see the next section). Anodic deposition of ceramics is also feasible, though limited, and reported to form PbO_2, MnO_2 and other oxides of high oxidation states. EP deposition of ceramic materials has also been demonstrated in particular for preparing relatively thick films.

In spite of all these studies, utilization of ED methods for coating medical implants by ceramic materials and oxides has been quite limited. The only mass of studies deals with the electrodeposition of HA and will be discussed below.

The application of ceramic materials for dental implants has been reported by Lacefield.[187] The most popular was HA. Others, including alumina were also mentioned. The application of EP (as well as other methods) as an alternative approach to plasma-spraying was discussed. Boccaccinni and Zhitomirsky[23] briefly reviewed the electrophoretic and electrolytic deposition techniques in ceramic processing for biomedical applications. They faced a problem of cracking in zirconia deposits that occurred upon drying. The addition of organic polymers, such as poly(diallyl-dimethylammonium chloride) provided better adhesion of zirconia and prevented cracking.[177] These studies paved the way for the ED of thick hybrid bioactive organic-inorganic films.[177-178]

The application of advanced materials made of metals and ceramics used for the bearing surfaces of total hip replacements (THRs), their standards, methods of manufacture and corrosion testing was reviewed by Dearnley.[188] Yet, ED of ceramic materials, i.e., ZrO_2 on hip prosthesis was reported only later by Yen et al.[189]. They deposited electrochemically ZrO_2 on a Co-Cr-Mo alloy at pH 2.2 by applying a cathodic current that caused the generation of hydroxyl ions by oxidizing water and oxygen. A monoclinic structure with preferred orientation parallel to the electrode plane was obtained at $623\text{ K} < T < 673\text{ K}$ and a tetragonal structure at higher temperatures.

More recently Liu et al. reviewed the work focusing on surface modification of titanium and its alloys for biomedical applications.[190] Among the different methods for surface modification electrochemical treatment was mentioned and in particular anodic oxidation. The latter is a well-established method to produce different types of protective oxide films on metals. The main technological advantage of anodizing titanium is improved adhesion and bonding. Micro arc deposition, MAD (see Methods of Deposition) is also described as a means of deposition ceramic coatings on metal surfaces, such as Al, Ti, Mg, Ta, W, Zn, and Zr and their alloys. Nevertheless, very few studies aimed at improving the biocompatibility or apply this technique to coat implantable devices. Nie et al.[39,191] described a hybrid treatment comprising MAD and EPD to improve the biocompatibility and durability of Ti–6Al–4V. A phosphate solution was used to produce a relatively thick and hard TiO_2 coating using MAD and EPD. Kern et al.[192] electrochemically deposited TiO_2 films with uniform thickness on AISI

316L stainless steel and Ti6Al4V substrates for potential use as color coded biocompatible coatings on biomedical implants. Deposition was carried out using $TiCl_4$ as a precursor and hydrogen peroxide. The reaction between the two led to the formation of a Ti peroxo species that was deposited under galvanostatic conditions by generating hydroxyl ions. They concluded that electrochemically deposited TiO_2 films were attractive as biocompatible colored coatings on non-anodizable biomedical alloys such as AISI 316L and CoCrMo as well as for Ti-alloys that are anodized for protective as well as coding reasons prior to implantation. The same group prepared composite bilayer coatings on Ti6Al4V substrates by EPD.[193] Biocompatible yttrium-stabilized zirconia (YSZ) in the form of nanoparticles and bioactive Bioglass (45S5) in the form of microparticles were used as coating materials. The bilayers were sintered at 900°C and yielded coatings with mechanical properties which were still insufficient for orthopedic applications.

As mentioned above, ceramic materials have been electrophoretically deposited on medical implants. The implications of EPD to nanotechnology including medical applications were reviewed by Boccaccini et al.[30] Such a medical application was described by Moritz et al. and involved coating of dental crowns and bridges with ceramics[194-195] made of Y_2O_3, SiO_2 and Al_2O_3. The dental crowns were coated with silver and were directly used as deposition electrodes.

It seems that the state-of-the-art in ED of ceramic materials for implants and medical application is the co-deposition of nano-objects, such as carbon nanotubes and nanoparticles in an inorganic matrix. Although we are not aware of any specific application yet, this will undoubtedly become a major field of research in the following years. Boccaccini reviewed recently[32] the up-to-date current research progress in the development of carbon nanotube (CNT)-ceramic nanocomposites by electrophoretic deposition. Ceramic particles of different size have been combined with single and multiwalled carbon nanotubes (CNTs) and electrophoretically deposited for a variety of functional, structural and biomedical applications. CNT-ceramic layers, e.g., SiO_2, TiO_2, MnO_2 and Fe_3O_4 and HA, with thickness varying between 1-50 μm were formed.[27,31,196] Systems reviewed include SiO_2/CNT, TiO_2/CNT, MnO_2/CNT, Fe_3O_4/CNT, HA/CNT and bioactive glass/CNT. Layered nanocomposites were formed by electrophoretic co-deposition from biphasic suspensions. Functionalization of the CNT was essential for their dispersion and electrophoretic deposition.

4. Hydroxyapatite and Calcium Phosphate

Hydroxyapatite (HA) sometimes called also hydroxylapatite is the hydroxylated form of the mineral apatite (calcium phosphate) $Ca_5(PO_4)_3OH$. Its hexagonal white crystal is made of two sub-units and therefore it is typically written as $Ca_{10}(PO_4)_6(OH)_2$.[197] The carbonated form of HA constitutes up to 50% of the bone. HA is also the primer mineral of tooth enamel. Substantial research has been devoted to the solubility of HA[198] and a large number of solubility constants have been reported. The solubility product of HA is ca. 10^{-37}, which makes it practically insoluble under physiological conditions. Hence, dissolution of HA is possible only in acidic conditions. On the other hand, dispersion of HA in particular in organic solvents gave micro and nanoparticles, which were found to be positively charged.[26]

The EPD of HA was first reported by Ducheyne[199] who carried out the experiments in isopropanol and aqueous solutions. Yet, Ducheyne et al.[200] claimed that adsorbed water in non-calcined powders prevented EPD. The EPD of HA on stainless steel and titanium was thoroughly studied by Zhitomirsky,[26,178,201-202] Eliaz[203-204] and many others.[49,205-218] The number of papers dealing with the electrochemical and electrophoretic deposition of HA exceeds 300 and is by far the most common electrochemical deposited material on medical devices and implants. The motivation for its deposition is clear as it can coat a metallic implant and used for replacement of lost or damaged bone tissue.[219] Moreover, HA is known to be an osteoinductive material, which accelerates the formation of new bone.[187]

Different procedures have been examined and developed for the electrochemical deposition (mostly electrophoretic deposition) of HA.[211-214,220-221] For example, Ban and Maruno reported[222] the hydrothermal-electrochemical deposition by dissolving NaCl, K_2HPO_4, $CaCl_2$, trishydroxyaminomethane, and HCl and applying a constant current at temperature 80-200°C. Yamashita et al.[223] used electrophoretic deposition followed by sintering to form composite multilayers of alumina and zirconia ceramics coated with a porous intermediate layer of HA and an adhesive calcium phosphate layer. Zhitomirsky[201] deposited via electrophoretic or/and electrochemical deposition different oxides as well as HA on individual carbon fibers, bundles and felts, which served as cathodic substrates. After burning out of carbon fibers the corresponding ceramic replicas in the shape of hollow fibers could be obtained. Huang et al.[224] studied the mechanism and kinetics of the electrochemical deposition of HA and post-hydrothermal synthesis. The electrochemical deposition was performed at various voltages ranging from 1.0–10 V. The depositing time varied from 1–3 hours and the temperature from 25–65°C. The results showed that the electrochemical deposition coatings were composed of $CaHPO_4 2H_2O$ crystals which were converted into needle-like HA crystals after post-hydrothermal treatment. The HA content of the coatings increased with the treatment temperature and time. The synthesis rate also increased with pH. Interestingly, Manso et al. reported[208] the electrodeposition of HA coatings under basic conditions and positive potentials (2–4 V). They proposed that the electrostatic attraction of hydroxyl ions in this case triggered the precipitation of HA nuclei at the interface. Nie and his coworkers[39,191] combined MAD) and EPD to obtain a double layer HA-TiO_2 coating on titanium alloys with HA as the top layer and a dense TiO_2 film as the inner layer. It was shown that a relatively thick and hard TiO_2 coating was produced by anodic MAD of titanium, and an HA coating incorporated on top of the TiO_2 layer can simultaneously be formed using a combination of plasma electrolysis and electrophoresis, with the suspension held at high pH. This dual layer gave high stability in corrosive environment of the human body and at the same time high biocompatibility. Stoch et al. showed[209] an improvement of HA coating when electrophoretically deposited on a joint intermediate layer of silica or calcium-silica. This was explained by the formation of silica and titania compounds in the interface region.

Not only Ti and its alloys were electrochemically coated with HA. Stainless steel, cobalt-chromium alloys[225] and even magnesium[226] have also been treated. Sridhar and coworkers[227] reviewed the EPD of HA on various metallic materials and the performance of these coatings in terms of corrosion resistance and biocompatibility. Corrosion inhibition has been the goal of many of studies where HA was electrophoretically deposited.[191,203,225-226,228-238] For example, Fathi and Azam claimed[232] to decrease the corrosion rate as a result of making a double-layer of HA on top of a tantalum coating.

The electrochemical co-deposition of HA and other monomers polymers has also been reported. Hu and coworkers[239] successfully electrochemically co-deposited HA and chitosan. The latter is a natural polysaccharide bearing amino groups that can easily be deposited under negative potentials. Deposition is a result of increasing the pH on the electrode surface, which causes deprotonation of chitosan. The composite material was found to be of different morphology than HA and possessed higher bonding strength. Wang[240] also found that the incorporation of chitosan by ED influenced calcium phosphate formation and crystallization. Hybrid coating exhibited an increased dissolution rate in both acidic and neutral simulated physiologic solution, whereas no significant difference on adhesive strength was found between the hybrid and original coatings. Most importantly, the calcium phosphate / chitosan coating proved to be a more favorable surface for goat bone marrow stromal cell attachment than an unincorporated coating.

Hu[241] also added vinyl acetate in the course of deposition of HA as a means of increasing the adhesion to the Ti substrate. Cheng et al.[242] electrochemically co-deposited calcium phosphate and the protein bovine serum albumin (BSA) on a HA coated Ti-6Al-4 V alloy. The co-precipitation of BSA/brushite ($CaHPO_4 \cdot 2H_2O$) coatings onto HA resulted in a 70-fold increase in BSA inclusion compared to simple adsorption, and was subsequently released by a slower mechanism (15% loss over 70 h).

Additional materials similar to HA were also electrochemically deposited. For example, Duan et al.[243] demonstrated the electrochemically deposition of calcium etidronate, which is a biphosphonate drug, on a Ti cathode under a constant potential (2.45 V). The concept was the same, increasing the pH by applying a negative potential caused deposition of the calcium salt. Fan et al.[244] successfully developed a uniform collagen fibril/octacalcium phosphate composite coating on silicon substrate by electrochemical deposition. Deposition was accomplished through applying a constant potential to the cathode in a three-electrode cell consisting of a mild acidic aqueous solution (pH 4.8–5.3) of collagen, Ca^{2+} and phosphate ions. The coating process involved self-assembly of collagen fibrils and the deposition of calcium phosphate minerals as a result of local pH increase. The porous composite layer consisted of a collagen fibril network on which clusters of calcium phosphate crystals nucleated and grew.

A related family of biomaterials, which has also been electrophoretically deposited, is bioglass, which is also termed bioactive glass.[245-247] These materials are commercially available glasses composed of relatively low concentration of SiO_2 (usually less than 60%), Na_2O, CaO and P_2O_5. The EPD of bioglasses has been reported[196,211,235,248] and reviewed by Garcia et al.[231] and Zhao et al.[249] In most, if not all, these studies the bioglass was introduced as a particulate powder of micrometer dimensions and was electrophoretically deposited on the anode due to the negative charge of the ceramic matrix. Oxygen generated in the anode was usually entrapped and lead to residual porosity and irregularities on the surface of the coating. Typical voltage was 5 V and coatings of micrometer thick were achieved within a few minutes. In several studies HA was co-deposited with bioglass.[211,235] Different substances were added to the deposition solution to tailor the coating properties. For example, the incorporation of carbon nanotubes[196] induced a nanostructured internal surface of the pores which was thought to be beneficial for osteoblast cell attachment and proliferation.

To summarize this part, it is evident that ED methods are highly attractive for coating medical devices and implants by HA and other bioceramics. Clearly, we will witness more sophisticated approaches in particular where EPD of HA will be combined with organic polymers, other ceramic materials and in particular nano-objects as a means of tailoring the coatings.

5. Metals

There are very few cases, which we are aware of where implants were electrochemically coated with a metallic surface. Most of the implants and the replacement components are made of metals, specifically of titanium, stainless steel and cobalt alloys[188] and therefore coating by another metal is quite rare.

Yet, there are many studies and patents where the metal was *electrochemically treated*, primarily by anodic oxidation to form a protecting oxide layer.[250-258] Titanium, in particular, is well known for its well controlled oxide grown coatings, which not only protect the underneath metal but can also enhance human osteoblast cell growth.[255] In general, rough surface morphology, high surface energy and low values of contact angles are important factors for better cell-materials interaction. Other studies reported[259-260] that neointimal hyperplasia and thrombogenicity were significantly decreased in electrochemically polished stainless steel stents, which were implanted in different animal models.

We found a few cases where biocompatibility was reported to increase due to electrochemical coating of metallic implants, e.g. coronary stents, by metals. Hehrlein et al.[261] coated 24 stainless steel stents by either electrochemical deposition of metals, such as Pt, Au and Cu, on the stent surface or with a metallic film of the same metals, which was implanted into the stent surface by argon ion bombardment. Coated and uncoated control stents were implanted in rabbit iliac arteries. Thrombus and neointima formation in arterial cross-sections of the coated stents were compared with the uncoated stents. The researchers found higher stent surface porosity and more surface cracks after stent expansion after electrochemical deposition than after ion implantation. Neointimal hyperplasia was increased in stents electrochemically coated compared with stents coated by ion implantation. In both study groups, the most electropositive coating (platinum or gold) induced markedly less neointima formation than the least electropositive (copper). The conclusions of this study were that stent surface texture was the most important factor determining biocompatibility of the coated stainless steel stents.

Very recently, another approach was reported by Weng et al.[262] They showed that the biocompatibility of Ti could be increased by electrochemically depositing Au nanoparticles on which arginine-glycine-asparagine-cysteine (RGDC) peptides were grafted by thiolate covalent coupling. The biological responses of the RGDC-grafted Ti substrates were evaluated using MG3 human osteoblast-like cells. Cell morphology showed that, on RGDC-immobilized titanium with Au particles, MG63 cells attached and spread more rapidly than on Ti substrates either without peptide or with direct loading of the peptide. Immuno-staining for focal adhesion kinase (FAK) demonstrated that RGDC enhanced cell attachment.

The electrochemical deposition of metals, in particular silver, has been used as a means of increasing antibacterial activity.[63,263-]

[268] For example, Ti surface was modified to simultaneously improve bone cell materials and antimicrobial activities.[269] Titanium surface was first anodized in sodium fluoride and sulfuric acid electrolytic solution to form titania nanotube on the surface to improve the biocompatibility of the surface. Then, silver oxide was electrodeposited on the titania nanotube surface at 5 V. Silver added titania nanotube surface was tested for compatibility with bone-cell materials interactions using human osteoblast bone cells. The antibacterial effect was studied and the results showed that silver-treated titania nanotube surface may provide antibacterial properties to prevent implants against postoperative infections without interference to the attachment and proliferation of bone tissue on titanium.

V. CONCLUSIONS

The combination of electrochemical deposition methods and medical implants is still far from being exhausted. These two worlds usually lie very far from each other and seem initially not to overlap. Yet and as has been shown in this review electrochemical methods offer many significant advantages as a coating technology specifically in the biosciences and medicine. The mild conditions, the ability to coat complex geometries, its low cost and flexibility make electrochemical deposition methods not only a transient episode for coating implants but can have a major impact and be implemented in commercial technologies. Understanding the basis of electrochemistry and at the same time learning the molecular level of the implant/tissue interface could lead to significant advancement in coating implants by electrochemical methods as well as to innovative applications.

REFERENCES

[1] *Fundamentals of Electrochemical* Deposition, Ed. by M. Paunovic, M. Schlesinger, Wiley and Sons, 2nd Edition, New Jersey 2006.
[2] J. Georgieva, S. Armyanov, *Journal of Solid State Electrochemistry* **11** (2007) 869.
[3] C. R. Shipley, *Plating and Surface Finishing* **71** (1984) 92.
[4] B. Viswanathan, *Current Science* **65** (1993) 537.

[5] G. W. Wen, Z. X. Guo, and C. K. L. Davies, *Materials Technology* **14** (1999) 210.
[6] J. W. Mullin, *Transactions of the Institution of Chemical Engineers and the Chemical Engineer* **45** (1967) C240.
[7] P. C. Searson, *Solar Energy Materials and Solar Cells* **27** (1992) 377.
[8] V. Danek, M. Chrenkova, and A. Silny, *Coordination Chemistry Reviews* **167** (1997) 1.
[9] P. Batail, K. Boubekeur, M. Fourmigue, and J. C. P. Gabriel, *Chemistry of Materials* **10** (1998) 3005.
[10] E. Budevski, G. Staikov, and W. J. Lorenz, *Electrochimica Acta* **45** (2000) 2559.
[11] M. E. Hyde, R. G. Compton, *Journal of Electroanalytical Chemistry* **549** (2003) 1.
[12] P. Allongue and F. Maroun, *Current Opinion in Solid State & Materials Science* **10** (2006) 173.
[13] V. Markovac and M. Cohen, *Journal of the Electrochemical Society* **114** (1967) 674.
[14] V. Markovac and M. Cohen, *Journal of the Electrochemical Society* **114** (1967) 678.
[15] M. C. A. Fantini and A. Gorenstein, *Solar Energy Materials* **16** (1987) 487.
[16] P. Oliva, J. Leonardi, J. F. Laurent, C. Delmas, J. J. Braconnier, M. Figlarz, F. Fievet, and A. Deguibert, *Journal of Power Sources* **8** (1982) 229.
[17] T. Pauporte, A. Goux, A. Kahn-Harari, N. de Tacconi, C. R. Chenthamarakshan, K. Rajeshwar, and D. Lincot, *Journal of Physics and Chemistry of Solids* **64** (2003) 1737.
[18] G. H. A. Therese and P. V. Kamath, *Chemistry of Materials* **12** (2000) 1195.
[19] I. Zhitomirsky, *Materials Letters* **33** (1998) 305.
[20] W. F. Pickard, *Journal of the Electrochemical Society* **115** (1968) C105.
[21] P. Sarkarand and P. S. Nicholson, *Journal of the American Ceramic Society* **79** (1996) 1987.
[22] O. O. Van der Biest and L. J. Vandeperre, *Annual Review of Materials Science* **29** (1999) 327.
[23] A. R. Boccaccini and I. Zhitomirsky, *Current Opinion in Solid State & Materials Science* **6** (2002) 251.
[24] I. Zhitomirsky, *Advances in Colloid and Interface Science* **97** (2002) 279.
[25] L. Besra and M. Liu, *Progress in Materials Science* **52** (2007) 1.
[26] I. Zhitomirsky and L. Galor, *Journal of Materials Science-Materials in Medicine* **8** (1997) 213.
[27] A. R. Boccaccini, J. Cho, J. A. Roether, B. J. C. Thomas, E. J. Minay, and M. S. P. Shaffer, *Carbon* **44** (2006) 3149.
[28] H. Imahori, *Journal of Materials Chemistry* **17** (2007) 31.
[29] T. Umeyama, M. Fujita, N. Tezuka, N. Kadota, Y. Matano, K. Yoshida, S. Isoda, and H. Imahori, *Journal of Physical Chemistry C* **111** (2007) 11484.
[30] I. Corni, M. P. Ryan and A. R. Boccaccini, *Journal of the European Ceramic Society* **28** (2008) 1353.
[31] J. Cho, M. Cannio and A. R. Boccaccini, *International Journal of Materials & Product Technology* **35** (2009) 260.
[32] A. R. Boccaccini and J. Cho, T. Subhani, C. Kaya, F. Kaya, *Journal of the European Ceramic Society* **30** (2010) 1115.
[33] P. S. Gordienko, S. V. Gnedenkov, S. L. Sinebryukhov, O. A. Khrisanfova and T. M. Skorobogatova, *Russian Electrochemistry* **29** (1993) 1232.

[34] O. A. Krisanfova, L. M. Volkova, S. V. Gnedenkov, T. A. Kaidalova, P. S. Gordienko, *Zhurnal Neorganicheskoi Khimii* **40** (1995) 558.
[35] S. V. Gnedenkov, S. L. Sinebryukhov, T. M. Skorobogatova, P. S. Gordienko, *Russian Journal of Electrochemistry* **34** (1998) 940.
[36] L. B. Sis, S. D. Brown, T. B. Van, G. P. Wirtz, *Journal of the American Ceramic Society* **57** (1974) 108.
[37] T. B. Van, G. P. Wirtz, S. D. Brown, *American Ceramic Society Bulletin* **55** (1976) 412.
[38] S. V. Gnedenkov, P. S. Gordienko, O. A. Khrisanfova, T. M. Scorobogatova, S. L. Sinebrukhov, *Journal of Materials Science* **37** (2002) 2263.
[39] X. Nie, A. Leyland, A. Matthews, *Surface & Coatings Technology* **125** (2000) 407.
[40] Y. Han, S. H. Hong, K. W. Xu, *Surface & Coatings Technology* **168** (2003) 249.
[41] W. H. Song, Y. K. Jun, Y. Han, S. H. Hong, *Biomaterials* **25** (2004) 3341.
[42] S. Won-Hoon, J. Youn-Ki, H. Yong, H. Seong-Hyeon, *Biomaterials* **25** (2004) 3341.
[43] J. Z. Chen, Y. L. Shi, L. Wang, F. Y. Yan, F. Q. Zhang, *Materials Letters* **60** (2006) 2538.
[44] Q. Dong, C. Z. Chen, D. G. Wang, Q. M. Ji, *Surface Review and Letters* **13** (2006) 35.
[45] S. H. Lee, H. W. Kim, E. J. Lee, L. F. Li, H. E. Kim, *Journal of Biomaterials Applications* **20** (2006) 195.
[46] M. S. Kim, J. J. Ryu, Y. M. Sung, *Electrochemistry Communications* **9** (2007) 1886.
[47] Y. Han, J. F. Sun, X. Huang, *Electrochemistry Communications* **10** (2008) 510.
[48] D. Y. Kim, M. Kim, H. E. Kim, Y. H. Koh, H. W. Kim, J. H. Jang, *Acta Biomaterialia* **5** (2009) 2196.
[49] S. R. Paital, N. B. Dahotre, *Materials Science & Engineering R-Reports* **66** (2009) 1.
[50] X. J. Tao, S. J. Li, C. Y. Zheng, J. Fu, Z. Guo, Y. L. Hao, R. Yang, Z. X. Guo, *Materials Science & Engineering C-Materials for Biological Applications* **29** (2009) 1923.
[51] J. L. Xu, F. Liu, F. P. Wang, D. Z. Yu, L. C. Zhao, *Current Applied Physics* **9** (2009) 663.
[52] H. J. Song, K. H. Shin, M. S. Kook, H. K. Oh, Y. J. Park, *Surface & Coatings Technology* **204** (2010) 2273.
[53] Z. Q. Yao, Y. Ivanisenko, T. Diemant, A. Caron, A. Chuvilin, J. Z. Jiang, R. Z. Valiev, M. Qi, H. J. Fecht, *Acta Biomater* **6** (2010) 2816.
[54] A. J. Heeger, S. Kivelson, J. R. Schrieffer, W. P. Su, *Reviews of Modern Physics* **60** (1988) 781.
[55] J. Roncali, *Chemical Reviews* **92** (1992) 711.
[56] W. J. Feast, J. Tsibouklis, K. L. Pouwer, L. Groenendaal, E. W. Meijer, *Polymer* **37** (1996) 5017.
[57] A. J. Heeger, *Angewandte Chemie-International Edition* **40** (2001) 2591.
[58] A. G. MacDiarmid, *Angewandte Chemie-International Edition* **40** (2001) 2581.
[59] J. L. Bredas, G. B. Street, *Accounts of Chemical Research* **18** (1985) 309.
[60] G. Inzelt, M. Pineri, J. W. Schultze, M. A. Vorotyntsev, *Electrochimica Acta* **45** (2000) 2403.
[61] G. Lecayon, Y. Bouizem, C. Legressus, C. Reynaud, C. Boiziau, C. Juret, *Chemical Physics Letters* **91** (1982) 506.

[62] S. Palacin, C. Bureau, J. Charlier, G. Deniau, B. Mouanda, P. Viel, *Chemphyschem* **5** (2004) 1469.
[63] S. Gabriel, R. Jerome, C. Jerome, *Progress in Polymer Science* **35** (2010) 113.
[64] A. Adenier, M. M. Chehimi, I. Gallardo, J. Pinson, N. Vila, *Langmuir* **20** (2004) 8243.
[65] M. Delamar, R. Hitmi, J. Pinson, J. M. Saveant, *Journal of the American Chemical Society* **114** (1992) 5883.
[66] J. Pinson, F. Podvorica, *Chemical Society Reviews* **34** (2005) 429.
[67] P. Allongue, M. Delamar, B. Desbat, O. Fagebaume, R. Hitmi, J. Pinson, J. M. Saveant, *Journal of the American Chemical Society* **119** (1997) 201.
[68] A. J. Downard, *Electroanalysis* **12** (2000) 1085.
[69] A. Adenier, M. C. Bernard, M. M. Chehimi, E. Cabet-Deliry, B. Desbat, O. Fagebaume, J. Pinson, F. Podvorica, *Journal of the American Chemical Society* **123** (2001) 4541.
[70] M. C. Bernard, A. Chausse, E. Cabet-Deliry, M. M. Chehimi, J. Pinson, F. Podvorica, C. Vautrin-Ul, *Chemistry of Materials* **15** (2003) 3450.
[71] D. Hodgins, J. M. Wasikiewicz, M. F. Grahn, D. Paul, N. Roohpour, P. Vadgama, A. M. Silmon, B. Cousins, B. Verdon, *Med Device Technol* **18** (2007) 30.
[72] B. Kasemo, J. Gold, *Adv Dent Res* **13** (1999) 8.
[73] A. G. Shard, P. E. Tomlins, *Regen Med* **1** (2006) 789.
[74] M. Zilberman, A. Kraitzer, O. Grinberg, J. J. Elsner, *Handb Exp Pharmacol* (2010) 299.
[75] J. A. Disegi, L. Eschbach, *Injury-Int. J. Care Inj.* **31** (2000) S2.
[76] G. Mani, M. D. Feldman, D. Patel, C. M. Agrawal, *Biomaterials* **28** (2007) 1689.
[77] P. J. Uggowitzer, R. Magdowski, M. O. Speidel, *Isij International* **36** (1996) 901.
[78] D. Eliezer, E. Aghion, F. H. Froes, *Advanced Performance Materials* **5** (1998) 201.
[79] Y. Kojima, in *Magnesium Alloys 2000*, Vol. 350-3 (Eds: Y. Kojima, T. Aizawa, S. Kamado), 2000, 3.
[80] S. Das, *Jom-Journal of the Minerals Metals & Materials Society* **55** (2003) A22.
[81] G. L. Makar, J. Kruger, *International Materials Reviews* **38** (1993) 138.
[82] G. Song, A. Atrens, D. Stjohn, J. Nairn, Y. Li, *Corrosion Science* **39** (1997) 855.
[83] G. L. Song, A. Atrens, *Advanced Engineering Materials* **1** (1999) 11.
[84] G. L. Song, A. Atrens, *Advanced Engineering Materials* **5** (2003) 837.
[85] E. Ghali, W. Dietzel, K. U. Kainer, *Journal of Materials Engineering and Performance* **13** (2004) 7.
[86] M. Peuster, P. Beerbaum, F. W. Bach, H. Hauser, *Cardiology in the Young* **16** (2006) 107.
[87] M. P. Staiger, A. M. Pietak, J. Huadmai, G. Dias, *Biomaterials* **27** (2006) 1728.
[88] G. L. Song, *Corrosion Science* **49** (2007) 1696.
[89] F. Witte, N. Hort, C. Vogt, S. Cohen, K. U. Kainer, R. Willumeit, F. Feyerabend, *Current Opinion in Solid State & Materials Science* **12** (2008) 63.
[90] T. Hanawa, *Journal of Artificial Organs* **12** (2009) 73.
[91] Y. H. Yun, Z. Y. Dong, N. Lee, Y. J. Liu, D. C. Xue, X. F. Guo, J. Kuhlmann, A. Doepke, H. B. Halsall, W. Heineman, S. Sundaramurthy, M. J. Schulz, Z. Z. Yin, V. Shanov, D. Hurd, P. Nagy, W. F. Li, C. Fox, *Materials Today* **12** (2009) 22.
[92] H. S. Brar, M. O. Platt, M. Sarntinoranont, P. I. Martin, M. V. Manuel, *Jom* **61** (2009) 31.

[93] *Superalloys: A Technical* Guide, Ed. by D. M. Donachie and S. J. Donachie, ASM International, 2nd edition, 2002.
[94] R. Michel, M. Nolte, M. Reich, F. Loer, *Archives of Orthopaedic and Trauma Surgery* **110** (1991) 61.
[95] *Cobalt-based alloys for biomedical* applications, Ed. by J. A. Disegi, R. L. Kennedy, R. Pilliar, ASTM International, 1999.
[96] A. Veronski, B. M. Surovska, *Metal Science and Heat Treatment* **34** (1992) 210.
[97] O. Ozturk, U. Turkan, A. E. Eroglu, *Surface & Coatings Technology* **200** (2006) 5687.
[98] G. Mani, M. D. Feldman, S. Oh, C. M. Agrawal, *Applied Surface Science* **255** (2009) 5961.
[99] D. Cox, D. Kereiakes, J. Hermiller, M. Midei, W. Bachinsky, D. Nutka, A. Lansky, S. Fink, L. Marin, T. Linnemeir, *Am. J. Cardiol.* **90** (2002) TCT34.
[100] M. Donachie, *Advanced Materials & Processes* **154** (1998) 63.
[101] Z. Zhang, R. Roy, F. J. Dugre, D. Tessier, L. H. Dao, *Journal of Biomedical Materials Research* **57** (2001) 63.
[102] J. Lahann, D. Klee, H. Thelen, H. Bienert, D. Vorwerk, H. Hocker, *Journal of Materials Science-Materials in Medicine* **10** (1999) 443.
[103] E. De Giglio, L. Sabbatini, P. G. Zambonin, *Journal of Biomaterials Science-Polymer Edition* **10** (1999) 845.
[104] E. De Giglio, M. R. Guascito, L. Sabbatini, G. Zambonin, *Biomaterials* **22** (2001) 2609.
[105] E. De Giglio, C. Stefania, C. D. Calvano, S. Luigia, G. Z. Pier, C. Silvia, D. B. Adriana, G. Colaianni, *Journal of Materials Science-Materials in Medicine* **18** (2007) 1781.
[106] J. Gunn, D. Cumberland, *European Heart Journal* **20** (1999) 1693.
[107] T. C. Woods, A. R. Marks, *Annual Review of Medicine* **55** (2004) 169.
[108] B. L. van der Hoeven, N. M. M. Pires, H. M. Warda, P. V. Oemrawsingh, B. J. M. van Vlijmen, P. H. A. Quax, M. J. Schalij, E. E. van der Wall, J. W. Jukema, *International Journal of Cardiology* **99** (2005) 9.
[109] O. F. Bertrand, R. Sipehia, R. Mongrain, J. R. Rodes, J. C. Tardif, L. Bilodeau, G. Cote, M. G. Bourassa, *Journal of the American College of Cardiology* **32** (1998) 562.
[110] Z. Weiss, D. Mandler, G. Shustak, A. J. Domb, *Journal of Polymer Science Part a-Polymer Chemistry* **42** (2004) 1658.
[111] R. Okner, A. J. Domb, D. Mandler, *Biomacromolecules* **8** (2007) 2928.
[112] D. J. Astley, J. A. Harrison, H. R. Thirsk, *Journal of Electroanalytical Chemistry* **19** (1968) 325.
[113] Armstron.Rd, J. A. Harrison, *Journal of the Electrochemical Society* **116** (1969) 328.
[114] M. Sluytersrehbach, J. Wijenberg, E. Bosco, J. H. Sluyters, *Journal of Electroanalytical Chemistry* **236** (1987) 1.
[115] R. Okner, M. Oron, N. Tal, A. Nyska, N. Kumar, D. Mandler, A. J. Domb, *Journal of Biomedical Materials Research Part A* **88A** (2009) 427.
[116] W. Khan, M. Kapoor, N. Kumar, *Acta Biomaterialia* **3** (2007) 541.
[117] R. Okner, Y. Shaulov, N. Tal, G. Favaro, A. J. Domb, D. Mandler, *Acs Applied Materials & Interfaces* **1** (2009) 758.
[118] J. Y. Lee, J. W. Lee, C. E. Schmidt, *Journal of the Royal Society Interface* **6** (2009) 801.

[119] P. M. George, A. W. Lyckman, D. A. LaVan, A. Hegde, Y. Leung, R. Avasare, C. Testa, P. M. Alexander, R. Langer, M. Sur, *Biomaterials* **26** (2005) 3511.
[120] X. Y. Cui, J. Wiler, M. Dzaman, R. A. Altschuler, D. C. Martin, *Biomaterials* **24** (2003) 777.
[121] K. Pihel, Q. D. Walker, R. M. Wightman, *Analytical Chemistry* **68** (1996) 2084.
[122] R. A. Green, N. H. Lovell, G. G. Wallace, L. A. Poole-Warren, *Biomaterials* **29** (2008) 3393.
[123] R. Wadhwa, C. F. Lagenaur, X. T. Cui, *Journal of Controlled Release* **110** (2006) 531.
[124] L. Leprince, A. Dogimont, D. Magnin, S. Demoustier-Champagne, *Journal of Materials Science-Materials in Medicine* **21** (2010) 925.
[125] B. Barbier, J. Pinson, G. Desarmot, M. Sanchez, *Journal of the Electrochemical Society* **137** (1990) 1757.
[126] Z. R. Scheibal, W. Xu, J. F. Audiffred, J. E. Henry, J. C. Flake, *Electrochemical and Solid State Letters* **11** (2008) K81.
[127] A. J. Downard, *Langmuir* **16** (2000) 9680.
[128] P. A. Brooksby, A. J. Downard, *Langmuir* **20** (2004) 5038.
[129] E. De Giglio, S. Cometa, L. Sabbatini, P. G. Zambonin, G. Spoto, *Analytical and Bioanalytical Chemistry* **381** (2005) 626.
[130] E. De Giglio, S. Cometa, N. Cioffi, L. Torsi, L. Sabbatini, *Analytical and Bioanalytical Chemistry* **389** (2007) 2055.
[131] J. R. Bertrand O, Gautier S, Maquet V, Detrembleur C, Jerome C 2002.
[132] C. Jerome, A. Aqil, S. Voccia, D. E. Labaye, V. Maquet, S. Gautier, O. F. Bertrand, R. Jerome, *Journal of Biomedical Materials Research Part A* **76A** (2006) 521.
[133] V. Langlois, E. Renard, G. Vergnol, P. Guerin, G. Loirand, F. Haroun, A. Seron, *Irbm* **31** (2010) 111.
[134] M. Ignatova, S. Voccia, B. Gilbert, N. Markova, D. Cossement, R. Gouttebaron, R. Jerome, C. Jerome, *Langmuir* **22** (2006) 255.
[135] M. Ignatova, S. Voccia, S. Gabriel, B. Gilbert, D. Cossement, R. Jerome, C. Jerome, *Langmuir* **25** (2009) 891.
[136] A. Adenier, C. Combellas, F. Kanoufi, J. Pinson, F. I. Podvorica, *Chemistry of Materials* **18** (2006) 2021.
[137] Y. Levy, N. Tal, G. Tzemach, J. Weinberger, A. J. Domb, D. Mandler, *Journal of Biomedical Materials Research Part B-Applied Biomaterials* **91B** (2009) 819.
[138] P. Kingshott, H. J. Griesser, *Current Opinion in Solid State & Materials Science* **4** (1999) 403.
[139] M. Morra, *Journal of Biomaterials Science-Polymer Edition* **11** (2000) 547.
[140] S. Gabriel, P. Dubruel, E. Schacht, A. M. Jonas, B. Gilbert, R. Jerome, C. Jerome, *Angewandte Chemie-International Edition* **44** (2005) 5505.
[141] Y. Tanaka, H. Doi, Y. Iwasaki, S. Hiromoto, T. Yoneyama, K. Asami, H. Imai, T. Hanawa, *Materials Science & Engineering C-Biomimetic and Supramolecular Systems* **27** (2007) 206.
[142] K. Oya, Y. Tanaka, H. Saito, K. Kurashima, K. Nogi, H. Tsutsumi, Y. Tsutsumi, H. I. Doi, N. Nomura, T. Hanawa, *Biomaterials* **30** (2009) 1281.
[143] I. Levy, S. Magdassi, D. Mandler, *Electrochimica Acta* (2010).
[144] C. J. Brinker, K. D. Keefer, D. W. Schaefer, C. S. Ashley, *Journal of Non-Crystalline Solids* **48** (1982) 47.
[145] J. Livage, M. Henry, C. Sanchez, *Progress in Solid State Chemistry* **18** (1988) 259.

[146] L. L. Hench, J. K. West, *Chemical Reviews* **90** (1990) 33.
[147] J. Y. Wen, G. L. Wilkes, *Chemistry of Materials* **8** (1996) 1667.
[148] C. J. Brinker, G. W. Scherer, Eds., *Sol-Gel Science*, Academic Press, INC, San Diego,CA 1990.
[149] C. J. Brinker, A. J. Hurd, G. C. Frye, K. J. Ward, C. S. Ashley, *Journal of Non-Crystalline Solids* **121** (1990) 294.
[150] D. Avnir, S. Braun, O. Lev, M. Ottolenghi, *Chemistry of Materials* **6** (1994) 1605.
[151] S. Braun, S. Rappoport, R. Zusman, D. Avnir, M. Ottolenghi, *Materials Letters* **10** (1990) 1.
[152] K. Amine, H. Tukamoto, H. Yasuda, Y. Fujita, *Journal of the Electrochemical Society* **143** (1996) 1607.
[153] E. Lindner, T. Schneller, F. Auer, H. A. Mayer, *Angewandte Chemie-International Edition* **38** (1999) 2155.
[154] M. M. Collinson, *Mikrochimica Acta* **129** (1998) 149.
[155] Y. Haruvy, S. E. Webber, *Chemistry of Materials* **3** (1991) 501.
[156] D. Avnir, T. Coradin, O. Lev, J. Livage, *Journal of Materials Chemistry* **16** (2006) 1013.
[157] P. Galliano, J. J. De Damborenea, M. J. Pascual, A. Duran, *Journal of Sol-Gel Science and Technology* **13** (1998) 723.
[158] R. Shacham, D. Avnir, D. Mandler, *Advanced Materials* **11** (1999) 384.
[159] R. Shacham, D. Avnir, D. Mandler, *Journal of Sol-Gel Science and Technology* **31** (2004) 329.
[160] R. Shacham, D. Mandler, D. Avnir, *Chemistry-a European Journal* **10** (2004) 1936.
[161] M. Sheffer, A. Groysman, D. Mandler, *Corrosion Science* **45** (2003) 2893.
[162] M. Sheffer, A. Groysman, D. Starosvetsky, N. Savchenko, D. Mandler, *Corrosion Science* **46** (2004) 2975.
[163] P. N. Deepa, M. Kanungo, G. Claycomb, P. M. A. Sherwood, M. M. Collinson, *Analytical Chemistry* **75** (2003) 5399.
[164] E. Doelsch, W. E. E. Stone, S. Petit, A. Masion, J. Rose, J. Y. Bottero, D. Nahon, *Langmuir* **17** (2001) 1399.
[165] K. Wapner, G. Grundmeier, *Surface & Coatings Technology* **200** (2005) 100.
[166] T. P. Chou, C. Chandrasekaran, S. J. Limmer, S. Seraji, Y. Wu, M. J. Forbess, C. Nguyen, G. Z. Cao, *Journal of Non-Crystalline Solids* **290** (2001) 153.
[167] X. L. Luo, J. J. Xu, Y. Du, H. Y. Chen, *Analytical Biochemistry* **334** (2004) 284.
[168] O. Nadzhafova, M. Etienne, A. Walcarius, *Electrochemistry Communications* **9** (2007) 1189.
[169] A. Walcarius, E. Sibottier, *Electroanalysis* **17** (2005) 1716.
[170] J. X. Liu, D. Z. Yang, F. Shi, Y. J. Cai, *Thin Solid Films* **429** (2003) 225.
[171] D. Grygier, W. Dudzinski, T. Wiktorczyk, K. Haimann, *Acta of Bioengineering and Biomechanics* **10** (2008) 27.
[172] R. Okner, A. J. Domb, M. Mandler, *New Journal of Chemistry* (2009).
[173] R. Okner, G. Favaro, A. Radko, A. J. Domb, D. Mandler, *submitted* (2010).
[174] J. A. Switzer, R. J. Phillips, R. P. Raffaelle, *Supramolecular Architecture* **499** (1992) 244.
[175] C. Fei, Z. Hai, C. Chen, Y. J. Xia, *Progress in Organic Coatings* **64** (2009) 264.
[176] Q. Jin, W. B. Xue, X. J. Li, Q. Z. Zhu, X. L. Wu, *Journal of Alloys and Compounds* **476** (2009) 356.
[177] I. Zhitomirsky, A. Petric, *Materials Letters* **46** (2000) 1.

[178] I. Zhitomirsky, A. Petric, *American Ceramic Society Bulletin* **80** (2001) 41.
[179] J. A. Switzer, *American Ceramic Society Bulletin* **66** (1987) 1521.
[180] J. A. Switzer, M. J. Shane, R. J. Phillips, *Science* **247** (1990) 444.
[181] R. Chaim, I. Silberman, L. Galor, *Journal of the Electrochemical Society* **138** (1991) 1942.
[182] R. Chaim, G. Stark, L. Galor, H. Bestgen, *Journal of Materials Science Letters* **13** (1994) 487.
[183] R. Chaim, G. Stark, L. Galor, H. Bestgen, *Journal of Materials Science* **29** (1994) 6241.
[184] I. Zhitomirsky, L. Galor, S. Klein, *Journal of Materials Science Letters* **14** (1995) 60.
[185] I. Zhitomirsky, L. Galor, A. Kohn, H. W. Hennicke, *Journal of Materials Science* **30** (1995) 5307.
[186] I. Zhitomirsky, L. GalOr, *Materials Letters* **31** (1997) 155.
[187] W. R. Lacefield, *Implant Dent* **7** (1998) 315.
[188] P. A. Dearnley, *Proceedings of the Institution of Mechanical Engineers Part H-Journal of Engineering in Medicine* **213** (1999) 107.
[189] C. C. Chang, S. K. Yen, *Surface & Coatings Technology* **182** (2004) 242.
[190] X. Y. Liu, P. K. Chu, C. X. Ding, *Materials Science & Engineering R-Reports* **47** (2004) 49.
[191] X. Nie, A. Leyland, A. Matthews, J. C. Jiang, E. I. Meletis, *Journal of Biomedical Materials Research* **57** (2001) 612.
[192] P. Kern, P. Schwaller, J. Michler, *Thin Solid Films* **494** (2006) 279.
[193] S. Radice, P. Kern, G. Burki, J. Michler, M. Textor, *Journal of Biomedical Materials Research Part a* **82A** (2007) 436.
[194] T. Moritz, W. Eiselt, K. Moritz, *Journal of Materials Science* **41** (2006) 8123.
[195] T. Moritz, D. Linaschke, W. Eiselt, *Electrophoretic Deposition: Fundamentals and Applications Ii* **314** (2006) 207.
[196] A. R. Boccaccini, F. Chicatun, J. Cho, O. Bretcanu, J. A. Roether, S. Novak, Q. Chen, *Advanced Functional Materials* **17** (2007) 2815.
[197] M. I. Kay, R. A. Young, A. S. Posner, *Nature* **204** (1964) 1050.
[198] T. S. B. Narasaraju, D. E. Phebe, *Journal of Materials Science* **31** (1996) 1.
[199] P. Ducheyne, W. Vanraemdonck, J. C. Heughebaert, M. Heughebaert, *Biomaterials* **7** (1986) 97.
[200] P. Ducheyne, S. Radin, M. Heughebaert, J. C. Heughebaert, *Biomaterials* **11** (1990) 244.
[201] I. Zhitomirsky, *Journal of the European Ceramic Society* **18** (1998) 849.
[202] K. Grandfield, I. Zhitomirsky, *Materials Characterization* **59** (2008) 61.
[203] N. Eliaz, T. M. Sridhar, U. K. Mudali, B. Raj, *Surface Engineering* **21** (2005) 238.
[204] N. Eliaz, M. Eliyahu, *Journal of Biomedical Materials Research Part a* **80A** (2007) 621.
[205] T. V. Vijayaraghavan, A. Bensalem, *Journal of Materials Science Letters* **13** (1994) 1782.
[206] H. Dasarathy, C. Riley, H. D. Coble, W. R. Lacefield, G. Maybee, *Journal of Biomedical Materials Research* **31** (1996) 81.
[207] T. M. Sridhar, T. K. Arumugam, S. Rajeswari, M. Subbaiyan, *Journal of Materials Science Letters* **16** (1997) 1964.
[208] M. Manso, C. Jimenez, C. Morant, P. Herrero, J. M. Martinez-Duart, *Biomaterials* **21** (2000) 1755.

[209] A. Stoch, A. Brozek, G. Kmita, J. Stoch, W. Jastrzebski, A. Rakowska, *Journal of Molecular Structure* **596** (2001) 191.
[210] M. C. Kuo, S. K. Yen, *Materials Science & Engineering C-Biomimetic and Supramolecular Systems* **20** (2002) 153.
[211] C. K. You, X. W. Meng, T. Y. Kwon, Y. Z. Yang, J. L. Ong, S. Kim, K. H. Kim, *Bioceramics 17* **284-286** (2005) 901.
[212] H. Mayr, M. Ordung, G. Ziegler, *Journal of Materials Science* **41** (2006) 8138.
[213] X. W. Meng, T. Y. Kwon, Y. Z. Yang, J. L. Ong, K. H. Kim, *Journal of Biomedical Materials Research Part B-Applied Biomaterials* **78B** (2006) 373.
[214] B. Plesingerova, G. Sucik, M. Maryska, D. Horkavcova, *Ceramics-Silikaty* **51** (2007) 15.
[215] S. P. Huang, K. C. Zhou, B. Y. Huang, Z. Y. Li, S. H. Zhu, G. H. Wang, *Journal of Materials Science-Materials in Medicine* **19** (2008) 437.
[216] Y. M. Liao, Z. D. Feng, S. W. Li, *Thin Solid Films* **516** (2008) 6145.
[217] D. J. Blackwood, K. H. W. Seah, *Materials Science & Engineering C-Biomimetic and Supramolecular Systems* **29** (2009) 1233.
[218] K. H. Kim, N. Ramaswamy, *Dental Materials Journal* **28** (2009) 20.
[219] K. De Groot, J. G. C. Wolke, J. A. Jansen, *Proceedings of the Institution of Mechanical Engineers Part H-Journal of Engineering in Medicine* **212** (1998) 137.
[220] Q. Y. Zhang, Y. Leng, *Biomaterials* **26** (2005) 3853.
[221] M. Saremi, B. M. Golshan, *Transactions of the Institute of Metal Finishing* **85** (2007) 99.
[222] S. Ban, S. Maruno, *Journal of Biomedical Materials Research* **42** (1998) 387.
[223] K. Yamashita, E. Yonehara, X. F. Ding, M. Nagai, T. Umegaki, M. Matsuda, *Journal of Biomedical Materials Research* **43** (1998) 46.
[224] L. Y. Huang, K. W. Xu, J. Lu, *Journal of Materials Science-Materials in Medicine* **11** (2000) 667.
[225] A. C. Lewis, P. J. Heard, *Journal of Biomedical Materials Research Part A* **75A** (2005) 365.
[226] Y. W. Song, D. Y. Shan, E. H. Han, *Materials Letters* **62** (2008) 3276.
[227] T. M. Sridhar, N. Eliaz, U. K. Mudali, B. Raj, *Corrosion Reviews* **20** (2002) 255.
[228] M. H. Fathi, M. Salehi, A. Saatchi, V. Mortazavi, S. B. Moosavi, *Surface Engineering* **17** (2001) 459.
[229] I. Gurappa, *Surface & Coatings Technology* **161** (2002) 70.
[230] T. M. Sridhar, U. K. Mudali, M. Subbaiyan, *Corrosion Science* **45** (2003) 237.
[231] C. Garcia, S. Cere, A. Duran, *Journal of Non-Crystalline Solids* **352** (2006) 3488.
[232] M. H. Fathi, F. Azam, *Materials Letters* **61** (2007) 1238.
[233] C. Arnould, J. Delhalle, Z. Mekhalif, *Electrochimica Acta* **53** (2008) 5632.
[234] N. Eliaz, T. M. Sridhar, *Crystal Growth & Design* **8** (2008) 3965.
[235] A. Balamurugan, G. Balossier, J. Michel, J. M. F. Ferreira, *Electrochimica Acta* **54** (2009) 1192.
[236] C. T. Kwok, P. K. Wong, F. T. Cheng, H. C. Man, *Applied Surface Science* **255** (2009) 6736.
[237] T. S. Wang, R. J. Zhang, Y. N. Yan, *Journal of Bioactive and Compatible Polymers* **24** (2009) 169.
[238] C. L. Wen, S. K. Guan, L. Peng, C. X. Ren, X. Wang, Z. H. Hu, *Applied Surface Science* **255** (2009) 6433.

[239] R. Hu, H. B. Hu, C. J. Lin, *Chemical Journal of Chinese Universities-Chinese* **23** (2002) 2142.
[240] J. Wang, J. De Boer, K. De Groot, *Journal of Dental Research* **83** (2004) 296.
[241] H. B. Hu, C. J. Lin, P. P. Y. Lui, Y. Leng, *Journal of Biomedical Materials Research Part a* **65A** (2003) 24.
[242] X. L. Cheng, M. Filiaggi, S. G. Roscoe, *Biomaterials* **25** (2004) 5395.
[243] K. Duan, Y. W. Fan, R. Z. Wang, *Journal of Biomedical Materials Research Part B-Applied Biomaterials* **72B** (2005) 43.
[244] Y. W. Fan, K. Duan, R. Z. Wang, *Biomaterials* **26** (2005) 1623.
[245] J. Wilson, G. H. Pigott, F. J. Schoen, L. L. Hench, *Journal of Biomedical Materials Research* **15** (1981) 805.
[246] T. Nakamura, T. Yamamuro, S. Higashi, T. Kokubo, S. Itoo, *Journal of Biomedical Materials Research* **19** (1985) 685.
[247] L. L. Hench, *Journal of the American Ceramic Society* **81** (1998) 1705.
[248] D. Krause, B. Thomas, C. Leinenbach, D. Eifler, E. J. Minay, A. R. Boccaccini, *Surface & Coatings Technology* **200** (2006) 4835.
[249] Y. F. Zhao, C. Z. Chen, D. G. Wang, *Surface Review and Letters* **12** (2005) 505.
[250] X. Cui, H. M. Kim, M. Kawashita, L. Wang, T. Xiong, T. Kokubo, T. Nakamura, *Dental Materials* **25** (2009) 80.
[251] G. Giavaresi, M. Fini, R. Chiesa, L. Rimondini, G. Rondelli, V. Borsari, L. Martini, N. N. Aldini, G. A. Guzzardella, R. Giardino, *International Journal of Artificial Organs* **25** (2002) 806.
[252] O. Zinger, P. F. Chauvy, D. Landolt, *Journal of the Electrochemical Society* **150** (2003) B495.
[253] V. M. Frauchiger, F. Schlottig, B. Gasser, M. Textor, *Biomaterials* **25** (2004) 593.
[254] C. Yao, V. Perla, J. L. McKenzie, E. B. Slamovich, T. J. Webster, *Journal of Biomedical Nanotechnology* **1** (2005) 68.
[255] K. Das, S. Bose, A. Bandyopadhyay, *Acta Biomaterialia* **3** (2007) 573.
[256] S. E. Kim, J. H. Lim, S. C. Lee, S. C. Nam, H. G. Kang, J. Choi, *Electrochimica Acta* **53** (2008) 4846.
[257] H. J. Oh, J. H. Lee, Y. J. Kim, S. J. Suh, C. S. Chi, *Materials Chemistry and Physics* **109** (2008) 10.
[258] Y. Yang, Y. Lai, Q. Zhang, K. Wu, L. Zhang, C. Lin, P. Tang, *Colloids Surf B Biointerfaces* **79** (2010) 309.
[259] I. De Scheerder, J. Sohier, K. Wang, E. Verbeken, X. R. Zhou, L. Froyen, J. Van Humbeeck, J. Piessens, F. Van de Werf, *Journal of Interventional Cardiology* **13** (2000) 179.
[260] I. De Scheerder, E. Verbeken, J. Van Humbeeck, *Semin Interv Cardiol* **3** (1998) 139.
[261] C. Hehrlein, M. Zimmermann, J. Metz, W. Ensinger, W. Kubler, *Coronary Artery Disease* **6** (1995) 581.
[262] H. A. Weng, C. C. Wu, C. G. Chen, C. C. Ho, S. J. Ding, *Journal of Materials Science-Materials in Medicine* **21** (2010) 1511.
[263] G. J. Chi, S. W. Yao, J. Fan, W. G. Zhang, H. Z. Wang, *Surface & Coatings Technology* **157** (2002) 162.
[264] S. Voccia, M. Ignatova, R. Jerome, C. Jerome, *Langmuir* **22** (2006) 8607.
[265] X. Pang, I. Zhitomirsky, *Surface & Coatings Technology* **202** (2008) 3815.

[266] L. Visai, L. Rimondini, C. Giordano, B. Del Curto, M. S. Sbarra, R. Franchini, C. Della Valle, R. Chiesa, *Journal of Applied Biomaterials & Biomechanics* **6** (2008) 170.
[267] W. Shao, Q. Zhao, *Colloids and Surfaces B-Biointerfaces* **76** (2010) 98.
[268] J. Wang, H. X. Li, L. Ren, A. S. Zhao, P. Li, Y. X. Leng, H. Sun, N. Huang, *Surface & Coatings Technology* **201** (2007) 6893.
[269] K. Das, S. Bose, A. Bandyopadhyay, B. Karandikar, B. L. Gibbins, *Journal of Biomedical Materials Research Part B-Applied Biomaterials* **87B** (2008) 455.

Regina Guslitzer-Okner

Regina Guslitzer-Okner received her B.Sc. in medicine chemistry from the Hebrew University in 2005. The research focused on thin films electrodeposition of conducting polymers on stainless steel and theirs biocompatibility. Regina continued this research and completed her Ph.D. entitled "Electrodeposition of Organic Polymers as Biocompatible, Drug Eluting Coatings for Metallic Medical Devices" in 2009 (supervised by Prof. A. J. Domb and Prof. D. Mandler) also at the Hebrew University. Dr. Guslitzer-Okner has published a number of peer-reviewed papers and participated in a few patents in a field of electropolymerization of conducting polymers and electrodeposition of sol-gel thin films on metallic implants.

Daniel Mandler

Daniel Mandler is a Professor of Chemistry in the Institute of Chemistry at the Hebrew University of Jerusalem. He received his B.Sc. from the Hebrew University in 1983 and completed his Ph.D. in 1988 (supervised by Prof. I. Willner) *summa cum laude* also at the Hebrew University. He spent two years (1988-1990) at the University of Texas with Prof. A. J. Bard where he carried his post-doctoral research and returned to the Hebrew University in 1990. Since 2003 he is a full professor of chemistry. His main interests are analytical and physical electrochemistry, surface science, forensic chemistry, and sol-gel technology. Prof. Mandler has published over 130 peer-reviewed papers, holds a few patents and has been in the organizing committees of numerous international meetings.

5

Electrochemical Techniques to Obtain Biofunctional Materials

Takao Hanawa

Institute of Biomaterials and Bioengineering, Tokyo Medical and Dental University, 2-3-10 Kanda-surugadai, Chiyoda-ku, Tokyo 101-0062, Japan

I. INTRODUCTION

Metals have a long history in medicine and dentistry. However, metals are artificial materials with no biofunctionality, which makes them less attractive as biomaterials for certain applications. On the other hand, biofunctionalities of ceramics and polymers are expected to show excellent properties as biomaterials; in fact, many devices consisting of metals have been substituted by those consisting of ceramics and polymers. In spite of this event, about 80% of implant devices in the medical field still consist of metals, and this share is currently maintained, because of their high strength, toughness, and durability. To-date, metallic biomaterials cannot be replaced with ceramics or polymers. The requirements for metallic medical devices are summarized in Table 1.

In order to add biofunctionality to metals, surface modification is necessary. Surface modification is a process that changes a material's surface composition, structure and morphology, leaving the bulk mechanical properties intact. In this chapter, some of

N. Eliaz (ed.), *Applications of Electrochemistry and Nanotechnology in Biology and Medicine I*, Modern Aspects of Electrochemistry 52, DOI 10.1007/978-1-4614-0347-0_5, © Springer Science+Business Media, LLC 2011

Table 1
Properties Required from Metals for Medical Devices and their Effects

Required property	Target medical devices	Effect
Elastic modulus	• bone fixation • spinal fixation	Prevention of bone absorption by stress shielding.
Superelasticity Shape memory effect	• multi-purpose	Improvement of mechanical compatibility.
Wear resistance	• artificial joint	Prevention of generation of wear debris. Improvement of durability.
Biodegradability	• stent • artificial bone • bone fixation	Elimination of materials after healing. Unnecessary of retrieval
Bone formation Bone bonding	• Stem of artificial hip joint dental implant	Fixation of devices in bone
Prevention of bone formation	• bone screw • bone nail	Prevention of assimilation
Adhesion of soft tissue	• dental implant • trans-skin device • external fixation • pacemaker housing	Fixation in soft tissue; prevention of inflectional disease
Inhibition of platelet adhesion	• devices contacting blood	Prevention of thrombus
Inhibition of biofilm formation	• all implant devices • treatment tools and apparatus	Prevention of infectious disease
Low magnetic susceptibility	• all implant devices • treatment tools and apparatus	No artifact in MRI

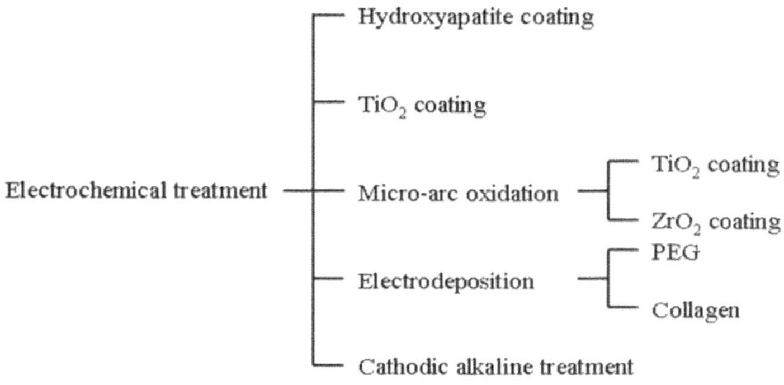

Figure 1. Types of electrochemical treatments.

the surface modification techniques to add biofunctionality to metals, especially electrochemical techniques, are reviewed.

II. OVERVIEW OF CURRENT ELECTROCHEMICAL TREATMENTS

Current electrochemical treatments are summarized in Fig. 1. In order to precipitate hydroxylapatite or calcium phosphate layer on titanium, electrochemical treatment is used commonly to form an hydroxyapatite (HA) layer on Ti.[1-4] Through an electrochemical process, carbonate-containing HA with a desirable shape such as plate, needle or particle could be precipitated on a Ti substrate, which is sometimes heated to obtain a better coating layer. Anodic polarization accelerates the formation of apatite layer on Ti. Beta-tricalcium phosphate (β-TCP) is cathodically coated on Ti for immobilization of collagen.[5] Cathodic polarization is also effective to precipitate HA and calcium phosphate, because cathode polarization generate alkaline condition on Ti surface. Therefore, alternative current that positive and negative charges are frequently repeated, is also effective. Low-voltage alternating current is applied for the precipitation of calcium phosphate on Ti.[6] This technique is useful for the treatment of thin wire and fiber without the dissolution of Ti. HA can also be electrodeposited by pulse current.[7]

The composition and pH of electrolyte for electrochemical treatment influence morphology of precipitated HA. Nano-grained calcium phosphate is electrochemically deposited on Ti using acidic electrolytes.[8] The coating layer contains dicalcium phosphate dihydrate (55–85 nm in grain size) with a small amount of HA (20–25 nm); the content of HA increases with the increase of the current density.[9] An electrochemical method of producing nano-crystalline HA coatings on Ti surface has been reported.[10,11] Also, HA has been coated by dynamic voltage during electrophoretic deposition.[12] Nano-size HA generally shows good osteoconductivity.

TiO_2 nanotube-type oxide film on Ti substrate has been fabricated using an electrochemical method. Anodizing to obtain a unique oxide layer on Ti in electrolytes of calcium glycerophosphate and calcium acetate has been attempted to obtain TiO_2 nanotube.[13] Self-organized porous nanotubular TiO_2 is anodically formed on Ti in Na_2SO_4 electrolyte containing NaF. The oxidation was carried out at 20 V, with the baths stirred using magnetic pellet and ultrasonic vibration. In this case, (101) and (200) poles were randomly oriented.[14] The formation and growth of a self-organized nanotube layer can be achieved directly by anodizing in NH_4-containing electrolytes. The diameter, length, and wall thickness of the nanotube are significantly influenced by anodizing conditions such as voltage, current density and anodizing time.[15] The TiO_2 nanotube on Ti material is expected to increase osteoconductivity, while the effect and role are not clear at present.

On the other hand, micro-arc oxidation (MAO, also named as plasma electrolytic oxidation or anodic spark oxidation) is a relatively convenient technique for forming oxide layers on metals. MAO is effective in formation of a porous or irregular-shaped TiO_2 layer on Ti substrate. The advantage of MAO is that coating layer is not only porous, but also uniformly coated on metal surfaces with complex geometry. Anodizing and micro-arc oxidation are not clearly distinguished. In the case of the formation of an oxide layer with connecting pore to the substrate metal by high voltage, this technique is usually categorized as micro-arc oxidation. In this sense, some of the electrochemical techniques explained above belong to MAO. MAO is currently used to obtain

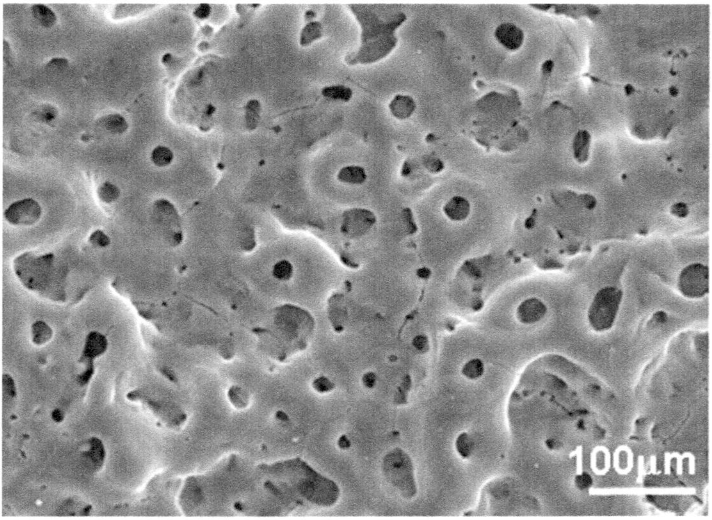

Figure 2. Scanning electron microphotograph of micro-arc oxidized zirconium.

thick and porous oxide or HA layer.[16-19] The most famous MAO product is well known as TiUnite®. Ultraviolet irradiation of micro-arc oxidation titania coating in distilled water has been found to enhance the bioactivity.[20] However, the effect of UV is not clear at present. This MAO technique is also applied to zirconium (Zr), forming ZrO_2 layer (Fig. 2).[21,22] ZrO_2-MAO technique could be useful in dental implants.

III. ELECTRODEPOSITION OF PEG

1. Electrodeposision Process and its Effects

Metals with biofunctionality have been required recently. For example, stents are placed at stenotic blood vessels for dilatation, and blood compatibility or prevention of adhesion of platelets is necessary. In guide wires and guiding catheters, lubrication in the blood vessels is important for proper sliding and insertion. In addi-

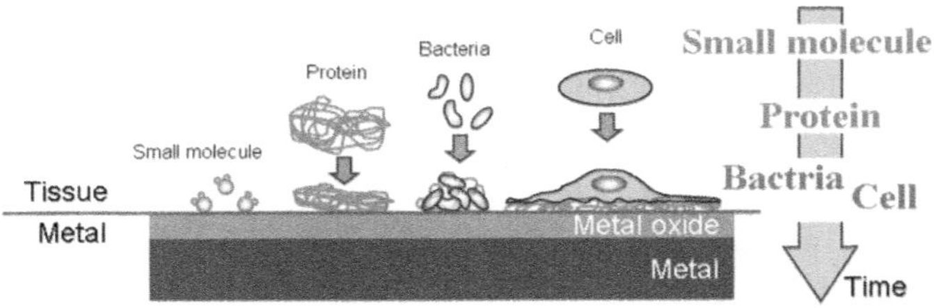

Figure 3. Initial surface reactions on a metal surface *in vivo*.

tion, if metals are used as sensing devices, the control of cell adhesion is necessary. In all implant materials, antibacterial property is required to avoid infection during the service. This property is also important in dental implants.

When a metal is implanted in the human body, the initial reaction between the metal surface and the biological environment is protein adsorption onto the surface. The protein adsorption at this initial stage governs the adhesion of cells and bacteria, as shown in Fig. 3, and even the tissue compatibility of the metal. A platelet aggregation on an implanted material is activated by adsorbed proteins, and a thrombus is formed.[23] The initial bacterial adhesion, followed by a biofilm formation, is also considered to be related with protein adsorption.[24] Thus, protein adsorption on the surface must be inhibited at the initial stage according to the purpose of the materials. For these purposes, the fundamental property is the inhibition of protein adsorption, because it is necessary for the adhesion of cells, platelets and bacteria.

Polyethylene glycol (PEG) is a biofunctional molecule on which the adsorption of proteins is inhibited. Therefore, the immobilization of PEG to a metal surface is an important step towards the biofunctionalization of the metal surface. Both terminals of PEG (MW = 1000) have been terminated with $-NH_2$ (NH_2–PEG–NH_2), but only one terminal was terminated with $-NH_2$(NH_2–PEG).[25] The cathodic potential was applied to Ti, starting from the open-circuit potential (OCP) and ending at –0.5 V vs. a saturated calomel electrode (SCE), and was maintained at this potential for 300 s. During charging, the terminated PEGs electrically migrated to and deposited on the Ti cathode, as shown in Fig. 4. Not only

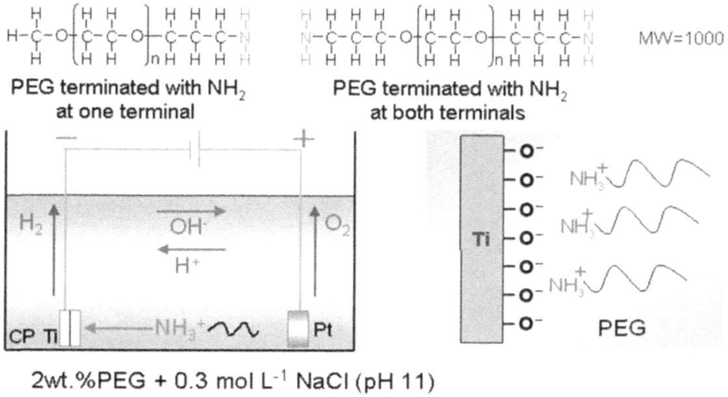

Figure 4. Principle of electrodeposition of PEG terminated with NH_2 at one or both terminals.

electrodeposition, but also immersion, led to the immobilization of PEG onto the Ti surface. However, more terminated amines combined with Ti oxide as an NH–O bond by electrodeposition, while more amines randomly existed as NH_3^+ in the PEG molecule by immersion (Fig. 5).[26,27] The amounts of the PEG layer immobilized

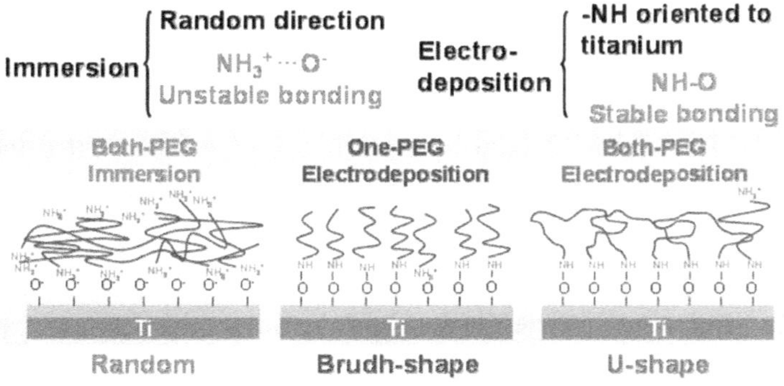

Figure 5. Schematic model of immobilized manners of PEG to titanium surface with immersion and electrodeposition.

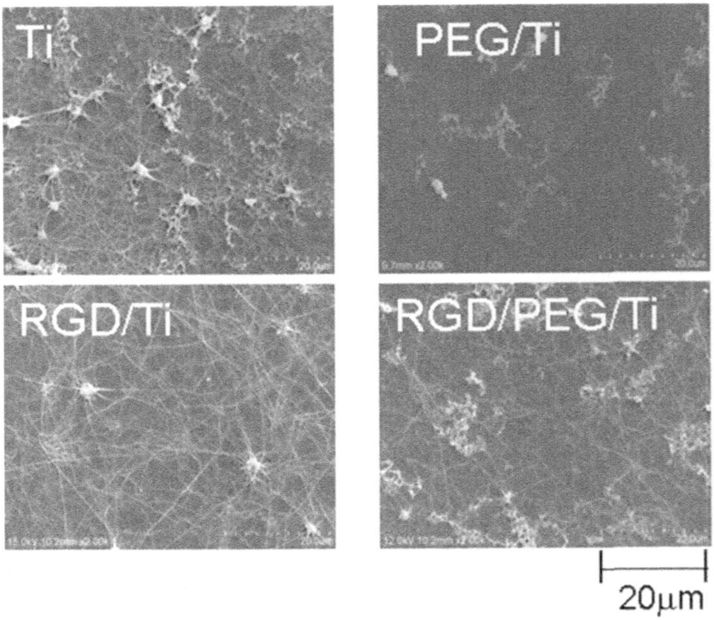

Figure 6. Adhesion of platelet and formation of fibrin network on different specimens.

onto the metals were governed by the concentrations of the active hydroxyl groups on each surface oxide in the case of electrodeposition, which was governed by the relative permittivity of the surface oxide in the case of immersion.[28] The PEG-immobilized surface inhibited the adsorption of proteins and cells, as well as the adhesion of platelets and bacteria (Fig. 6),[27] indicating that this electrodeposition technique is useful for the biofunctionalization of metal surfaces; it is also useful for all electroconductive and morphological materials.

2. Evaluation of the Thickness of the Immobilized Layer with Ellipsometry

The thickness of the PEG layer deposited on titanium has been determined by an ellipsometer in air. Ellipsometry measures the

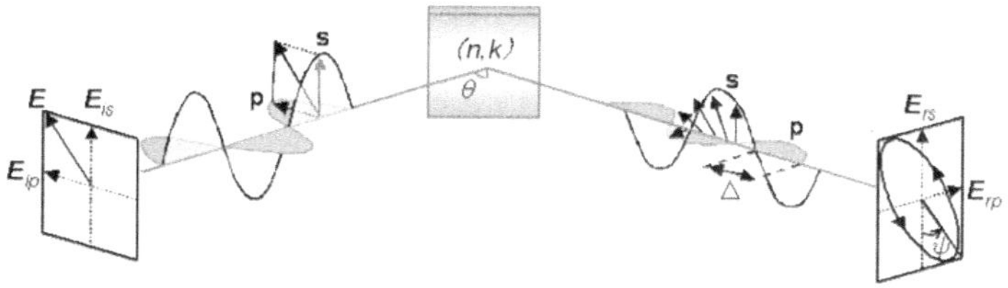

Figure 7. Principle of ellipsometry.

change of polarization upon reflection or transmission. Typically, ellipsometry is done only in the reflection setup. The exact nature of the polarization change is determined by the sample's properties (thickness, complex refractive index or dielectric function tensor). Although optical techniques are inherently diffraction limited, ellipsometry exploits phase information and the polarization state of light, and can achieve angstrom resolution. In its simplest form, the technique is applicable to thin films with thickness less than a nanometer to several micrometers. The sample must be composed of a small number of discrete, well-defined layers that are optically homogeneous and isotropic. Violation of these assumptions will invalidate the standard ellipsometric modeling procedure, and more advanced variants of the technique must be applied (Fig. 7). According to this principle, it is possible to identify 0.1-nm thickness.

The use of an ellipsometer has resulted in the underestimation of the thickness compared to that in solution. The light source was a He-Ne laser with a wavelength of 632.8 nm, and the incident angle to the titanium surface was 70°. The immobilized PEG layer is not uniform and condensed; therefore, the thickness is an apparent one that is influenced by chain length of PEG and immobilized density per unit area. In addition, the thicknesses are measured in air; therefore, the real thickness in solutions is larger than these values. The thickness of immobilized PEG layer on Ti with electrodeposition and immersion is shown in Fig.8.

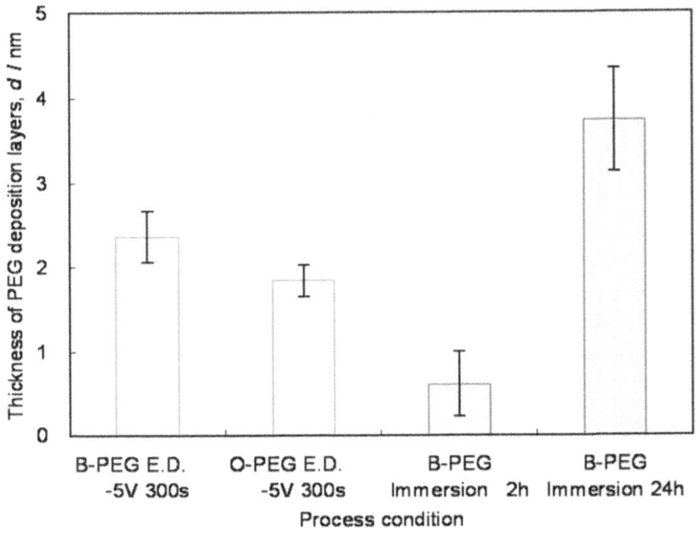

Figure 8. Thickness of immobilized PEG layer on Ti with electrodeposition and immersion.

IV. ELECTRODEPOSITION OF PEG TWITTER IONS TO ACCELERATE CELL ADHESION

1. The Isoelectric Points of PRG Twitter Ions According to pH

Cell growth and differentiation on a material are necessary events for hard and soft tissue compatibility of the material[29] they are generally controlled by the adsorption of proteins on the material. Peptides consisting of arginine (R)-glycine (G)-asparaginic acid (D) sequence, RGD peptide, accelerate cell attachment and extension of bone cells on Ti.[30] RGD is a peptide known to involve cell adhesion, which is involved in many extracellular matrix proteins.[31] Bone formation is accelerated by immobilizing RGD on a Ti surface.[32] Peptides with terminal cysteine residues were immobilized on maleimide-activated oxides.[33-35]

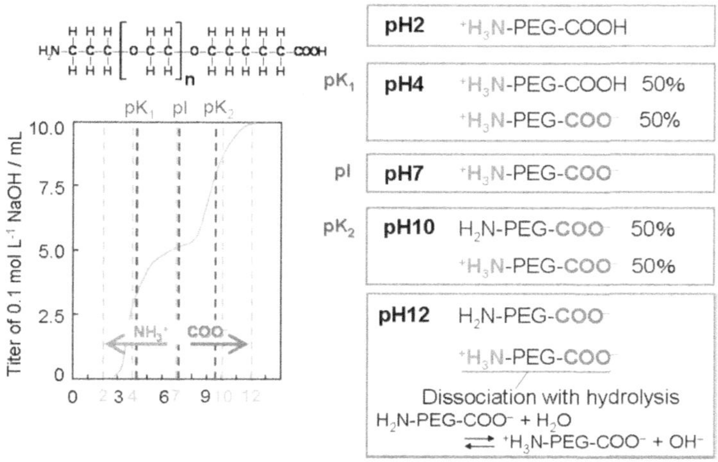

Figure 9. Titration curve of PEG and dissociation condition of PEG according to pH."pI" represents the isoelectric point.

The isoelectric points (IEPs) of two zwitterions, glycine and both-terminal-terminated polyethylene glycol (NH_2-PEG-COOH), were determined from the titration curves, and the thickness of zwitterion layers immobilized on Ti with immersion and electrodeposition at various pH based on IEPs were evaluated with ellipsometry to investigate the effect of pH and the immobilization technique on the interactions between the zwitterions and Ti surface.[36] From the titration curves, pK_1, pK_2, and the IEP of glycine were determined as 2.8, 8.9, and 5.9, respectively, and pK_1, pK_2, and the IEP of NH_2-PEG-COOH were determined as 2.1, 11.7, and 6.9, respectively. At a certain specific pH, $^{+}H_3N$-CH_2-COO^- or $^{+}H_3N$-PEG-COO^- were formed by hydrolysis of glycine or NH_2-PEG-COOH (Fig. 9). In addition, the Ti surface was negatively charged at this pH. As a result, for immersion, the electrostatic reactivity between terminal groups of zwitterion and hydroxyl groups on Ti surface was the highest and the thickness of the immobilized layer was significantly the largest at pH 12. For electrodeposition, glycine, with its smaller molecular weight, was more easily attracted to Ti surface than NH_2-PEG-COOH, which

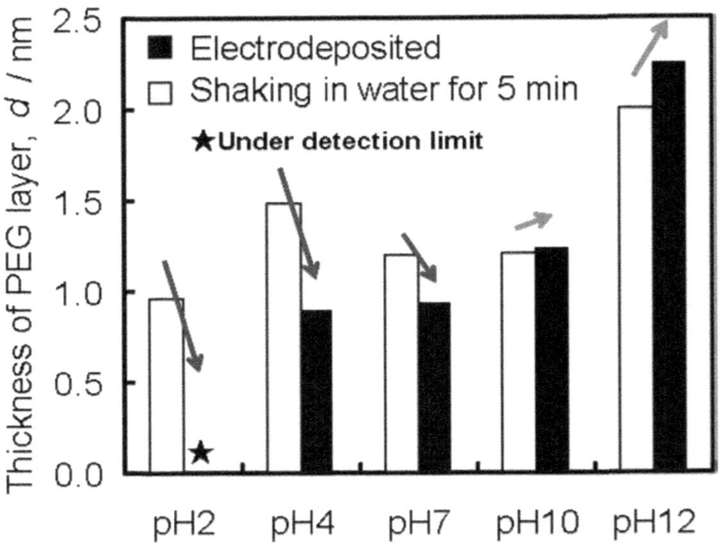

Figure 10. Thickness of PEG-immobilized layer electrodeposited at various pH and those after shaking in water to evaluate durability of the immobilized layer.

has a larger molecular weight, while the thickness of the immobilized layer was the largest at pH 12 in both zwitterions. The thickness of the residual layer after shaking in distilled water is shown in Fig. 10. In addition, the negatively charged hydroxyl groups (OH^-), rather than the positively charged ones (OH_2^+ or Ti^+), attracted ^+H_3N-H_2-COO^- and ^+H_3N-PEG-COO^-. The negatively, rather than the positively, charged hydroxyl groups attracted more zwitterions. In the case of electrodeposition, glycine, with its smaller molecular weight, was more easily attracted to the surface than NH_2-PEG-COOH, which has a large molecular weight, while the thickness of the immobilized layer was the largest at pH 12 in both zwitterions. In addition, the degree of interactions of glycine and NH_2-PEG-COOH with Ti surface influenced the durability of the immobilized layer. A Ti surface electrodeposited with NH_2-PEG-COOH at pH 10 or 12 is useful for the inhibition of protein adsorption; on the other hand, it is also useful for the immobilization of other biofunctional molecules to accelerate bone formation.

2. The Effect of RGD Peptide Immobilized through PEG on Calcification by MC3T3-E1

The effect of a cell-adhesive peptide containing RGD immobilized through PEG twitter ion on Ti on calcification by MC3T3-E1 cells was investigated in an attempt to develop a new surface modification technique using biofunctional molecules. RGD was immobilized on Ti through PEG, both terminals of which were terminated with -NH_2 and -COOH to combine with the Ti surface and RGD, as shown in Fig. 11.[37] PEG was immobilized on Ti by electrodepo-

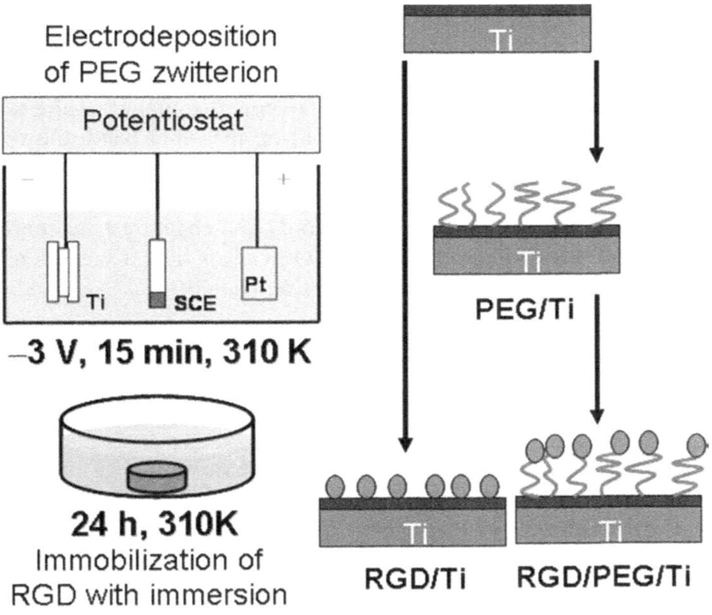

Figure 11. Immobilization of RGD peptide to Ti surface through electrodeposited PEG twitter ion.

sition, and RGD – by immersion. For comparison, glycine was employed because it is the simplest molecule containing both $-NH_2$ and -COOH at its terminals. The specimens prepared are shown in Fig. 12. MC3T3-E1 cells were cultured and differentiation-induced on each specimen, and the cell calcification properties were investigated. As a result, there was no significant difference in the morphology and extension of MC3T3-E1 cells cultured on each specimen, while the number of cells cultured on RGD/PEG/Ti was the largest. After differentiation-induction, there was no significant difference in the ALP activity among all specimens. On the other hand, the level of cell calcification on RGD/PEG/Ti was the highest. Therefore, the hard tissue compatibility of Ti is improved by immobilizing RGD through functional molecules which have a long molecular chain.

The calcification level on RGD/PEG/Ti was the largest among all specimens. The RGD peptide on RGD/PEG/Ti was the most separated from Ti because PEG is the long straight-chain molecule. Therefore, the RGD peptide on RGD/PEG/Ti is not affected by Ti, i.e., the surface oxide film and the Ti substrate. The relative permittivity of the surface oxide film governs the change in the conformation of proteins adsorbed on it; on the other hand, the conformation of proteins always changes when they are adsorbed on a solid surface, and the conformation plays an important role in the function of the proteins. In other words, the change in the conformation of proteins adsorbed on RGD/PEG/Ti is little. Consequently, adsorbed cells on the proteins may be activated.

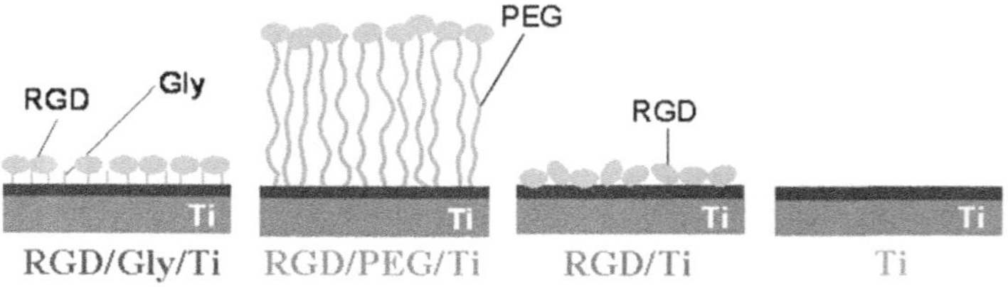

Figure 12. Specimens for evaluation of calcification by MC3T3-E1 cells on their surfaces.

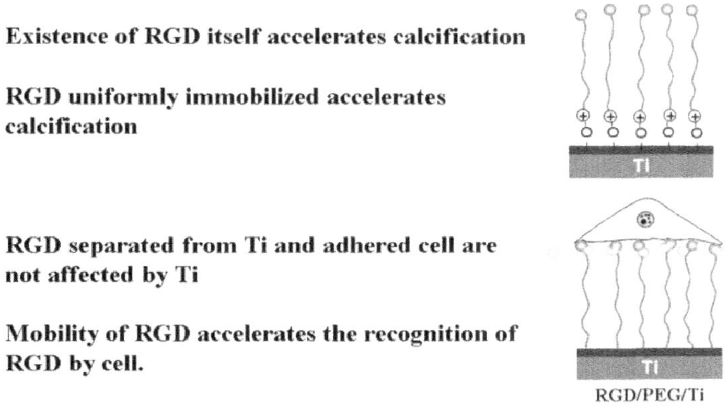

Figure 13. Mechanism of good calcification by MC3T3-E1 cell on RGD/PEG/Ti.

The culture environment of osteoblastic cells in RGD/PEG/Ti could be better for cell attachment and proliferation than the environment in RGD/Gly/Ti, RGD/Ti, and Ti. The molecular length of PEG was much longer than that of Gly, and the mobility of RGD immobilized at the terminal of the PEG increased compared with that of RGD/Gly/Ti. Consequently, osteoblastic cells on RGD/PEG/Ti easily recognized the RGD peptide, and the environment of the differentiation might have been improved. Furthermore, the high mobility of PEG caused the local control of the RGD density. The possible mechanism is illustrated in Fig. 13.

V. NICKEL-FREE SURFACE OXIDE FILM ON TI-NI ALLOY

The Ni-Ti alloy is a promising metallic material used in many fields due to its special properties of shape memory effect and superelasticity. Because of its high corrosion resistance[38] and good biocompatibility,[39] this alloy has been increasingly applied to medical and dental devices, e.g., orthodontic wires,[40] root canal instruments,[41] coronary stents,[42] and catheters.[43] However, it was reported that more amount of released Ti ion was detected from Ni-Ti alloy than from unalloyed titanium or Ti-6Al-4V alloy, which suggested that the corrosion resistance of Ni-Ti alloy was not so good as the other two metals.[44] Ni contained in Ni-Ti alloy may possibly cause adverse reactions like metallic allergy,[45,46] toxicity,[47,48] and carcinogenecity.[48] Therefore, it is necessary to reduce the Ni ion release from the alloy to improve its corrosion resistance and biocompatibility.

Improvement of the corrosion resistance of Ni-Ti alloy by electrolytic treatment, by using different acidic electrolyte compositions, has been made. Specimens electrolyzed with lactic acid, water and glycerol were found to show higher corrosion potential and to release lower amount of titanium and nickel ions than mechanically-polished specimens.[49] With the electrolytic treatment, nickel concentration in the surface oxide layer of Ni-Ti alloy decreased, and the thickness of the surface oxide layer increased. It was shown that electrolytic treatment with the suitable electrolyte could improve the corrosion resistance of Ni-Ti, which is effective to produce medical and dental devices that utilize shape memory effect or superelasticity with better biocompatibility. Electrolytic treatment was effective in improving the corrosion resistance of Ni-Ti. Titanium in the surface oxide layer of Ni-Ti was oxidized to Ti^{4+} by electrolytic treatment, and thus became more stable. The thickness of the surface oxide layer on Ni-Ti increased after being subjected to electrolytic treatment. The nickel concentration in the surface oxide layer of Ni-Ti decreased after being subjected to electrolytic treatment. Characteristics of the surface oxide layer of electrolyzed Ni-Ti varied according to the electrolyte composition. During electrolysis, Ni is preferentially released from the surface oxide film and Ti is preferentially oxidized to form new oxide film: The Ni-free surface oxide film is finally formed as shown in Fig. 14. This resultant surface may show low toxicity.

VI. CATHODIC POLARIZATION FOR CHARACTERIZATION OF METAL SURFACES

1. Changes of Surface Composition

When a material is implanted into a human body, an immediate reaction occurs between its surface and the living tissues. In other words, the immediate reaction at this initial stage determines and

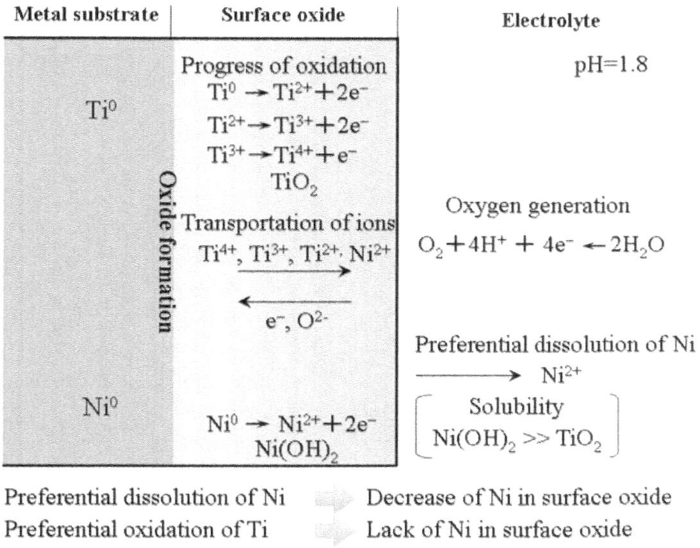

Figure 14. Mechanism of the formation of nickel-free surface oxide film on Ni-Ti alloy by electrolysis.

defines a material's tissue compatibility right away. Since conventional metallic biomaterials are usually covered with metal oxides, the surface oxide films on metals play an important role not only against corrosion but also regarding tissue compatibility. The composition of the surface oxide film varies according to environmental changes, although the film is macroscopically stable. Passive surfaces co-exist in close contact with electrolytes, undergoing a continuous process of partial dissolution and reprecipitation from the microscopic viewpoint.[50] In this sense, the surface composition constantly changes according to the environment. Table 2 represents compositional changes of surface oxide films on metallic biomaterials.

Table 2
Change in the Composition of the Surface Oxide on Metals Used for Medical Devices in Living Body and Simulated Body Fluids.

Metal	Changing matter	Ref.
CP-Ti	When titanium which has been surgically implanted into the human jaw is characterized using AES, its surface oxide film reveals constituents of calcium, phosphorus and sulfur. By immersing titanium and its alloys in Hanks' solution (Fig. 9) and other solutions,[3-5,15] preferential adsorption of phosphate ions occurs — leading to formation of calcium phosphate on their surfaces.[16]	52-62
316L stainless steel	In pins and wires made of 316L austenitic stainless steel, calcium and phosphorus are present in the surface oxide. Chromium combined with sulfur, and/or iron combined with phosphorus (where the latter contains calcium and chlorine). For specimens immersed in Hanks' solution and in cell culture medium, as well as incubated with culture cells, calcium phosphate was formed. Sulfate was adsorbed by the surface oxide film and reduced to sulfite in cell culture medium and with the cultured cells. The results in this study suggest that nickel and manganese are depleted in the oxide film. The surface oxide changes into iron and chromium oxides, where a small amount of molybdenum oxide will be present in the human body.	63-65
Co-Cr-Mo ASTM F799-95	After dissolution, the surface oxide consisted of chromium oxide (Cr^{3+}) which contained molybdenum oxide (Mo^{4+}, Mo^{5+} and Mo^{6+}). Results from angle-resolved XPS revealed that chromium and molybdenum were more widely distributed in the inner layer than in the outer layer of the oxide film. In body fluids, cobalt is completely dissolved. The surface oxide changes into chromium oxide containing a small amount of molybdenum oxide. Calcium phosphate is also formed on the top surface.	66
Zr Ti-Zr	Only phosphate, without calcium, is formed on Ti-Zr alloys that contain over 50 wt.% Zr.[5]	67, 68

(*i*) *Titanium*

The film on Ti consists of amorphous or low-crystalline and non-stoichiometric TiO_2.[50] The surface oxide film on Ti just after polishing in water contains not only Ti^{4+} but also Ti^{3+} and Ti^{2+}.[50,51] When Ti that has been surgically implanted into the human jaw is characterized by Auger electron spectroscopy (AES), its surface oxide film reveals three constituents, namely, calcium, phosphorus and sulfur.[52,53] Calcium phosphates are formed on Ti and Ti alloys by immersion in Hanks' solution and other simulated body fluids and by culturing cells on it.[52,54-62] The above phenomena are characteristic of Ti and Ti alloys.[54,57] Sulfite or sulfide is also formed on titanium during cell culturing on it.[60] The ability of Ti to form calcium phosphate on itself is one of the reasons for its better hard-tissue compatibility than those of other metals.

(*ii*) *Stainless Steel*

In pins and wires made of 316L austenitic stainless steel, calcium and phosphorus are present in the surface oxide.[63] The corrosion product of 316L steel implanted in the femur as part of an artificial hip joint consists of Cr combined with S, and/or Fe combined with P (where the latter contains Ca and Cl).[64]

For 316L steel polished in deionized water, the surface oxide film consisted of iron and chromium oxides that contained small amount of nickel, molybdenum and manganese oxides.[65] The surface oxide also contained a large amount of OH^-. For specimens immersed in Hanks' solution and in cell culture medium, as well as incubated with culture cells, calcium phosphate was formed. Sulfate was adsorbed onto the surface oxide film and reduced to sulfite in cell culture medium and with the cultured cells. The results in this study suggest that Ni and Mn are depleted in the oxide film. The surface oxide changes into iron and chromium oxides, where a small amount of molybdenum oxide will be present in the human body.

(*iii*) *Cobalt-Chromium-Molybdenum Alloy*

In the case of Co-Cr-Mo alloy, Co was dissolved during immersion in Hanks' solution and in cell culture medium, as well as during incubation in a cell culture.[66] After dissolution, the surface oxide consisted of chromium oxide (Cr^{3+}) which contained molybdenum oxide (Mo^{4+}, Mo^{5+} and Mo^{6+}). Results from angle-resolved X-ray photoelectron spectroscopy (XPS) revealed that Cr and Mo were more widely distributed in the inner layer than in the outer layer of the oxide film.

In body fluids, Co is completely dissolved. The surface oxide changes into chromium oxide containing a small amount of molybdenum oxide. Calcium phosphate is also formed on the top surface.

(*iv*) *Zirconium*

Zr does not form calcium phosphate in Hanks' solution, but it forms zirconium phosphate.[67,68] When Zr is added to titanium as an alloying element, the ability of calcium phosphate formation on titanium is reduced, and only phosphate without calcium is formed on Ti-Zr alloys that contain more than 50 wt.% Zr.[67] This property of Zr could be applied to internal fixations used in bone marrow, such as bone screws and bone nails, to avoid assimilation with bone,[69] because bone sometimes refractures when devices consisting of titanium alloy are retrieved.[70,71]

2. Difference between Calcium Phosphate Formation on Titanium and Zirconium

The most important question is the mechanism responsible for the difference in calcium phosphate formation on Ti and Zr. In the case of alkali-heat-treated Ti, calcium and phosphate are orderly deposited, and calcium deposition is the pre-requisite for phosphate deposition.[72] In this regard, chemical precipitation of calcium salts can occur on 316L stainless steel, as reported for steel in sea water with a cathodic protection system. The precipitation of mineral scales rapidly occurs when the steels are cathodically polarized because the cathodic reaction products of OH^- increase the pH at a suitable level for the precipitation of calcareous deposits.[73,74]

Therefore, the cathodic reaction is one of the key factors for the elucidation of this mechanism.

Ti and Zr were immersed in Hanks' solution, with or without calcium and phosphate ions, and the surfaces were characterized with XPS. The compositions of solutions used for immersion and immersion stages are listed in Table 3:

(1) not immersed in any electrolyte,
(2) immersed in Hanks' for 1 d,
(3) immersed in Ca-free for 1 d,
(4) first immersed in Ca-free and then immersed in P-free for 1 d,
(5) immersed in P-free for 1 d, and
(6) first immersed in P-free and then immersed in Ca-free for 1 d.

They became thick after one day immersion for the detection of their whole region by XPS. After the above processes, the specimens were rinsed with deionized water to remove chemical species that were physically adsorbed but not chemically incorporated in the surface oxide film. After rinsing, the specimens were immediately stored in an auto-dried desiccator, before XPS analysis.

Table 3
Compositions and pH Values of Electrolytes Used in this Study.

Ion	Concentration (mol L^{-1})			
	0.9% NaCl	Hanks'	Ca-free Hanks'	P-free Hanks'
Na^+	1.54×10^1	1.42×10^1	1.42×10^1	1.41×10^1
K^+	---	5.81×10^3	5.81×10^3	5.37×10^3
Mg^{2+}	---	8.11×10^4	8.11×10^4	8.11×10^4
Ca^{2+}	---	1.26×10^3	---	1.26×10^3
Cl^-	1.54×10^1	1.45×10^1	1.42×10^1	1.45×10^1
PO_4^{3-}	---	7.78×10^4	7.78×10^4	---
SO_4^{2-}	---	8.11×10^4	8.11×10^4	8.11×10^4
CO_3^{2-}	---	4.17×10^3	4.17×10^3	4.17×10^3
pH	5.4	7.4	7.2	8.1

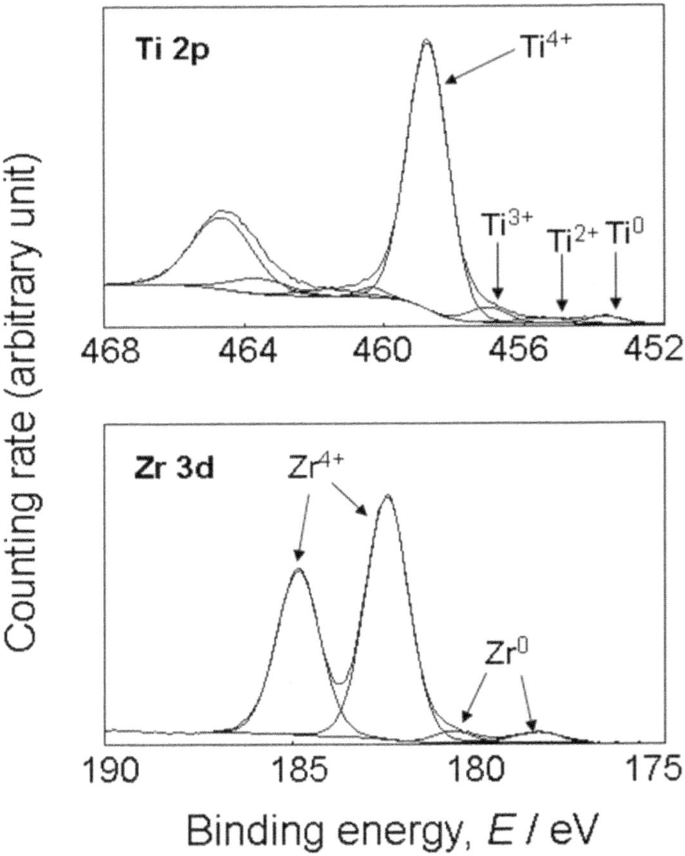

Figure 15. XPS spectra of the Ti 2p and Zr 3d electron energy regions.

Only titanium and oxygen for Ti and only zirconium and oxygen for Zr were detected with XPS before immersion. The corresponding spectra of the Ti 2p and Zr 3d electron energy regions are shown in Fig. 15. The spectrum of the Ti 2p region gave four doublets according to the valence, Ti^0, Ti^{2+}, Ti^{3+}, and Ti^{4+}, while that of Zr 3d gave only two, Zr^{4+} and Zr^0. The decomposition spectrum

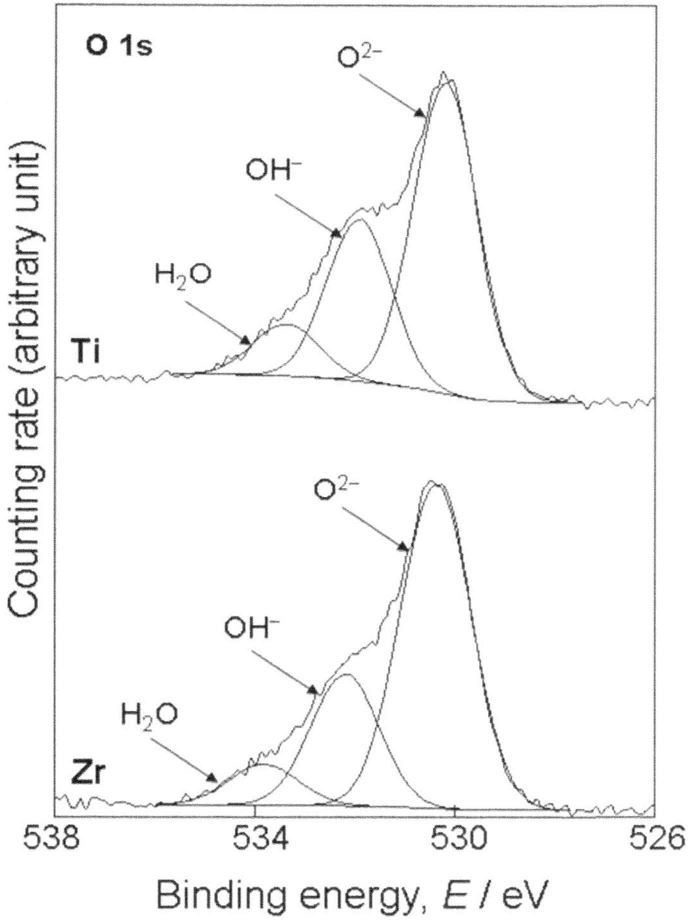

Figure 16. XPS spectra of the O 1s electron energy regions.

shows that Ti^{2+} oxide exists inside the surface oxide layer; however, Ti^{2+} formation is always thermodynamically less favorable than Ti^{3+} formation at the surface. The corresponding spectra of the O 1s electron energy regions from Ti and Zr are shown in Fig. 16. The XPS spectrum of the O 1s region contained three peaks originating from oxide, O^{2-}, hydroxide or hydroxyl groups, OH^-, and hydrate and/or adsorbed water, H_2O.[75] The compositions and thicknesses of surface oxide films on Ti and Zr are listed in Table 4. The table also contains the proportions of oxygen in each chemical state. The surface oxide film on Zr contains only Zr^{4+}, and that on Ti contains Ti^{2+} and Ti^{3+} as well as Ti^{4+}. In other words, the surface oxide film on Zr is stoichiometrically oxidized and stable, while that on Ti is not stoichiometric and still has room for oxidation. Therefore, the surface oxide film on Zr is more stable than that on Ti. Moreover, less OH^- is contained in surface oxide film on Zr than on Ti, indicating that the film on Zr is less hydrated. The dehydration proceeds inside the surface film, while the dehydration process takes place only partly for Ti^{4+} oxide. Since a considerable portion of oxidized Ti stays as Ti^{2+} and Ti^{3+} in the surface film, the oxidation process may proceed to the end just at the uppermost part of the surface film.[76]

The phenomena involved in the formation of the calcium and/or phosphate layer are schematically summarized in Fig. 17. In the case of Ti, calcium phosphate was formed when Ti was immersed in Hanks' solution. Calcium was incorporated onto Ti by immersion in P-free solution; calcium did not form $CaTiO_3$, but formed CaO or $Ca(OH)_2$ because the binding energy of Ca $2p_{3/2}$ from this specimen was 347.6 eV, which is similar to that of CaO or $Ca(OH)_2$, while that of $CaTiO_3$ is 346.6 eV. Then, calcium phosphate was formed on it by immersion in Ca-free solution. On the other hand, phosphate was formed on Ti by immersion in Ca-free solution, but was completely substituted with calcium after immersion in P-free solution. This indicates that calcium or phos-

Table 4
Composition, Proportion of Chemical States of Oxygen, and Thickness of the Surface Oxide Film on Ti and Zr, as Estimated by XPS

Specimen	Concentrations (at.%)			Proportion of chemical states of oxygen			d (nm)
	Ti	Zr	O	O^{2-}	OH^-	H_2O	
Ti	20.4	---	79.4	0.59	0.31	0.10	5.2
Zr	---	25.8	79.2	0.66	0.26	0.08	4.3

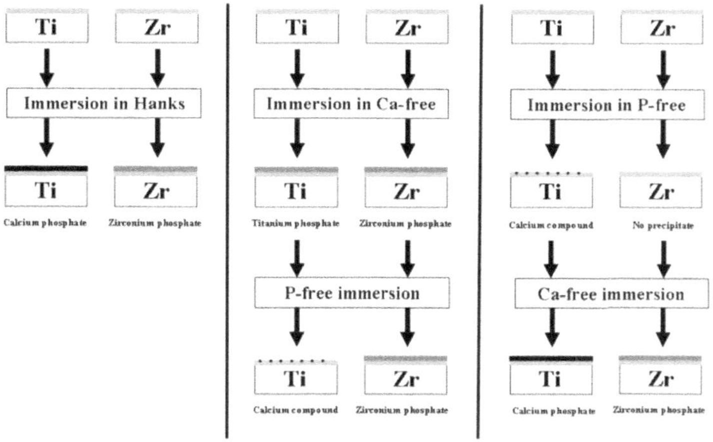

Figure 17. Illustrations of changes in surface layers on Ti and Zr immersed in each electrolyte, as determined by XPS.

phate could not exist stably alone on Ti, and eventually formed calcium phosphate on it in solutions containing both calcium and phosphate ions. In the case of Zr, zirconium phosphate was formed in Hanks' solution, and calcium was not incorporated. Calcium was also never incorporated by immersion in P-free solution, and the original surface oxide film remained. On the other hand, phosphate was formed on Zr by immersion in Ca-free solution, and this phosphate film remained even after immersion in P-free solution. Therefore, Zr does not react with calcium because phosphate is more stable with Zr, so that zirconium phosphate is formed.

(*i*) *Cathodic Polarization*

Ti and Zr were cathodically polarized in Hanks' solution, with or without calcium and phosphate ions. Figure 18 shows the cathodic polarization curves of Ti, Zr and Au in 0.9% NaCl. Au showed a common shape of polarization curve in the potential-current density diagram. The plateau region that indicates cathodic current density was independent of potential change and

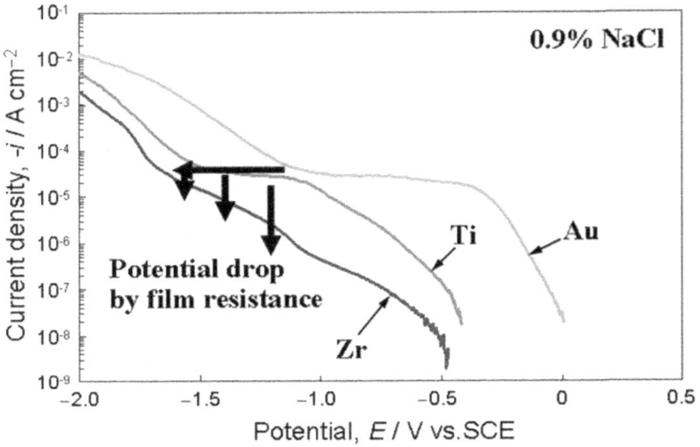

Figure 18. Cathodic polarization curves of Ti, Zr and Au in 0.9% NaCl.

was observed around –0.4 to –1.0 V_{SCE}. The primary electrochemical reaction that occurred at this potential was reduction of dissolved oxygen. The rate-limiting step of this region was diffusion of the oxygen from the bulk solution to the electrode surface. The constant current density is often observed in a neutral solution and is so-called limiting current density, i_L; it is commonly about –20 μA cm^{-2}. The i_L value could be translated as indicator of barrier effect on the surface layer that prevents mass transport. From the results for Au, there was no barrier layer on Au obstructing the diffusion of oxygen, i.e., there was no oxide or hydroxide layer on it.

The polarization curve of Ti showed a similar tendency to that of Au – the curve had flat region, and i_L corresponded to the value of Au. On the other hand, the cathodic current density for Zr was much smaller than that for Ti and did not show the obvious i_L of dissolved oxygen. In other words, the density (or compactness) of the passive film on Zr was much higher than that for Ti. From the viewpoint of the mixed-potential theory, the suppression of a cathodic reaction with similar values of rest potentials means that the anodic reaction of Zr could also be suppressed. That is, the surface oxide film on Zr is electrochemically stronger and more protective against metal dissolution than that on Ti. This difference between the surface oxide films on Ti and Zr may be caused by their compositions, as determined with XPS (see Fig. 15 and Table 4).

Figure 19 shows polarization curves of Ti and Zr in Hanks' and P-free Hanks' solutions. The cathodic reactions in the case of Ti and Zr were accelerated in P-free solution compared to the common Hanks' solution. Therefore, the calcium compounds that formed on Ti and Zr were unstable in P-free solution. This indicated that the presence of the phosphate ion was important for the protectiveness and reactivity of both metals. In fact, calcium phosphate on Ti and zirconium phosphate on Zr were formed after immersion in Ca-free solution to stabilize chemically their surfaces.

The above results revealed that the calcium phosphate layer formed on Ti in Hanks' solution is relatively protective against mass transport throughout the layer. However, the zirconium phosphate layer formed on Zr is much more protective and stable than that on Ti. Therefore, calcium ions were not incorporated. The difference in calcium phosphate formation on Ti and Zr is thus attributed to the reactivity of the surface oxide films.

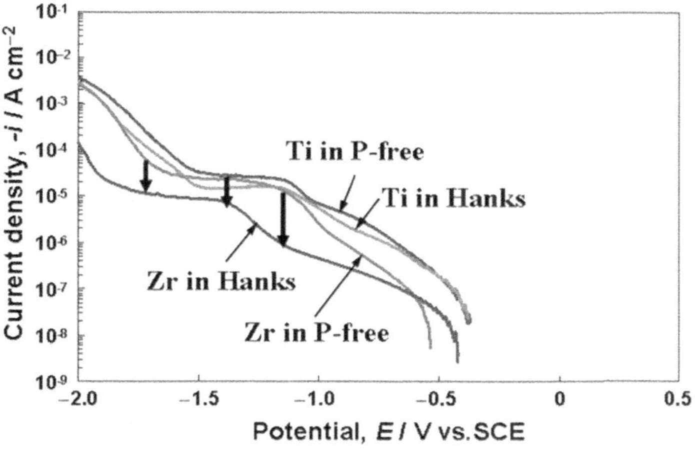

Figure 19. Cathodic polarization curves of Ti and Zr in Hanks and P-free Hanks solutions.

(ii) Characteristics of the Surface Oxide Film

Surface oxide films as passive films on Ti and Zr are amorphous and non-stoichiometric oxides containing a small amount of thin crystals,[50] because the formation of passive films is rapid while time-consuming crystallization does not occur. This amorphous structure is effective for electrochemical protection because of the absence of grain boundaries. In addition, the surface oxide film is very thin (approximately 2–5 nm) and its properties are influenced by the metal substrate under it.[50,57,60] Therefore, the surface oxide films are different from crystallized titanium oxide and zirconium oxide bulk ceramics with regard to those chemical properties. As a result, the surface oxide film on a substrate metal exhibits unique properties, according to the composition, structure, etc.

The surface oxide film on Ti is not completely oxidized and is relatively reactive, while that on Zr is stably oxidized; that on Zr is more passive and protective than that on Ti. Neither calcium nor phosphate stably exists alone on Ti, and calcium phosphate is formed in biological environments; calcium phosphate formed on Ti is stable and protective. On the other hand, calcium is never incorporated onto Zr, while phosphate zirconium formed on Zr is highly stable and forms a protective layer; therefore, no calcium reacts with the layer.

VII. ELECTROCHEMICAL MEASUREMENTS

Various electrochemical measurements were performed on titanium with and without culturing murine fibroblasts L929 in order to characterize the effects of cells on the interfacial electrochemical properties between titanium and cells. To this aim, an originally developed electrochemical cell [77] was used, as shown in Fig. 20. The open-circuit potential of titanium was decreased in the presence of L929 cells, which was caused by the shift of equilibrium potential between cathodic and anodic reactions indicated by the decrease in the cathodic current density in the presence of L929 cells. In the cathodic potential step test, the decrease in current density following the peak current density was delayed due to L929 cells, indicating that the diffusivity of molecules and ions

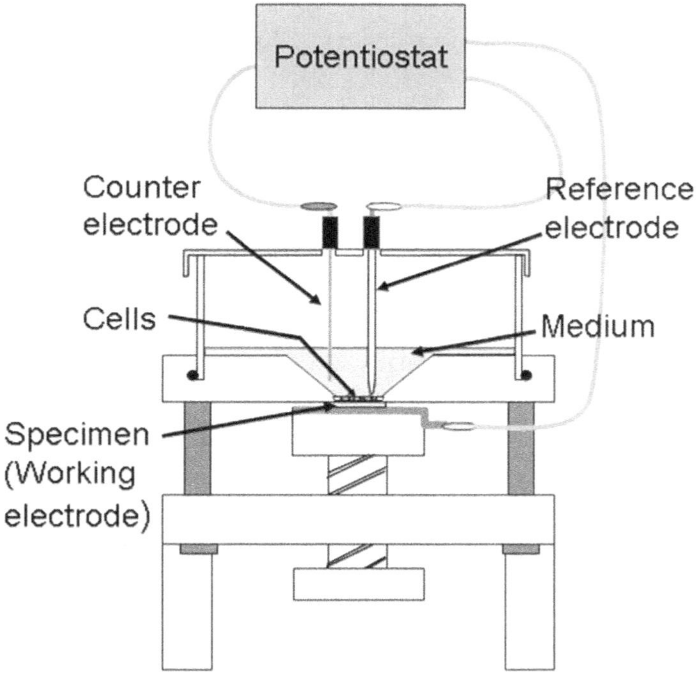

Figure 20. Electrochemical cell for measurement under cell culture.

decreased in the presence of cells. In addition, alternating current impedance measurement and data approximation to the electrical equivalent circuit model revealed that the circuit element for diffusion resistance of biomolecule adsorption layer increased in the presence of L929 cells. Consequently, the effect of cells on the interface property is the retardation of diffusion through the biomolecule adsorption layer due to the increase in biomolecule density with extracellular matrix consisting with proteins and glycosaminoglycans generated by the cells.

In addition, potentiodynamic polarization and impedance tests were carried out on 316L stainless steel with culturing murine fibroblast L929 cells, to elucidate the corrosion behaviour of 316L steel in the presence of L929 cells and to understand the electrochemical interface between 316L steel and cells, respectively.[78] Potential step test was carried out on 316L steel with type I collagen coating and culturing L929 cells to compare the effects of collagen and L929 cells. The open-circuit potential of 316L steel slightly shifted in the negative direction and the passive current density increased due to cells, indicating a decrease in the protective ability of passive oxide film. The pitting potential decreased in the presence of cells, indicating a decrease in the pitting corrosion resistance. In addition, a decrease in diffusivity at the interface was indicated from the decrease in the cathodic current density and the increase in the diffusion resistance parameter in the impedance test. The anodic peak current in the potential step test decreased in the presence of cells and collagen. Consequently, the corrosion resistance of 316L steel decreased in the presence of L929 cells. In addition, collagen coating would provide an environment for an anodic reaction similar to that with culturing cells.

VIII. QUARTZ CRYSTAL MICROBALANCE (QCM)

Quartz crystal microbalance (QCM) is an excellent tool to monitor the adsorption of molecules onto metal surfaces. In addition, the combination of QCM with electrochemical analysis (EQCM) is a powerful tool to elucidate the interaction between biomolecules and metals.

The adsorption of the simplest amino acid, glycine, on gold and titanium surfaces in Hanks' solution was monitored using a QCM.[79] The changes in mass and open-circuit potential at rest potential and in mass and current at constant potentials were measured before and after the injection of glycine. Results showed a momentary adsorption of glycine, followed by the desorption of part of the adsorbed glycine. The adsorbed amount of glycine increased with the increase of glycine concentration. The time before stabilizing the adsorption-desorption phenomena also increased with the concentration. More glycine was adsorbed on titanium than on gold because of the difference in the surface electric charges, the concentration of electric charges on titanium is much larger than that on gold.

On the other hand, the adsorption of Bovine Serum Albumin (BSA) on gold and titanium surfaces in Hanks' solution was monitored using an electrochemical quartz crystal microbalance (EQCM).[80] The changes in mass and open-circuit potential at rest potential and in mass and current at constant potentials were measured before and after the introduction of BSA. The mass of BSA adsorbed onto titanium reached a steady value within one hour, while that of the albumin adsorbed on gold continued to increase. The amount of BSA adsorbed on titanium increased with concentration and stabilized at 0.8 mg/mL while that on gold did not stabilize. This indicates a monolayer of the protein on titanium and multi-layer on gold. The time before stabilizing the adsorption-desorption phenomena also increased with the BSA concentration.

REFERENCES

[1]S. Ban and S. Maruno, *Biomaterials* **19** (1998) 1245.
[2]S. Bodhak, S. Bose, and A. Bandyopadhyay, *Acta Biomater.* **6** (2010) 641.
[3]A. Yuda, S. Ban, and Y. Izumi, *Dent. Mater. J.* **24** (2005) 588.
[4]K. Kuroda, M. Moriyama, R. Ichino, M. Okido, and A. Seki, *Mater. Trans.* **49** (2008) 1434.
[5]M. Hosaka, Y. Shibata, and T. Miyazaki, *J. Biomed. Mater. Res.* **78B** (2006) 237.
[6]Y. Tanaka, E. Kobayashi, S. Hiromoto, K. Asami, H. Imai, and T. Hanawa, *J. Mater. Sci. Mater. Med.* **18** (2007) 797.
[7]T. Hayakawa, M. Kawasaki, and G. H. Takaoka, *J. Ceram. Soc. Jpn.* **116** (2008) 68.
[8]R. Narayanan, S. K. Seshadri, T. Y. Kwon, and K. H. Kim, *Scr. Mater.* **56** (2007) 229.
[9]R. Narayanan, T. Y. Kwon, and K. H. Kim, *J. Biomed. Mater. Res.* **85B** (2008) 231.
[10]R. Narayanan, T. Y. Kwon, and K. H. Kim, *Mater. Sci. Eng.* **C28** (2008) 1265.
[11]R. Narayanan, S. Y. Kim, T. Y. Kwon, and K. H. Kim, *J. Biomed. Mater. Res.* **87A** (2008) 1053.
[12]X. W. Meng, T. Y. Kwon, and K. H. Kim, *Dent. Mater. J.* **27** (2008): 666.
[13]X. I. Zhu, K. H. Kim, and Y. S. Jeong, *Biomaterials* **22** (2001) 2199.
[14]R. Narayanan, J. Y. Ha, T. Y. Kwon, K. H. Kim, *Metal. Mater. Trans.* **39B** (2008) 493.
[15]J. M. Jang, S. J. Park, G. S. Choi, T. Y. Kwon, and K. H. Kim, *Metal. Mater. Int.* **14** (2008) 457.
[16]H. Ishizawa and M. Ogino, *J. Biomed. Mater. Res.* **29** (1995) 1071.
[17]Y. Han, S. H. Hong, and K. W. Xu, *Surf.Coat. Technol.* **168** (2003) 249.
[18]F. Liu, F. P. Wang, T. Shimizu, K. Igarashi, and L. C. Zhao, *Surf. Coat. Technol.* **199** (2005) 220.
[19]W. Ma, J. H. Wei, Y. Z. Li, X. M. Wang, H. Y. Shi, and S. Tsutsumi, *J. Biomed. Mater. Res.* **86B** (2008) 162.
[20]Y. Han, D. Chen, J. Sun, Y. Zhang, and K. Xu, *Acta Biomater.* **4** (2008) 1518.

[21] Y. Yan and Y. Han, *Surf. Coat. Technol.* **201** (2007) 5692.
[22] Y. Han, Y. Yan, and C. Lu, *Thin Solid Film.* **517** (2009) 1577.
[23] S. Offermanns, *Circ. Res.* **99** (2006) 1293.
[24] H. Zhang, M. Ahmad, G. Gronowicz, *Biomaterials* **24** (2003) 2013.
[25] Y. Tanaka, H. Doi, Y. Iwasaki, S. Hiromoto, T. Yoneyama, K. Asami, and T. Hanawa, *Mater. Sci. Eng.* **C27** (2007) 206.
[26] Y. Tanaka, H. Doi, E. Kobayashi, T. Yoneyama, and T. Hanawa, *Mater. Trans.* **48** (2007) 287.
[27] Y. Tanaka, Y. Matsuo, T. Komiya, Y. Tsutsumi, H. Doi, T. Yoneyama, and T. Hanawa, *J. Biomed. Mater. Res.* **92A** (2010) 350.
[28] Y. Tanaka, H. Saito, Y. Tsutsumi, H. Doi, H. Imai, and T. Hanawa, *Mater.. Trans.* **49** (2008) 805.
[29] J. E. Davies, B. Lowenberg, and A. Shiga, *J. Biomed. Mater. Res.* **24** (1990) 1289.
[30] Rezania, C. H. Thomas, A. B. Branger, C. M. Waters, and K. E. Healy, J. *Biomed. Mater. Res.* **37** (1997) 9.
[31] M. D. Pierschbacher and E. Ruoslahti, *Nature* **309** (1984) 30.
[32] H. Schliephake, D. Scharnweber, M. Dard, S. Rössler, and A. Sewing, Clin. Oral Implant Res. 13 (2002) 312.
[33] S. J. Xiao, M. Textor, N. D. Spencer, and H. Sigrist, *Langmuir* **114** (1998) 5507.
[34] S. J. Xiao, M. Textor, N. D. Spencer, M. Wieland, B. Keller, and H. Sigrist., *J Mater. Sci. Mater. Med.* **8** (1997) 867.
[35] Rezania, R. Johnson, A. R. Lefkow ,and K. E. Healy, *Langmuir* **15** (1999) 6931.
[36] Y. Tanaka, H. Saito, Y. Tsutsumi, H. Doi, N. Nomura, H. Imai, T. Hanawa, *J. Colloid. Interface Sci.* **330** (2009) 138.
[37] K. Oya, Y. Tanaka, H. Saito, K. Kurashima, K. Nogi, H. Tsutsumi, Y. Tsutsumi, H. Doi, N. Nomura, and T. Hanawa, *Biomaterials* **30** (2009) 1281.
[38] K. M. Speck and A. C. Fraker, *J. Dent. Res.* **59** (1980) 1590.
[39] L. S. Castleman, S.M. Motzkin, F. P. Alicandri, V.L Bonawit, and A. A. Johnson, *J. Biomed. Mater. Res.* **10** (1976) 695.
[40] F. Miura, M. Mogi, Y. Ohura and H. Hamanaka, *Am. J. Orthod. Dentofacial Orthop.* **90** (1986) 1.
[41] H. M. Walia, W. A. Brantley and H. Gerstein, *J. Endodont.* **14** (1988) 346.
[42] Roguin, E. Grenadier, S. Linn, W. Markiewicz and R. Beyar, *Am. Heart J.* **138** (1999) 326.
[43] D. Rosenthal, M. B. Herring, T. G. Odonovan, D. F. Cikrit, A. I. Comerota, and J. D. Corson, *J. Vasc. Surg.* **16** (1992) 453.
[44] T. Yoneyama, H. Doi, E. Kobayashi and H. Hamanaka, M. Yamamoto, T. and Kuroda, *J. Biomed. Mater. Res.* **27** (1993) 399.
[45] C. Lidén, T. Menné, and D. Burrows, *Brit. J. Dermatol.* **134** (1996) 193.
[46] Schnuch, J. Geier, W. Uter, P. J. Frosch, W. Lehmacher, W. Aberer, M. Agathos, R. Arnold, T. Fuchs, B. Laubstein, G. Lischka, P. M. Pietrzyk, J. Rakoski, G. Richter, and F. Rueff. *Contact Dermatitis* **37** (1997) 200.
[47] G.C. McKay, R. Macnair, C. MacDonald, and M. H. Grant, *Biomaterials* **17** (1996) 1339.
[48] E. Denkhaus, and K. Salnikow, *Crit. Rev. Oncol. Hemat.* **42** (2002) 35.
[49] O. Fukushima, T. Yoneyama, H. Doi, and T. Hanawa, *Dent. Mater. J.* **25** (2006) 151
[50] E. J. Kelly, *Modern Aspects of Electrochemistry,* Vol. 14, Springer, Berlin, Germany, 1982, p. 319.
[51] T. R. Beck, *Electrochem. Acta* **18** (1973) 807.

[52] P. Li, and P. Ducheyne, *J. Biomed. Mater. Res.* **41** (1998) 341.
[53] L. Frauchiger, M. Taborelli, B. O. Aronsson and P. Descouts, *Appl. Surf. Sci.* 143 (1999) 67.
[54] T. Hanawa, in *The Bone-Biomaterial Interface*, Ed. By in: J. E. Davies, University of Toronto Press, Toronto, 1991, p. 49.
[55] T. Hanawa and M. Ota, *Biomaterials* **12** (1991) 767.
[56] D. J. Wever, A. G. Veldhuizen, J. deVries, H. J. Busscher, D. R. A. Uges and J. R. van Horn, *Biomaterials* **19** (1998) 761.
[57] T. Hanawa and M. Ota, *Appl. Surf. Sci.* **55** (1992) 269.
[58] T. Hanawa, O. Okuno, and H. Hamanaka, *J. Jpn. Inst. Met.* **56** (1992) 1168.
[59] J. L. Ong, L. C. Lucas, G. N. Raikar, R. Connatser, and J. C. Gregory, *J. Mater. Sci. Mater. Med,* **6** (1995) 113.
[60] S. Hiromoto, T. Hanawa, and K. Asami, *Biomaterials* **25** (2004) 979.
[61] K. E. Healy and P. D. Ducheyne, *Biomaterials* **13** (1992) 553.
[62] P. Serro, A. C. Fernandes, B. Saramago, J. Lima, and M. A. Barbosa, *Biomaterials* **18** (1997) 963.
[63] J. E. Sundgren, P. Bodo, I. Lundstrom, A. Berggren, and S. Hellem, *J. Biomed. Mater. Res.* 19 (1985) 663.
[64] J. Walczak, F. Shahgaldi, and F. Heatley, *Biomaterials* 19 (1998) 229.
[65] T. Hanawa, S. Hiromoto, A. Yamamoto, D. Kuroda, and K. Asami, *Mater. Trans.* **43** (2002) 3088.
[66] T. Hanawa, S. Hiromoto, and K. Asami, *Appl. Surf. Sci.* **183** (2001) 68.
[67] T. Hanawa, O. Okuno, and H. Hamanaka, *J. Jpn. Inst. Met.* **56** (1992) 1168.
[68] T. Hanawa, S. Hiromoto, K. Asami, O. Okuno, and K. Asaoka, *Mater. Trans.* **43** (2002) 3000.
[69] E. Kobayashi, M. Ando, Y. Tsutsumi, H. Doi, T. Yoneyama, M. Kobayashi, and T. Hanawa, Mater. Trans. 48 (2007) 301.
[70] P. L. Sanderson, W. Ryan, and P. G. Turner, *Injury* **23** (1992) 29.
[71] D. A. Labosky, M. B. Cermak, and C. A. Waggy, *J. Hand. Surg.* **15A** (1990) 294.
[72] B. C. Yang, J. Weng, X. D. Li, and X. D. Zhang, *J. Biomed. Mater. Res.* 47 (1999) 213.
[73] J. S. Peaker and J. T. Czernuszka, *Thin Solid Film.* **287** (1996) 174.
[74] Nevile, T. Hodgkiess, and A. P. Morizot, *British Corros. J.* **32** (1997) 277.
[75] K. Asami, S. C. Chen, H. Habazaki, and K. Hashimoto, *Corros. Sci.* **35** (1993) 43.
[76] T. Hanawa, K. Asami, and K. Asaoka, *J. Biomed. Mater. Res.* **40** (1998) 530.
[77] S. Hiromoto, K. Noda, and T. Hanawa, *Electrochem. Act.* **48** (2002) 387.
[78] S. Hiromoto and T. Hanawa, *J. Royal Soc. Interface* **3** (2006) 495.
[79] J. A. Lori and T. Hanawa, *Corros. Sci.* **43** (2001) 2111.
[80] J. A. Lori and T. Hanawa, *Spectrosc. Int. J.* **18** (2004) 545.

Takao Hanawa

Dr. Takao Hanawa is a full professor at the Department of Metals, Institute of Biomaterials and Bioengineering, Tokyo Medical and Dental University (since July 2004) as well as at the Graduate School of Engineering, The University of Tokyo (since April 2009). He received a B.Sc. in Metallurgical Engineering from Hokkaido University in 1981, a D.D.Sc. from Hokkaido University in 1989, and a Ph.D. form Tohoku University in 1998. His professional experience includes a Deputy-Director-General position at the Biomaterials Research Center of the National Institute for Materials Science. His research interests include metallic biomaterials, dental materials, bio-functionalization of metals, and metal-tissue interactions. Dr. Hanawa has contributed nearly 250 journal papers and 37 book chapters. He has won many awards, including The Young Investigator award of the Japanese Society for Biomaterials (1995), The Minister Award of the Japanese Science and Technology Agency (1999), Award for Distinguished Service by Japan Institute of Metals (2001), Award for Developmental Engineering by Japan Institute of Metals (2004), and an award by the Japanese Society for Biomaterials (2009).

6

Bioactive Metals Prepared by Surface Modification: Preparation and Properties

Tadashi Kokubo and Seiji Yamaguchi

Department of Biomedical Sciences, College of Life and Health Sciences, Chubu University, 1200 Matsumoto-cho, Kasugai, 487-8501 Japan

1. INTRODUCTION

In general, synthetic materials implanted into bone defects are encapsulated by a fibrous tissue that isolates them from the surrounding bone, and hence do not fix to the surrounding bone well. Several types of ceramics based on calcium phosphate, such as Bioglass in the Na_2O–CaO–SiO_2–P_2O_5 system, glass-ceramic A-W in the MgO–CaO–SiO_2–P_2O_5 system, sintered hydroxyapatite, sintered β-tricalcium phosphate, and a combination of these two phases, have been found to bond to bone directly without forming any fibrous tissue around them.[1] They are called bioactive ceramics, and are already used clinically as important bone substitutes. However, they are brittle and have poor fracture toughness, and hence cannot be used under load-bearing conditions.

Metallic materials, such as stainless steel, Co–Cr alloys, and titanium (Ti) metal and its alloys, are used under load-bearing conditions. Among these, Ti metal and its alloys exhibit better compatibility with living tissues than other metals do. Ti metal and

N. Eliaz (ed.), *Applications of Electrochemistry and Nanotechnology in Biology and Medicine I*, Modern Aspects of Electrochemistry 52, DOI 10.1007/978-1-4614-0347-0_6,

its alloys can be in direct contact with living bone when implanted with a rough surface produced by grit blasting or acid etching[2], although they can be encapsulated by a fibrous tissue when implanted with a polished or abraded surface.[3] Therefore, in many cases, orthopaedic and dental implants are subjected to grit blasting and/or acid etching.[4] However, this direct contact does not imply the formation of a bond between the bone and the metal.

To provide the metal with a bone-bonding ability, hydroxyapatite is sometimes coated on the metal using a suitable method[5] such as plasma spraying, sputtering, sol-gel, or alternative soaking.[6] However, the resulting hydroxyapatite coating is not stable in a body environment for long periods, because of the weak bonding of the coated hydroxyapatite to the metallic substrate, the presence of strain at the interface between the hydroxyapatite coating and the metallic substrate, and the instability of the hydroxyapatite coating in an aqueous body environment.[5]

The development of bioactive metals that are able to bond to living bone is desired in orthopaedic and dental fields. The preparation of bioactive metals by surface modification and their properties are reviewed in this chapter.

II. REQUIREMENTS FOR BIOACTIVE MATERIALS

All the bioactive ceramics described above, except for β-tricalcium phosphate, form a bone-like apatite layer on their surfaces in the living body, and bond to living bone through this apatite layer. This indicates that any material that forms a bone-like apatite on its surface in the living body bonds to living bone through this apatite layer.[7]

It has been shown that the apatite formed on bioactive ceramics in the living body can be reproduced on their surfaces in an acellular simulated body fluid (SBF) with ion concentrations nearly equal to those of human blood plasma.[7] This means that any material that is able to form a bone-like apatite layer on its surface in an SBF can bond to living bone through this apatite layer formed on its surface in the living body.[7]

It has been found that even pure gels of SiO_2,[8] TiO_2,[9] ZrO_2,[10] Nb_2O_5,[11] and Ta_2O_5,[12] prepared using a sol-gel process, form bone-like apatite on their surfaces in an SBF, as shown in Fig. 1.

Apatite
Silica gel
10μm
(a)
Titania gel
Apatite
10μm
(b)
Zirconia gel
Apatite
10μm
(c)
Apatite
Niobium oxide gel
10μm
(d)
Tantalum oxide gel
Apatite
10μm
(e)

This means that some functional groups, such as Si–OH, Ti–OH, Zr–OH, Nb–OH, and Ta–OH, which are abundant on the surface of these gels, induce apatite nucleation in an SBF. Once the apatite nuclei form, they can grow spontaneously, because an SBF is supersaturated with respect to the apatite.[13] Both, –COOH[14] and $–SO_4H$[15] groups, have also been shown to be effective for apatite nucleation. This indicates that any type of ceramic, metal, or organic polymer can bond to living bone through the apatite layer formed on their surface, if some functional groups effective for apatite nucleation form on their surface.

III. BIOACTIVE TI METAL AND ITS ALLOYS

1. Surface Modification with Sodium Titanate

It is assumed that apatite formation on a titania gel in an SBF[9] is induced by the Ti–OH groups on its surface. Ti metal is generally covered with a thin passive titanium oxide layer. It is expected that if sodium ions are incorporated into the surface, the sodium titanate can form on its surface, and this can form Ti–OH groups on the surface in a body environment by exchanging Na^+ ions with H_3O^+ ions in the body fluid to induce apatite formation. Sodium titanate has been shown to form by simple chemical and heat treatments.[16,17]

If Ti metal is soaked in a 5 M NaOH solution at 60°C for a period of 24 h, then sodium and oxygen ions are incorporated into the surface of the Ti metal to a depth of 1 μm, as shown in Fig. 2.[18] A layer with a feather-like structure composed of sodium hydrogen

Figure 1. SEM photographs of surfaces of (a) SiO_2, (b) TiO_2, (c) ZrO_2, (d) Nb_2O_5 and (e) Ta_2O_5 gel after soaking in SBF for 14 days. Figs.(a) and (b) are reprinted from Ref. 9, Copyright (1994) with permission from John Wiley and Sons; Fig. (c) is reprinted from Ref.10, Copyright (2001) with permission from Wiley-Blackwell; Fig. (d) is reprinted from Ref.11, Copyright (2001) with permission from The Ceramic Society of Japan; and Fig. (e) is reprinted from Ref.12, Copyright (2001) with permission from Springer.

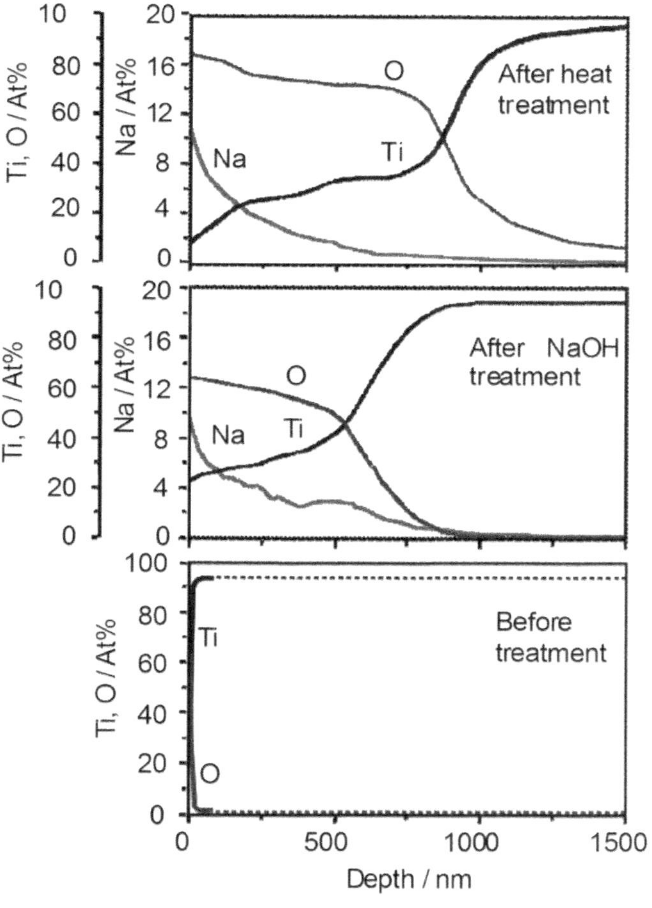

Figure 2. Auger electron spectroscopy depth profiles of surfaces of untreated Ti metals untreated and those subjected to NaOH and heat treatments.Reprinted from Ref. 18, Copyright (1999) with permission from John Wiley and Sons.

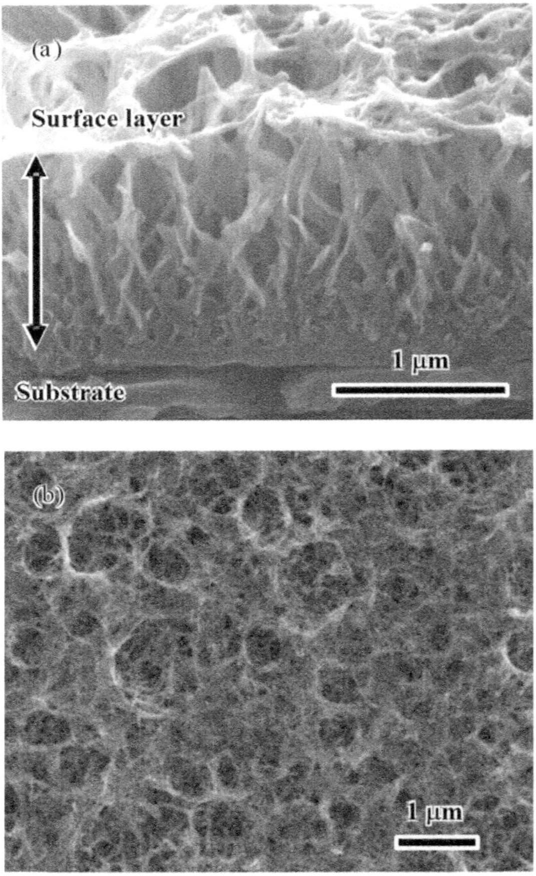

Figure 3. SEM photographs of (a) cross-section and (b) surface of Ti metals subjected to NaOH treatment. Reprinted from Ref. 20, copyright (2009) with permission from The Ceramic Society of Japan.

titanate ($Na_xH_{2-x}Ti_yO_{2y+1}$, $0 < x < 2$)[19] elongated perpendicular to the substrate is formed on the surface of the Ti metal by the NaOH treatment, as shown in Fig. 3(a).[20] This surface layer exhibits a fine network structure on the nanometre scale, as shown in Fig. 3(b).[20] Such a sodium hydrogen titanate layer forms apatite on its surface in

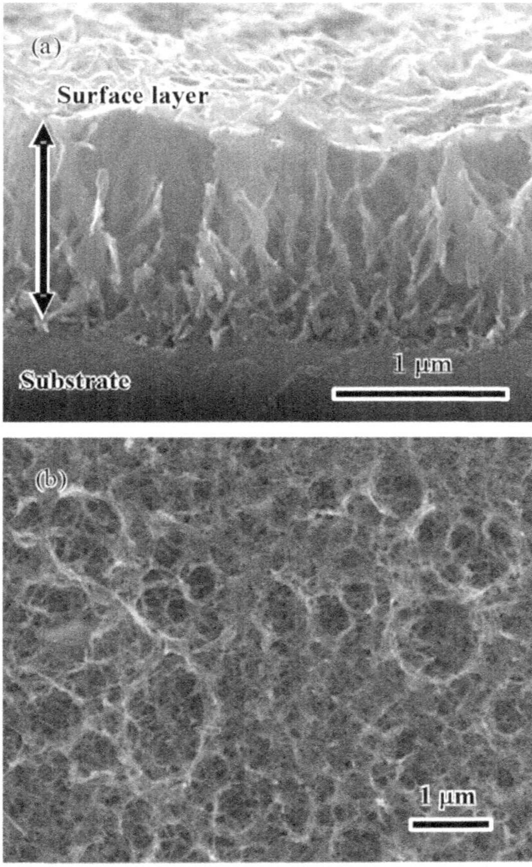

Figure 4. SEM photographs of (a) cross-section and (b) surface of Ti metals subjected to NaOH and heat treatments.Reprinted from Ref. 20, Copyright (2009) with permission from The Ceramic Society of Japan.

an SBF.[18, 19] However, it has a low apatite-forming ability and scratch resistance.[19]

If the NaOH-treated Ti metal is heat-treated at 600°C for a period of 1 h in air, then only oxygen ions penetrate deep into the Ti metal, as shown in Fig. 2,[18] and the sodium hydrogen titanate is transformed into sodium titanate ($Na_2Ti_yO_{2y+1}$, y = 5, 6...) and rutile[19] to give a dense feather-like structure, as shown in Fig. 4.[20] Structural changes of the surface of Ti metal due to the NaOH and heat treatments are schematically shown in Fig. 5. It should be noted that the surface layer consisting of the sodium titanate and rutile gradually changes, passing into the Ti metal substrate without exhibiting a distinct boundary between the materials. The heat-treated surface layer shows a scratch resistance as high as 50 mN.[19, 22, 23]

When the NaOH- and heat-treated Ti metal is soaked in an SBF, apatite begins to precipitate in the deep regions of the interspaces of the feather-like titanate phase, grows by filling these interspaces to integrate with the titanate layer to form a dense composite composed of apatite and titanate, and spreads over the entire surface, as shown in Fig. 6.[18] According to transmission electron microscopic (TEM) observations, the apatite precipitated on the NaOH- and heat-treated Ti metal in an SBF assumes a fine needle-like form with a composition having a Ca/P atomic ratio of Ca/P = 1.65, accompanied by a small amounts of Na and Mg, as shown in Fig. 7, similar to bone mineral.[24]

These NaOH and heat treatments are also effective for inducing an apatite-forming ability in an SBF for conventional Ti-based alloys, such as Ti–6Al–4V, Ti–6Al–2Nb–Ta, and Ti–15Mo–5Zr–3Al.[17] A graded surface structure, in which sodium titanate and rutile gradually change into titanium metal, is also formed on these alloys by an NaOH treatment and a heat treatment.[25,26] During the NaOH treatment, the alloying elements, such as Al, V, Mo, and Zr, are selectively released from the surface.

According to TEM observations,[24] X-ray photoelectron spectroscopy (XPS) data,[27] and zeta-potential measurements,[28] the mechanism of the formation of apatite on NaOH- and heat-treated Ti metal in an SBF can be interpreted in terms of the electrostatic interaction of the Ti metal with the ions in the SBF, as shown in Fig. 8. The sodium titanate on the surface of the Ti metal releases Na^+ ions via an exchange with the H_3O^+ ions in the SBF to form Ti–OH groups on the surface, as expected. However, these Ti–OH groups do not induce apatite nucleation directly, as the Ti–OH groups are negatively charged, because the pH of the surrounding fluid is increased by the release of Na^+ ions[29, 30], and hence, these initially only combine with the positively charged Ca^{2+} ions[29, 30] in the SBF to form calcium titanate. As the Ca^{2+} ions begin to accumulate, the

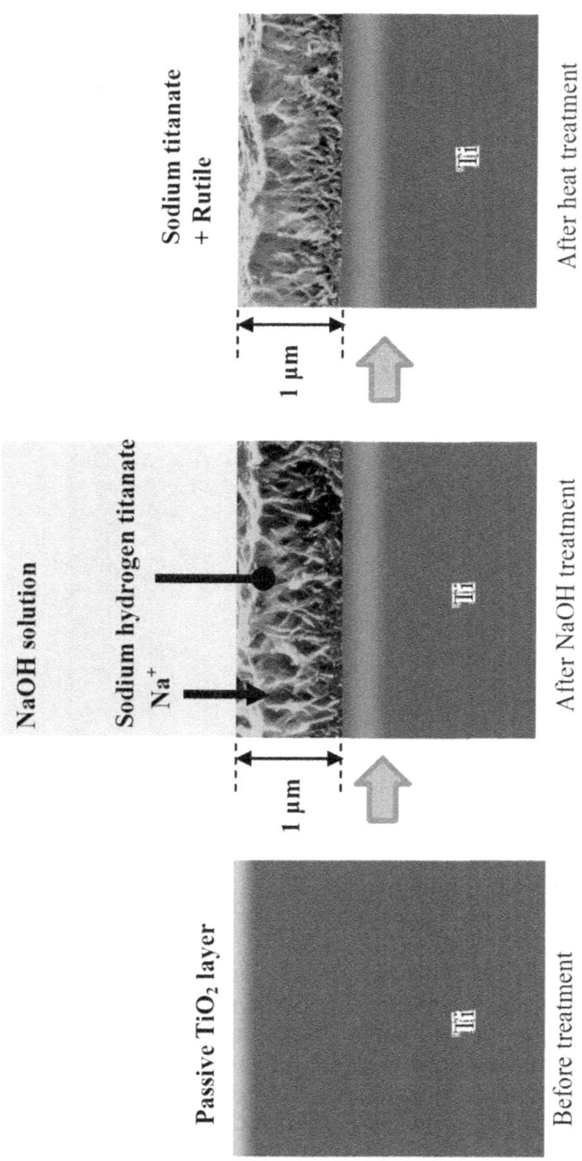

Figure 5. Structural change of the surface of Ti metal subjected to NaOH and heat treatments.

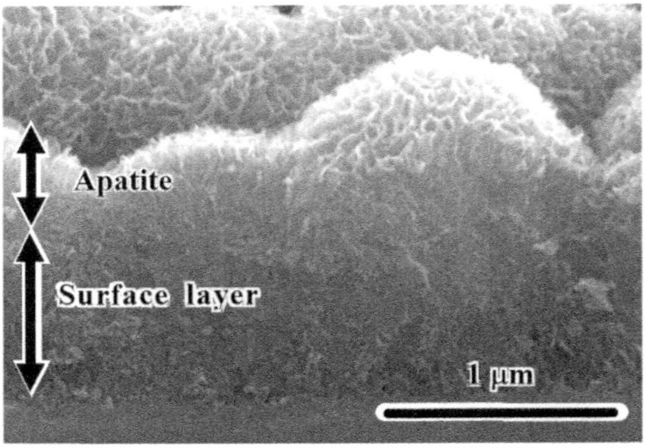

Figure 6. SEM photographs of cross-section and surface of NaOH- and heat-treated Ti metal after soaking in SBF for 1 day. Reprinted from Ref. 20, Copyright (2009) with permission from The Ceramic Society of Japan.

surface of the titanate becomes positively charged and then it combines with the negatively charged phosphate ions to form an amorphous calcium phosphate phase. The calcium phosphate formed is metastable, and it eventually transforms into a stable crystalline bone-like apatite layer. Conforto et al.[31] investigated the mechanism of apatite formation on Ti metal that was subjected only to an NaOH treatment using TEM observations, and confirmed this apatite-formation process. Takadama et al.[32,33] also reported a sim-

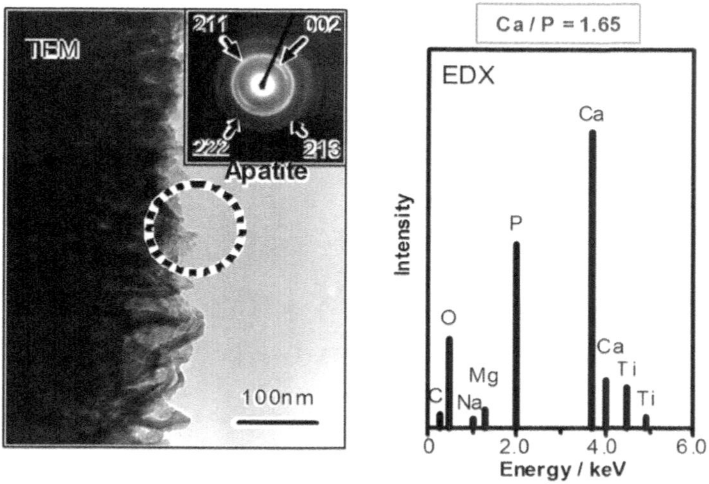

Figure 7. TEM photograph and EDX spectrum of the surface of the NaOH- and heat-treated Ti metal after soaking in SBF for 5 days (dotted circle: area of electron diffraction and EDX analysis). Reprinted from Ref. 24, Copyright (2001) with permission from John Wiley and Sons.

ilar mechanism of apatite formation on Ti–6Al–4V alloy subjected to NaOH and heat treatments.

According to Strnad et al.[34,35], Ti metal subjected to only an NaOH treatment shows uniform apatite formation in an SBF if it is subjected to a preliminary acid treatment. Jonášová et al.[36] showed that the apatite-forming ability of Ti metal subjected to an NaOH treatment is not decreased in an SBF, even if the sodium content of the surface layer is considerably decreased. However, the apatite-forming ability and scratch resistance of the NaOH-treated Ti metal were low before the heat treatment. Kawai et al.[19] showed that these properties are improved markedly by a subsequent heat treatment, that the apatite-forming ability was increased by a heat treatment, and that the samples were stable even in a humid environment if the Na^+ ions in the surface layer of the Ti metal were partially removed using a water treatment before the heat treatment. This is an important property for practical applications of such

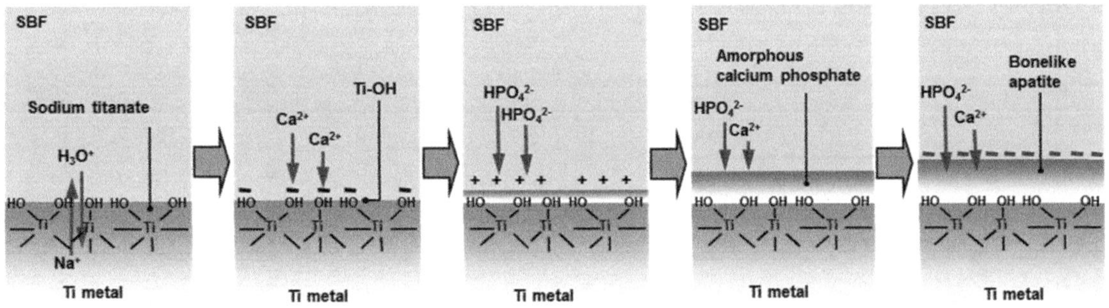

Figure 8. Process of apatite formation on NaOH- and heat-treated Ti metal in SBF. Reprinted from Ref. 28, Copyright (2003) with permission from John Wiley and Sons)

treated Ti metal, as implants are sometimes stored in a humid environment for long periods before implantation. Such NaOH- and heat-treated Ti metal is liable to exhibit a decrease in its apatite-forming ability in a humid environment, because the Na^+ ions of the sodium titanate would be slowly released via exchange with H_3O^+ ions in a humid environment. However, if the Na^+ ions in the upper layer of the surface of the Ti metal are removed by a water treatment before the heat treatment, the Ti metal would precipitate anatase on its top surface layer after a heat treatment, and such a decrease in apatite-forming ability in a humid environment would be prevented. Krupa et al. showed that the corrosion resistance of NaOH-treated Ti metal is also increased by a heat treatment.[37] Lakstein et al. reported that Ti-6Al-4V electrochemically deposited with hydroxyapatite on its surface after the NaOH treatment showed better bone formation than the same alloy coated with the hydroxyapatite by plasma spray method.[38]

Using the NaOH and heat treatments, the formation of an apatite layer can be induced easily and uniformly both on the flat surface of Ti metal and on the irregular inner surfaces of porous Ti metal.[39] During the fabrication of a porous body and the subsequent NaOH and heat treatments, oxygen ions are incorporated into the surface of the Ti metal. However, this oxygen contamination does have an adverse effect on the apatite-forming ability of the resulting porous body.[40]

Cell cultures on NaOH- and heat-treated Ti metal show that such treated Ti metal promotes the differentiation of osteoblasts,[41] as well as the formation of bone nodules[42] on its surface. These properties indicate that NaOH- and heat-treated Ti metal forms bone-like apatite on its surface in the living body, and bonds to the surrounding living bone through this apatite layer when it is implanted into a bone defect. When a rectangular sample of such treated Ti metal was implanted into a bone defect of a rabbit tibia, it was confirmed to form an apatite layer on its surface and to bond to living bone, as shown in Fig. 9.[3,43,44] When a tensile stress was applied to the interface, a considerably larger load was required to produce a fracture at the interface than at the interface of untreated Ti metal with the bone, and the difference between these samples increased with increasing period of time after implantation, as

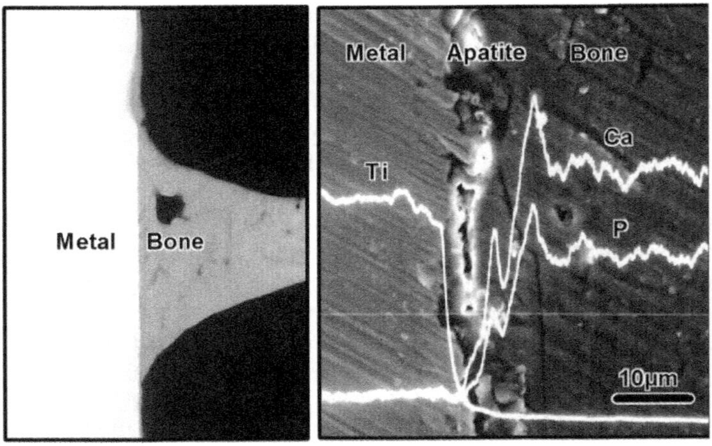

Figure 9. Contact microradiograph and SEM-EDX photographs at the interface between Ti metal which was subjected to NaOH and heat treatments and rabbit tibial bone, 8 weeks after implantation. Reprinted from Ref. 3, Copyright (1997) with permission from John Wiley and Sons)

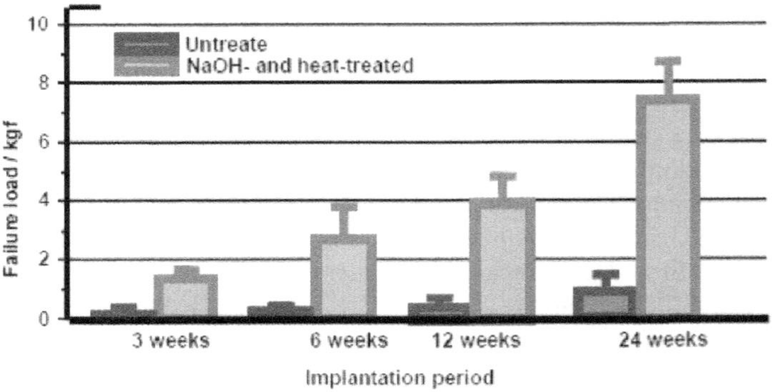

Figure 10. Detaching failure loads of the untreated and the NaOH- and heat-treated Ti metals as a function of implantation period in rabbit tibia. Reprinted from Ref. 3 Copyright (1997) with permission from John Wiley and Sons.

shown in Fig. 10.[3] Ti-based alloys, such as Ti–6Al–4V, Ti–6Al–2 Nb–Ta, and Ti–15Mo–5Zr–3Al, which were subjected to the same NaOH and heat treatments also displayed high failure loads under tensile stress when they were implanted into the tibia of a rabbit.[45] When implanted into a canine femur, such Ti metal and its alloys subjected to the same treatment showed a high bonding strength, even under a shear stress applied by a push-out test.[46] When a rod sample of such treated Ti metal was implanted into the medullar canal of a rabbit femur, it formed a bone-like apatite layer on its surface within a period of three weeks, and was fully surrounded by newly grown bone within a period of 12 weeks, as shown in Fig. 11.[47] At twelve weeks after implantation, the treated Ti metal rod could not be pulled out from the surrounding bone without being accompanied by a bone fragment, as shown in Fig. 12.[46,48]

Porous Ti metal subjected to NaOH and heat treatments was fully penetrated and had bonded with bone that had newly grown from the surrounding bone.[49] Based on these animal experiments, the NaOH and heat treatments were applied to the porous Ti metal layer on the outer cap and on part of the stem of a Ti–6Al–2Nb–Ta alloy of a total artificial hip joint. Following successful clinical trials, the resulting bioactive artificial joint has been used clinically in Japan since 2007, as shown in Fig. 13.[50,51]

2. Surface Modification with Calcium Titanate

NaOH treatments and heat treatments are simple methods for inducing an apatite-forming ability for Ti metal, and are effective for pure Ti metal and also for conventional Ti-based alloys, such as Ti–6Al–4V, Ti–6Al–2Nb–Ta, and Ti–15Mo–5Zr–3Al. However, they are not effective for inducing an apatite-forming ability for new types of Ti–Zr–Nb–Ta alloys that are free of cytotoxic elements.

The apatite-forming ability of Ti metal induced by the NaOH and heat treatments is liable to decrease when the treated Ti metal is stored in a humid environment for long periods, as described above. Although this instability in the apatite-forming ability is improved by the partial release of the sodium ions from the surface layer by a water treatment, a simpler method for obtaining stable bioactive Ti metal and its alloys is desired to be developed.

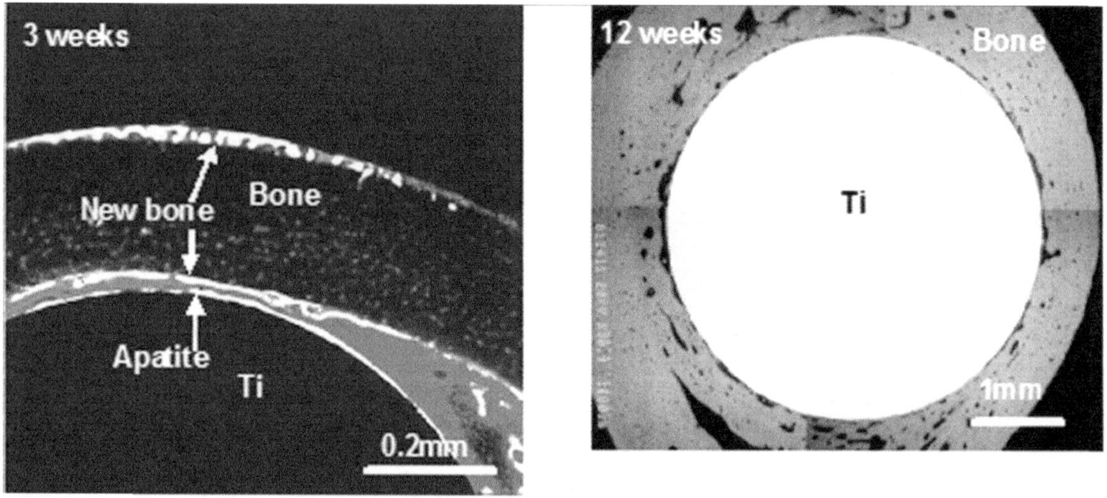

Figure 11. Confocal laser scanning microscopic photograph (left-hand side) and SEM photograph (right-hand side) of the cross-section of the NaOH- and heat-treated Ti metal rod implanted into rabbit femur for 3 and 12 weeks, respectively. Reprinted from Ref. 47, Copyright (2003) with permission from John Wiley and Sons.

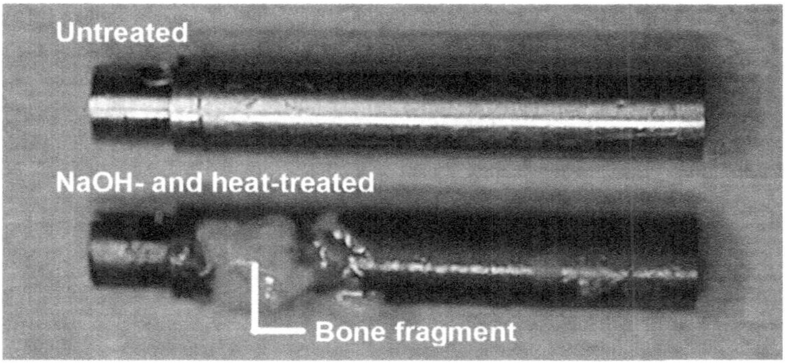

Figure 12. Surfaces of the untreated and the NaOH- and heat-treated Ti metal rods after the pull-out tests at 12 weeks after implantation. Reprinted from Ref. 48, Copyright (2004) with permission from Springer.

It is expected that a calcium titanate surface layer could give such Ti metals a higher and more stable apatite-forming ability than the sodium titanate surface layer does, because the calcium ions released from the surface layer in a body environment would increase the ionic activity product of the apatite in the surrounding body environment more effectively than sodium ions can, and the calcium ions in the surface layer would be more difficult to release in a humid environment than sodium ions are. However, a calcium titanate layer cannot form on Ti-based metals using a simple $Ca(OH)_2$ solution and a heat treatment, as the solubility of $Ca(OH)_2$ in aqueous solutions is low.

Calcium ions were first attempted to be incorporated into the surface of Ti metal using an ion implantation technique. Hanawa et al.,[52] Armitage et al.,[53] and Nayab et al.[54] showed that calcium ions can be incorporated into the surface of Ti metal to a depth of 100 nm at a dose of 1×10^{17} ions cm^{-2} under an acceleration voltage of 40 keV, and that Ti metal implanted with calcium ions provides a larger bone–Ti metal contact area. However, this technique requires a special apparatus for ion implantation, which is expensive.

Calcium ions can also be incorporated into the surface of Ti metal using micro-arc oxidation in a calcium-containing electrolyte. Sul et al.[55] showed that about 11 at.% Ca ions are incorporated into a 1,300-nm-thick TiO_2 layer on the surface of Ti metal using this

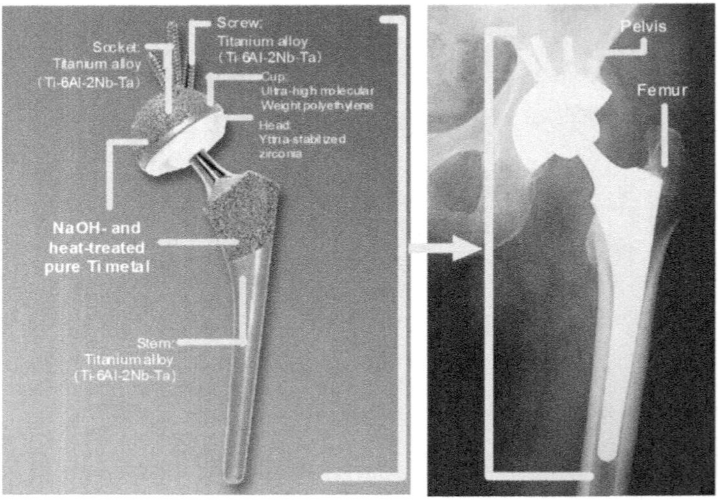

Figure 13. Clinical application of NaOH- and heat-treated Ti metal in hip joint. Reprinted from Ref. 51, Copyright (2005) with permission from Elsevier.

method, and that a dental implant of such treated Ti metal showed a higher removal torque than untreated Ti metal did. However, this higher torque cannot be simply attributed to the bone-bonding ability of the treated Ti metal because the micro-arc oxidation forms a porous structure with a number of craters on the surface of the Ti metal. Fröjd et al.[56] observed good contact with bone by Ti metal incorporated with calcium ions in its surface oxide layer after a micro-arc oxidation treatment. Wu et al.[57] showed that a Ti–24Nb–4Zr–7.9Sn alloy forms a Ca- and P-containing titanium oxide layer on its surface after micro-arc oxidation in a Ca- and P-containing aqueous solution, and that this showed a high removal torque when it was implanted into a rabbit's tibia. According to Song et al.,[58] Ti metal incorporated with calcium ions in its surface oxide layer after micro-arc oxidation does not form apatite on its surface in an SBF, but the same Ti metal that formed an amorphous $Ca(OH)_2$ layer after a subsequent hydrothermal treatment with water at 250°C for a period of 2 h formed an apatite layer. However, these micro-arc oxidation methods also require a special apparatus for applying the electric field on medical devices.

Calcium ions can also be incorporated into the surface of Ti metal by a hydrothermal treatment in calcium-containing aqueous solutions. Nakagawa et al.[59] first showed that a small amount of calcium ions are incorporated into a shallow region of the surface of Ti metal by a hydrothermal treatment using aqueous $CaCl_2$ at 200°C for a period of 24 h, and that such treated Ti metal forms apatite on its surface in an SBF. Park et al.[60] later showed that larger amounts of calcium ions are incorporated into the surface of Ti–6Al–4V alloy by forming $CaTiO_3$ from a hydrothermal treatment using a mixed NaOH and CaO solution at 180°C for a period of 24 h. This alloy subjected to the treatment above also forms apatite on its surface in Hank's solution, and shows increased bone–implant contact. Ueda et al.[61] showed that apatite formed in an SBF on Ti metal that was first treated using an H_2O_2/HNO_3 solution at 80°C for a period of 50 min, and then hydrothermally treated with a $Ca(OH)_2$ solution at 180°C for a period of 12 h. Chen et al.[62] showed that calcium ions are incorporated into the surface of porous Ti metal by a hydrothermal treatment using a 0.2 M $Ca(OH)_2$ solution at 250°C for a period of 8 h. It was also reported by Park et al.[63] that the removal torque of a Ti metal dental implant increased by the formation of P-incorporated titanium oxide on the Ti metal after a hydrothermal treatment in an H_3PO_4 solution at 180°C for a period of 2 h. However, these hydrothermal treatments also require a special high-pressure apparatus in an aqueous environment at high temperatures for application to medical devices.

The sodium hydrogen titanate formed on Ti metal by the NaOH treatment described in the previous section has the layered structure[64] shown in Fig. 14,[65] and therefore, its sodium ions can be easily replaced with calcium ions in aqueous calcium-containing solutions. Rakngarm et al.[66] soaked Ti metal and Ti–6Al–4V alloy in a $Ca(OH)_2$ solution after an NaOH treatment, and reported that such treated metals also form apatite on their surfaces in an SBF. However, they did not mention the state of the calcium ions in the treated metals.

Kizuki et al.[67] soaked Ti metal in a 100 mM $CaCl_2$ solution at 40°C for a period of 24 h after an NaOH treatment, and showed that the sodium hydrogen titanate formed on Ti metal after the NaOH

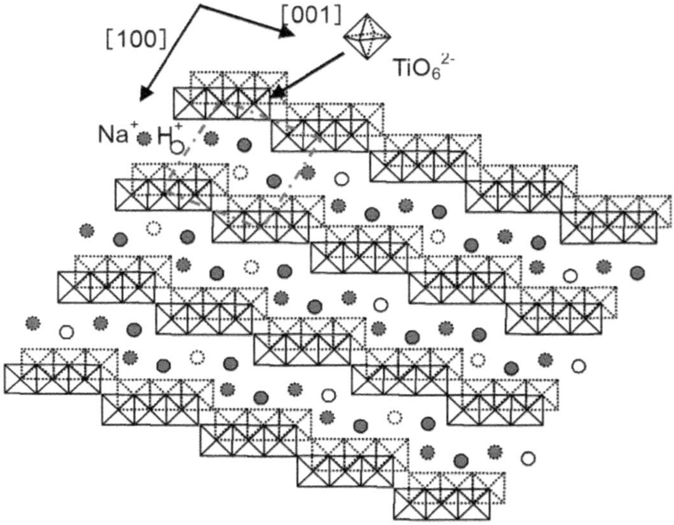

Figure 14. Structure of sodium hydrogen titanate. Reprinted from Ref. 65, Copyright (2010) with Molecular Diversity Preservation International.

treatment transformed into isomorphous calcium hydrogen titanate after a $CaCl_2$ treatment. The calcium hydrogen titanate layer formed showed an apatite-forming ability in an SBF. However, it had a poor scratch resistance. Then, the NaOH- and $CaCl_2$-treated Ti metal was heat-treated at 600°C for a period of 1 h, and the calcium hydrogen titanate was transformed into calcium titanate and rutile by this heat treatment, as shown in Fig. 15.[68] The scratch resistance of the surface layer increased to 190 mN after the heat treatment. However, the apatite-forming ability of the surface layer in an SBF was lost using this heat treatment. This may be attributed to a lower mobility of the calcium ions in the calcium titanate. The apatite-forming ability of the surface layer was recovered using a subsequent water treatment at 80°C for a period of 24 h, as shown in Fig. 16.[67] The calcium ions in the calcium titanate were partially replaced with H_3O^+ ions during the water treatment, as shown in Fig. 17.[67] As a result, the mobility of the calcium ions in the calcium titanate increased, and therefore, the apatite-forming ability of the calcium

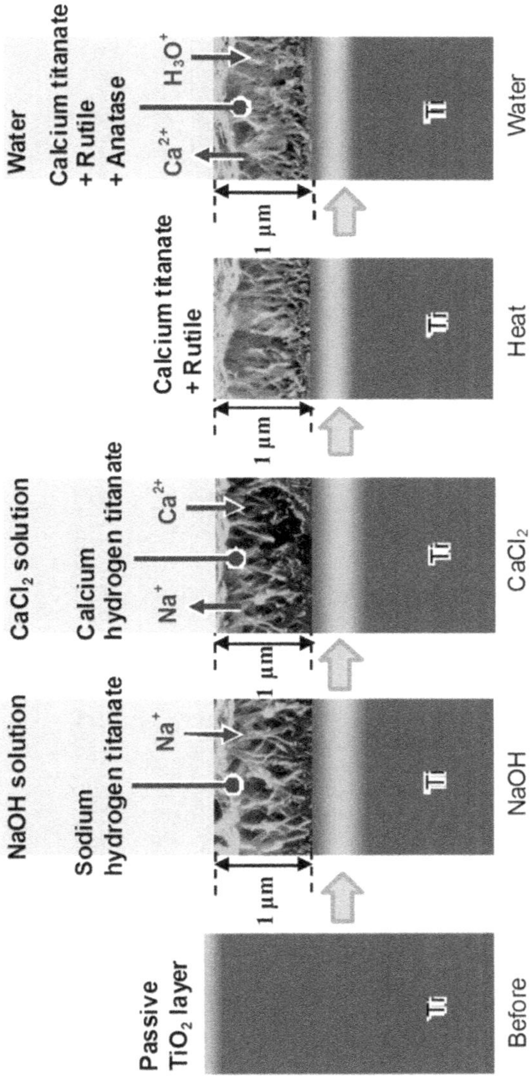

Figure 15. Structural change of the surface of Ti metal due to NaOH, $CaCl_2$, heat and water treatments. Reprinted from Ref. 68, Copyright (2010) with permission from Springer.

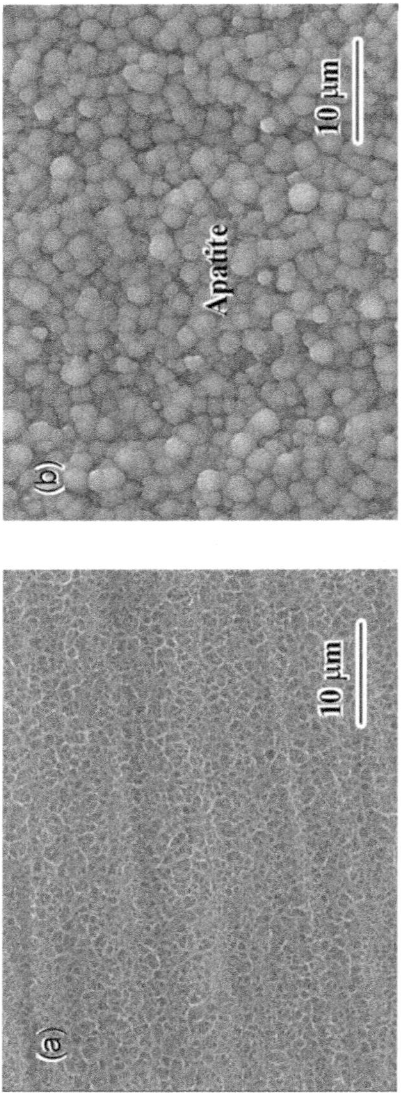

Figure 16. SEM photographs of surfaces of Ti metal soaked in SBF for 1 day (a) after NaOH, $CaCl_2$, heat treatments and (b) subsequent water treatment. Reprinted from Ref. 67, Copyright (2010) with permission from Elsevier.

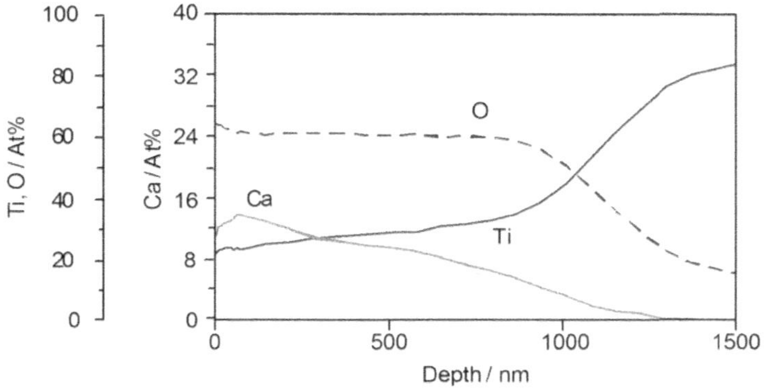

Figure 17. Auger electron spectroscopy depth profiles of the surface of Ti metal subjected to NaOH, $CaCl_2$, heat and water treatments. Reprinted from Ref. 67, Copyright (2010) with permission from Elsevier.

titanate increased. The final water treatment can be replaced by an autoclave treatment for sterilization before implantation.

The mechanism of the formation of apatite on such treated Ti metal in an SBF can also be interpreted in terms of an electrostatic interaction of the Ti metal with the ions in the SBF, similar to that of Ti metal subjected to the NaOH and heat treatments. The calcium ions in the calcium titanate are released as Ca^{2+} ions via an exchange with the H_3O^+ ions in the SBF to form Ti–OH groups on the surface. This increases the ionic activity product of the apatite in the surrounding SBF. The Ti–OH groups formed are negatively charged because of the increase in pH from the released calcium ions. These negatively charged Ti–OH groups induce apatite formation from an electrostatic interaction with the ions in the SBF similar to the Ti–OH groups formed on sodium titanate.

The apatite-forming ability of such treated Ti metal is not noticeably decreased, even when the treated Ti metal is stored in a humid environment for long periods. This is attributed to the lower mobility of Ca^{2+} ions in calcium titanate than that of Na^+ ions in sodium titanate. These NaOH, $CaCl_2$, heat, and water treatments are also effective for inducing an apatite-forming ability in the new types of Ti–Zr–Nb–Ta alloys, such as Ti–15Zr–4Nb–4Ta[68], Ti–37Nb–2Ta–3Zr–0.3O[69], and Ti–29Nb–13Ta–4.6Zr.[70] This may be attributed to the higher ionic activity product of the apatite in the surrounding SBF, caused by the release of Ca^{2+} ions. The Ti metal and Ti–15Zr–4Nb–4Ta alloy formed with the calcium titanate on their surfaces using these NaOH, $CaCl_2$, heat, and water treatments were confirmed to be tightly bonded to living bone in animal experiments using rabbits. Alloys in the system Ti-Zr-Nb-Ta can exhibit unique mechanical properties such as low elastic modulus[71] and gum property.[72] They could find new applications as orthopedic and dental implants, if they are given bioactivity by chemical and heat treatments.

3. Surface Modification with Titanium Oxide

The sodium titanate and calcium titanate described above release either Na^+ or Ca^{2+} ions, respectively, into the living body, and hence, are liable to have an unfavourable effect on living cells from the increase in pH of the surrounding environment, especially in the narrow spaces of porous implants. Titanium oxide releases no soluble ions. It is known that even pure titanium oxide can induce apatite formation on its surface in a body environment, as shown by the titania gel in Fig. 1. Ti metal is usually covered with a passive thin titanium oxide layer, as described earlier. This oxide layer grows to form a thick rutile layer after heat treatment in air. However, this oxide layer does not form apatite on the surfaces in a body environment.[73] Various attempts at forming a titanium oxide layer that can induce apatite formation on Ti metal and its alloys have been made.

Sul et al.[74] showed that a thick layer of titanium oxide consisting of anatase and rutile phases forms on Ti metal after anodic oxidation in a 0.1 M CH_3COOH solution, and that such treated Ti metal dental implants show a high removal torque when implanted into a rabbit's tibia. Wang et al.[75] reported that Ti metal forms an anatase layer on its surface after treatment using an H_2O_2/HCl solution and subsequent heat treatment around 400–500°C, and that such treated Ti metal shows apatite formation on its surface in an SBF. Wang et al.[76] also reported that Ti metal forms a rutile layer on its surface after treatment in an HF/HNO_3 solution and subsequent heat treatment at 400–500°C, and that such treated Ti metal shows apatite formation in an SBF, but only in a narrow gap. Wu et al.[77] reported that Ti metal forms a titanium oxide layer with a mixture of anatase and rutile after treatment using an $H_2O_2/TaCl_2$ solution at 80°C after an HF/HNO_3 mixed acid treatment, and that such treated Ti metal also forms apatite on its surface in an SBF.

Yang et al.[78] and Cui et al.[79] showed that an anatase–rutile mixed layer is formed on Ti metal after anodic oxidation using a spark discharge in H_2SO_4 and Na_2SO_4 solutions, and that such treated Ti metal forms apatite on its surface in an SBF. Lu et al.[80] reported that Ti metal forms apatite on its surface in an SBF after it had been treated with a strong nitric acid solution at 60°C for a period of 600 min, although an oxide layer was not detected on its surface.

Zhou et al.[81] reported that a layer of nanosized anatase formed on Ti metal after magnetron spattering in a mixed Ar/O_2 plasma formed apatite on its surface in an SBF. Dong et al.[82] showed that Ti metal forms a layer of self-assembled nanowires of hydrogen titanate on its surface after a hydrothermal treatment in NaOH solution at 100–250°C for a period of 2–10 h and subsequent rinsing with water, and that such treated Ti metal was fully covered by bone tissue in mice. Lu et al.[83] reported that Ti metal forms a rutile layer on its surface after treatment with an H_2SO_4/HCl solution and subsequent heat treatment at 500°C, and that such treated Ti metal forms apatite on its surface in an SBF. Oh et al.,[84] Popat et al.,[85] and Bjursten et al.[86] reported that Ti metal forms a layer of well-aligned nanotubes of anatase on its surface by anodic oxidation in a 0.5% HF solution and subsequent heat treatment around 500°C, and that such treated Ti metal accelerates the growth of osteoblasts, bone formation on its surface, and bonding to bone.

Sugino et al.[87] found that a rutile layer formed on Ti–15Zr–4Ta–4Nb alloy after heat treatment at 500°C forms apatite on its surface in an SBF, but only in the internal surfaces of microgrooves. Karthega et al.[88] showed that Ti–6Al–4V alloy forms an anatase layer on its surface after treatment with 15 and 25 wt% H_2O_2 and subsequent heat treatment at 400°C, and that such treated alloy forms apatite on its surface in an SBF.

These results show that apatite formation on titanium oxide in a body environment does not depend upon the type of crystalline phase present. Depending on their preparation conditions, both anatase and rutile can form apatite on their surfaces. From a literature survey, it seems that the acid and heat treatments of Ti metal described above give favourable conditions for apatite formation on titanium oxide. However, it is difficult to derive a general principle governing the apatite-forming ability of titanium oxide from the processes described above.

The sodium hydrogen titanate formed on the surface of the Ti metal by treatment with 5 M NaOH solution at 60°C for a period of 24 h has a layered structure, as described above, and hence, its sodium ions can be completely replaced with H_3O^+ ions to be transformed into hydrogen titanate if this NaOH-treated Ti metal is soaked in water or acidic solutions, such as HCl, HNO_3, or H_2SO_4, as shown in Fig. 18. The apatite-forming ability of sodium hydrogen titanate in an SBF is lost after its transformation into hydrogen titanate, but is increased by its transformation into anatase and rutile by a subsequent heat treatment.[89,90] The apatite-forming ability increases with decreasing pH of the acid solution, irrespective of the type of acid solution, as shown in Fig. 19.[91] This is assumed to be attributable to the increased positive surface charge of the titanium oxide with decreasing pH of the acid solution.

When Ti metal is heat-treated at various temperatures after treatment with an H_2SO_4/HCl mixed solution without treatment with NaOH, apatite is only formed in an SBF on the surface of Ti metal heat-treated at 500–650°C, as shown in Fig. 20.[73] Titanium hydride is formed on Ti metal after the acid treatment, and this transforms into rutile at 500°C, and its content increases with increasing heat-treatment temperature. The apatite-forming ability of Ti metal does not depend on the amount of rutile. A surface roughness on the micrometre scale was produced by the acid treatment, and this gradually decreases above 700°C. The apatite-forming ability of Ti metal also does not depend on the surface roughness. The zeta potential was almost zero for Ti metal that was heat-treated below 500°C and above 700°C, but was only positive for Ti metal samples that were heat-treated at 500–650°C, as shown in Fig. 21.[73] This indicates that the formation of apatite on Ti metal subjected to acid treatments and heat treatments is attributable to the positive surface charge of the Ti metal.

When the surface of Ti metal is positively charged, it first combines with negatively charged phosphate ions, as shown in Fig. 22.[73] As the phosphate ions accumulate, the surface becomes

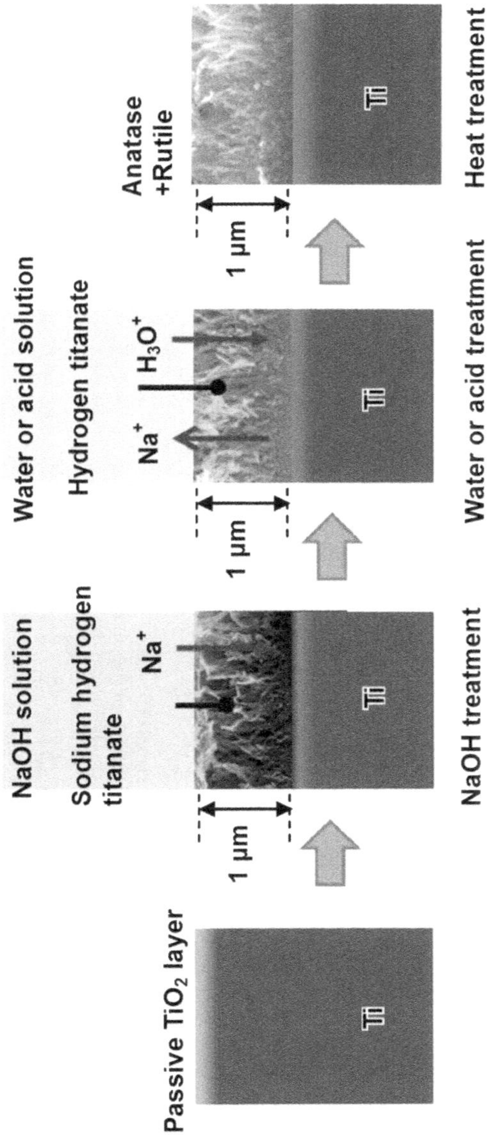

Figure 18. Surface structural change of Ti metal by NaOH, water or acid solution and heat treatments.

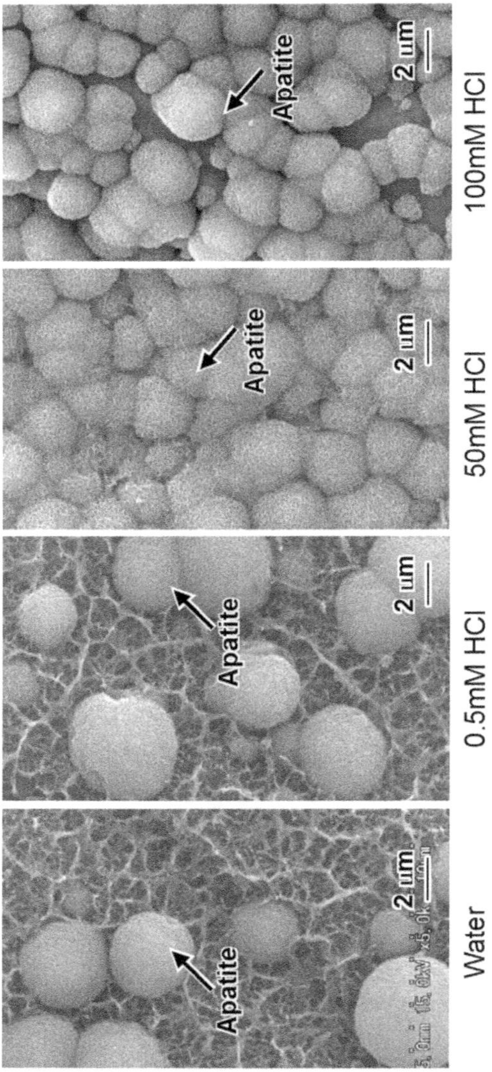

Figure 19. SEM photographs of surfaces of Ti metals soaked in SBF for 1 day, after NaOH, water or HCl and heat treatments. Reprinted from Ref. 90, Copyright (2009) with permission from Springer.

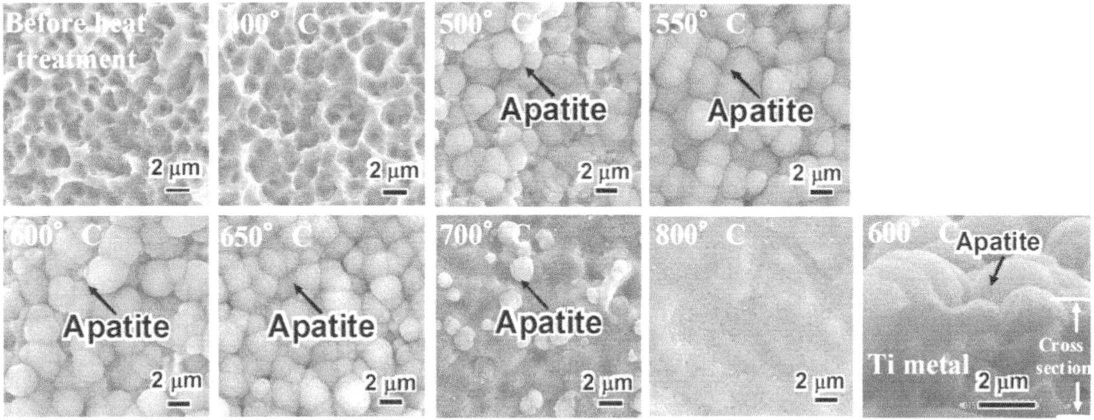

Figure 20. SEM photographs of surfaces of Ti metals soaked in SBF for 1 day after heat-treated at various temperatures following the H_2SO_4/HCl mixed acid treatment 70°C for 1 h. Reprinted from Ref. 73, Copyright (2010) with permission from The Royal Society.

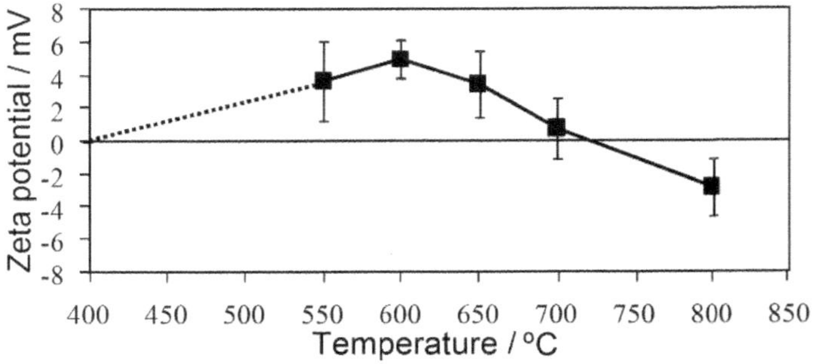

Figure 21. Zeta potentials of surfaces of Ti metals heat-treated at various temperatures after the mixed acid treatment.(Reprinted from ref. 73 with permission from The Royal Society)

negatively charged, and combines with positively charged calcium ions to form an amorphous calcium phosphate phase. This calcium phosphate is metastable, and is eventually transformed into crystalline bone-like apatite. This apatite-formation process was confirmed from XPS spectra of Ti metal heat-treated at 600°C after an acid treatment as a function of soaking time in an SBF. This is in contrast to the apatite-formation process on a negatively charged surface of Ti metal formed with sodium titanate, which is shown in Fig. 8.

The reason the acid- and heat-treated Ti metal shows a positive charge on its surface in a body environment is interpreted in terms of the adsorbed acid groups, such as Cl^-, NO_3^-, and SO_4^{2-}. These acid groups are adsorbed onto, and incorporated into, a layer of titanium hydride on the Ti metal's surface during the acid treatment, and remain, even after the transformation of the titanium hydride into titanium oxide by the heat treatment, as long as the temperature of the heat treatment is not too high, as shown in Fig. 23.[73] These dissociate in a body environment to give the same surface as Ti metal in an acid environment. In general, titanium oxide is positively charged in an acid environment, as shown in Fig. 24.[29,30] As a result, Ti metal heat-treated at 500–650°C after an acid treatment shows a positively charged surface in a body environment. Ti metal heat-treated below 400°C after an acid treatment does not have a

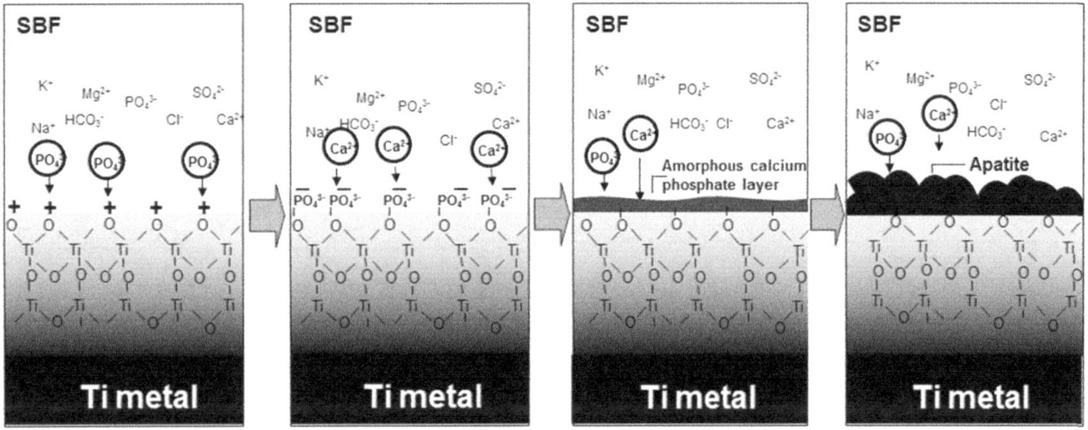

Figure 22. Process of formation of apatite on positively charged Ti metal in SBF. Reprinted from Ref. 73, Copyright (2010) with permission from The Royal Society)

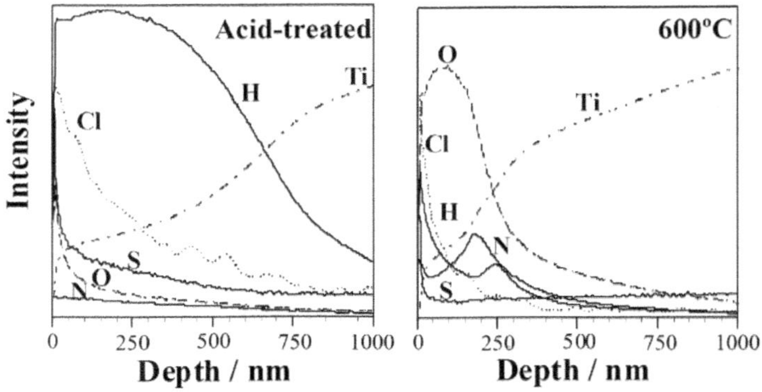

Figure 23. Depth profiles of radio-frequency glow discharge optical emission spectroscopy spectra of Ti metals (a) acid-treated, and (b) subsequently heat-treated at 600°C. Reprinted from Ref. 73, Copyright (2010) with permission from The Royal Society.

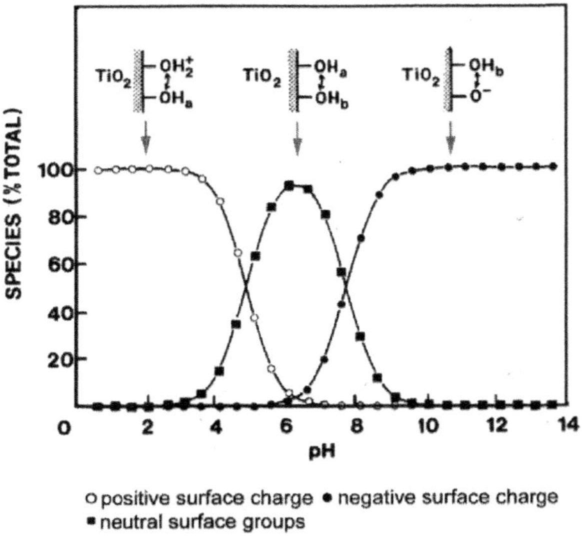

Figure 24. Diagram indicating the concentration of the different charged and uncharged hydroxyrelated species at a titanium oxide surface as a function of pH of the aqueous solution. $-OH_a$ and $-OH_b$ denote acidic and basic hydroxyl group. Open circles: relative surface concentrations of positive species; filled circles: negative species; squares: neutral species. Reprinted from Ref. 30, Copyright (2001) with permission from Springer.

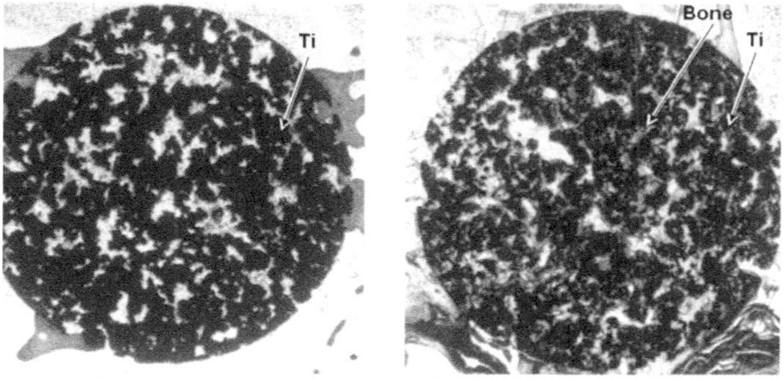

Figure 25. Bone formation in porous Ti metals subjected to no treatment and NaOH, HCl and heat treatments, 26 weeks after implantation into rabbit femur. Reprinted from Ref. 65, Copyright (2010) with permission from Molecular Diversity Preservation International.

charged surface, because the titanium hydride on the surfaces is electrically conductive. Ti metal heat-treated above 700°C after an acid treatment also does not show a charged surface, because the acid groups on the surface are decomposed at high temperatures.

Ti samples subjected to acid and heat treatments after an NaOH treatment, as well as those subjected to acid and heat treatments without an NaOH treatment, showed a stable apatite-forming ability, even in a humid environment. Both of these Ti samples were confirmed to be tightly bonded to living bone in animal experiments using a rabbit's tibia.[92-94] However, there is a difference in their surface roughness. The former has a surface roughness on the nanometre scale, whereas the latter has a surface roughness on the micrometre scale. Bioactive porous titanium metal prepared by the former treatment exhibited bone formation at bone defect sites, as shown in Fig. 25,[65,95-97] and also ectopically in the muscle tissue of a dog, as shown in Fig 26.[98,99] It would be interesting to know whether porous Ti metal prepared by the latter treatment also exhibits a similar osteoinduction, because osteoinduction is reported to be highly influenced by the surface roughness.[100] Because of its high bone-forming ability, the porous Ti metal described above is being

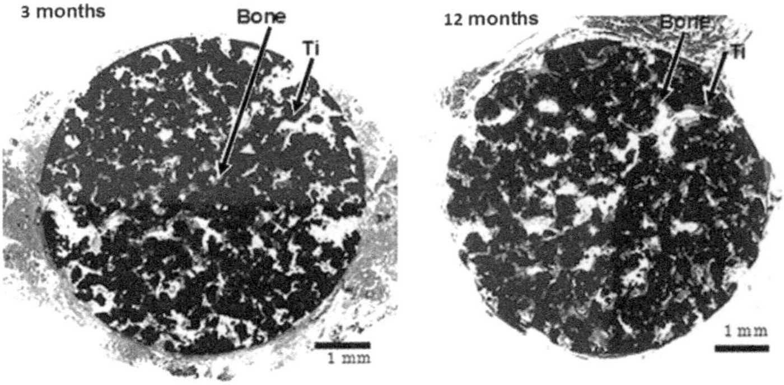

Figure 26. Bone formation in porous Ti metals subjected to no treatment and NaOH, HCl and heat treatments, 3 and 12 months after implantation into muscle of beagle dog. Reprinted from Ref. 99, Copyright (2006) with permission from Elsevier.

subjected to clinical trials after animal experiments[101] for application in spinal fusion devices, as shown in Fig. 27.

These results show that when Ti metal is heat-treated after an acid treatment, titanium oxide with adsorbed acid groups is formed on its surface, and these induce apatite formation in a body environment from their positive charges. In view of this finding, most of the formation of apatite on titanium oxide described in the literature cited in the first part of this section can also be interpreted in terms of the positive surface charge on the titanium oxide. Zhao et al.[102,103] reported that a plasma-sprayed TiO_2 coated layer with a rutile phase showed a high apatite-forming ability in an SBF when treated with an acid solution, such as H_2SO_4 and HNO_3. Kokubo et al.[104] reported that a titania gel layer coated on polyethylene terephthalate using a sol-gel method formed apatite on its surface in an SBF, and bonded tightly to living bone[105] when it was treated with an HCl solution at 80°C for a period of 8 d. These results can also be interpreted in terms of a positive surface charge on the titanium oxide produced by the acid treatment.

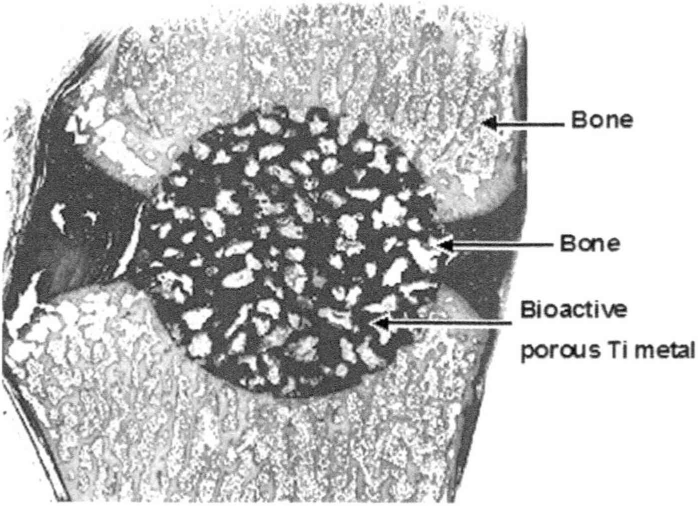

Figure 27. Fusion of spine of dog by a bioactive porous Ti metal. Reprinted from Ref. 101, Copyright (2007) with permission from American Association of Neurosurgeons.

4. Principles for Preparing Bioactive Ti Metal and its Alloys by Surface Modification

It is known that Ti metal and its alloys generally exhibit good compatibility with living tissue, but do not bond to living bone. This is interpreted in terms of the surface charge of the metal, as follows. These metals are usually covered with a passive thin layer of titanium oxide, which shows almost zero surface charge in a body environment with a pH = 7, because the number of negatively charged sites is balanced with the number of positively charged sites, as shown in Fig. 24. The negatively charged sites adsorb the positively charged calcium ions, whereas the positively charged sites adsorb the negatively charged phosphate ions simultaneously[73] in a body environment to form a calcium phosphate layer on the surface of the titanium oxide. As a result, these sites are soon electrically neutral, and the adsorption of calcium phosphate stops. Because of the presence of the thin layer of calcium phosphate on the surface of the titanium oxide, Ti metal and its alloys usually exhibit good compatibility with living tissue, but because of the lack of hydroxyapatite on their surface, they do not bond to living bone.

However, if the titanium oxide is exposed to a strong alkali or acid environment, then it becomes strongly charged, either negatively or positively, on its surface, as shown in Fig. 24, and can adsorb a large amount of calcium or phosphate ions selectively at first, and then later on, the corresponding counter ions. As a result, a thick apatite layer is formed on the surface of the titanium oxide, as shown in Figs. 8 and 22. In these cases, Ti metal and its alloys formed with titanium oxide on their surfaces bond to living bone through an apatite layer. A strong alkaline environment is produced on the surface of the titanium oxide if the Ti metal and its alloys are formed with an alkali or alkaline earth titanate, such as sodium titanate or calcium titanate on their surface from treatment with an alkali and/or alkaline earth solution and heat treatment, because the alkali or alkaline earth ions of the titanates are released in a body environment. On the other hand, a strong acid environment is produced on the surface of the titanium oxide if the Ti metal and its alloys are formed with titanium oxide on their surface by treatment with an acid solution and heat treatment, because the acid groups adsorbed on the surface of the titanium oxide are dissociated in a body environment.

The reason why calcium or phosphate ions are preferentially adsorbed on the charged metal in a body environment, where various types of charged ions and organic substances coexist, is attributed to the high degree of supersaturation in a body environment with respect to apatite. It can be said from these considerations that bioactive Ti metal and its alloys can be prepared using alkaline or acidic solutions and a subsequent heat treatment.

IV. BIOACTIVE Zr METAL

Some zirconium oxide gels prepared using a sol-gel method have also been found to form apatite on their surface in an SBF.[10,106] This indicates that zirconium metal and its alloys can exhibit bioactivity after a surface treatment. However, the simple NaOH and heat treatments only confer a slight apatite-forming ability in an SBF to Zr metal.[107] According to Chen et al., Zr metal forms apatite on its surface in an SBF when soaked in a 10 M NaOH solution at 60°C for a period of 24 h and then heat-treated at 600°C for a period of 1 h in a vacuum.[108] Hang et al. reported that Zr metal forms apatite on its surface in an SBF when subjected to micro-arc oxidation in an aqueous electrolyte containing calcium acetate monohydrate and β-glycerophosphate disodium salt pentahydrate under a potential of 500 V for a period of 5 min, to form zirconia-containing CaO and a phosphate layer on its surface.[109]

V. BIOACTIVE Nb METAL

Niobium oxide gels prepared using a sol-gel method also form apatite on their surface in an SBF.[11] This indicates that niobium and its alloys can exhibit bioactivity after a surface treatment. However, Nb metal does not form apatite on its surface in an SBF, even when sodium niobate forms on its surface after soaking in a 1 M NaOH solution at 60°C for a period of 24 h.[11] According to Wang et al., Nb metal forms apatite on its surface in an SBF when it is soaked in a 0.5 M NaOH solution at 80°C for a period of 24 h and then heat-treated at 600°C for a period of 1 h in a vacuum.[110]

VI. BIOACTIVE Ta METAL

Tantalum oxide gels prepared using a sol-gel method also form apatite on their surface in an SBF,[12] and their apatite-forming ability increases with increasing sodium content of the gel.[12] Tantalum metal forms apatite on its surface in an SBF without any chemical treatment. However, it takes a long period for the apatite to form.[111] If Ta is soaked in a 0.2–0.5 mM NaOH solution at 50°C for a period of 24 h to form a sodium tantalate hydrogel on its surface, then the induction period for apatite formation is decreased considerably,[111] and if the sample is then further heat-treated at 300°C for a period of 1 h, the bonding strength of the surface layer to the substrate increases without any decrease in its apatite-forming ability.[112,113] The mechanism of apatite formation on NaOH- and heat-treated Ta metal can be interpreted in terms of the Ta–OH groups formed on its surface from the exchange of Na^+ ions with the H_3O^+ ions in the

SBF. When Ta metal is soaked in a 100 mM $CaCl_2$ solution at 80°C for a period of 24 h after an NaOH treatment and then heat-treated at 500°C for a period of 1 h, its apatite-forming ability is increased markedly.[114] Ta metal subjected to NaOH and heat treatments has been confirmed to bond tightly to the living tibial bone of a rabbit within 16 weeks of implantation.[115]

VII. SUMMARY

In this chapter, it has been shown that various types of bone-bonding bioactive metals with different characteristics are obtained using surface modifications employing various methods, such as ion implantation, micro-arc treatment, hydrothermal treatment, sputtering, and chemical and heat treatment. Among these techniques, chemical and heat treatments are simple and versatile methods for the surface modification of metals. Various ions can be incorporated into the surface layer of a metal to give different functions after the chemical treatments. Various bioactive metals with different functions can be developed using these methods in the future.

The bone-bonding ability of metals with modified surfaces is attributable to the formation of apatite on their surface in a body environment. The formation of apatite can be interpreted in terms of the electrostatic interaction of the metal surface with the calcium or phosphate ions in a body fluid. Metal surfaces that are either negatively or positively charged in a body environment can induce apatite formation. These findings provide us with a useful guiding principle for developing new bioactive materials.

Some bioactive metals with modified surfaces after chemical and heat treatments are already used clinically as implants, and these can find various applications as orthopaedic and dental implants in the future.

REFERENCES

[1]T. Kokubo, *Bioceramics and their clinical applications,* Woodhead Publishing, Cambridge, 2008.
[2]S. A. Hacking, M. Tanzer, E. Harvey, J. J. Krygier, and J. D. Bobyn, *Clin. Or-*

thopaed. Relat. Res. **405** (2002) 24.
[3]W.-Q. Yan, T. Nakamura, M. Kobayashi, H.-M. Kim, F. Miyaji, and T. Kokubo, *J. Biomed. Mater. Res.* **37** (1997) 267.
[4]P. G. Coelho, J. M. Granjeiro, G. E. Romanos, M. Suzuki, N. R. F. Silva, G. Cardaropoli, V. P. Thompson, and J. E. Lemons, *J. Biomed. Mater. Res.* **88B** (2009) 579.
[5]S. C. G. Leeuwenburgh, J. G. C. Wolke, J. A. Jansen and K. de Groot, in *Bioceramics and their clinical applications,* Ed. by T. Kokubo, Woodhead Publishing, Cambridge, 2008, p. 464.
[6]T. Taguchi, Y. Muraoka, H. Matsuyama, A. Kishida, and M. Akashi, *Biomaterials* **22** (2001) 53.
[7]T. Kokubo, and H. Takadama, *Biomaterials* **15** (2006) 2907.
[8]P. Li, C. Ohtsuki, T. Kokubo, K. Nakanishi, N. Soga, T. Nakamura, and T. Yamamuro, *J. Am. Ceram. Soc.* **75** (1992) 2094.
[9]P. Li, C. Ohtsuki, T. Kokubo, K. Nakanishi, N. Soga, and K. de Groot, *J. Biomed. Mater. Res.* **28** (1994) 7.
[10]M. Uchida, H.-M. Kim, T. and Kokubo, *J. Am. Ceram. Soc.* **84** (2001) 2041.
[11]T. Miyazaki, H.-M. Kim, T. Kokubo, C. Ohtsuki, and T. Nakamura, *J. Ceram. Soc. Japan* **109** (2001) 929.
[12]T. Miyazaki, H.-M. Kim, and T. Kokubo, *J. Sol-Gel Sci. Technol.* **21** (2001) 83.
[13]C. Ohtsuki, T. Kokubo, and T. Yamamuro, *J. Non-Cryst. Solids.*, **143** (1992) 84.
[14]M. Tanahashi, and T. Matsuda, *J. Biomed. Mater. Res.* **34** (1997) 305.
[15]T. Kawai, C. Ohtsuki, M. Kamitakahara, T. Miyazaki, M. Tanihara, Y. Sakaguchi, and S. Konagaya, *Biomaterials* **25** (2004) 4529.
[16]T. Kokubo, F. Miyaji, H.-M. Kim, and T. Nakamura, *J. Am. Ceram. Soc.* **79** (1996) 1127.
[17]H.-M. Kim, F. Miyaji, T. Kokubo, and T. Nakamura, *J. Biomed. Mater. Res.* **32** (1996) 409.
[18]H.-M. Kim, F. Miyaji, T. Kokubo, S. Nishiguchi, and T. Nakamura, *J. Biomed. Mater. Res.* **45** (1999) 100.
[19]T. Kawai, T. Kizuki, H. Takadama, T. Matsushita, T. Kokubo, H. Unuma and T. Nakamura, *J. Ceram. Soc. Japan* **118** (2010) 19.
[20] S. Yamaguchi, H. Takadama, T. Matsushita, T. Nakamura, and T. Kokubo, *J. Ceram. Soc. Japan* **117** (2009) 1126.
[21]H.-M. Kim, F. Miyaji, T. Kokubo, and T. Nakamura, *J. Ceram. Soc. Japan* **105** (1997) 111.
[22]H.-M. Kim, F. Miyaji, T. Kokubo, and T. Nakamura, *J. Mater. Sci.: Mater. Med.* **8** (1997) 341.
[23]H.-M. Kim, F. Miyaji, T. Kokubo, and T. Nakamura, *J. Biomed. Mater. Res. (Appl. Biomater.)* **38** (1997) 121.
[24]H. Takadama, H.-M. Kim, T. Kokubo, and T. Nakamura, *J. Biomed. Mater. Res.* **57** (2001) 441.
[25]H.-M. Kim, H. Takadama, F. Miyaji, T. Kokubo, S. Nishiguchi, and T. Nakamura, *J. Mater. Sci.: Mater. Med.* **11** (2000) 555.
[26]H.-M. Kim, H. Takadama, T. Kokubo, S. Nishiguchi, and T. Nakamura, 2000b. *Biomaterials* **21** (2000) 353.
[27]H. Takadama, H.-M. Kim, T. Kokubo, and T. Nakamura, *J. Biomed. Mater. Res.* **57** (2001) 185.
[28]H.-M. Kim, T. Himeno, M. Kawashita, J.-H. Lee, T. Kokubo, and T. Nakamura, *J. Biomed. Mater. Res.* **67A** (2003) 1305.

[29]J. M. Gold, M. Schmidt, and S. G. Steinemann, *Helv. Phys. Acta* **62** (1989) 246.
[30]M. Textor, C. Sitting, V. Franchiger, S. Tosatti and D. M. Brunette, In *Tianium in Medicine*, Ed. by D. M. Brunette, P. Tengrall, M. Textor, and P. Thomsen, Germany, Springer, Verlag, 2001, p. 172.
S. C. G. Leeuwenburgh, J. G. C. Wolke, J. A. Jansen and K. de Groot, in *Bioceramics and their clinical applications,* Ed. by T. Kokubo, Woodhead Publishing, Cambridge, 2008, p. 464.
[31]E. Conforto, D. Caillard, L. Müller, and F. A. Müller, *Acta Biomater.* **4** (2008) 1934.
[32]H. Takadama, M. Mizuno, H.-M. Kim, T. Kokubo and T. Nakamura, in *Ceramic Engineering and Science Proceedings,* Vol. 23, Ed. by H.-T. Lin, and M. Single, Am. Ceram. Soc., Ohio, 2002, p. 753.
[33]H. Takadama, H.-M. Kim, T. Kokubo, and T. Nakamura, *Sci. and Technol. Advanc. Mater.* **2** (2001) 389.
[34]J. Strnad, J. Protivínský, D. Mazur, K. Veltruská, Z. Strnad, A. Helebrant, and J. Šesták, *J. Therm. Anal. Calor.* **76** (2004) 17.
[35]L. Jonášová, A. F. Müller, A. Hlebrant, J. Strnad, and P. Greil, *Biomaterials.* **25** (2004) 1187.
[36]L. Jonášová, A. F. Müller, A. Hlebrant, J. Strnad, and P. Greil, *Biomaterials.* **23** (2002) 3905.
[37]D. Krupa, J. Baszkiewicz, J. Mizera, T. Borowski, A. Barcz, J. W. Sobczak, A. Biliński, M. Lewandowska-Szumiel, and M. Wojewódzka, *J. Biomed. Mater. Res.* **88A** (2008) 589.
[38] D. Lakstein, W. Kopelovitch, Z. Barkay, M. Bahaa, D. Hendel, and N. Eliaz, *Acta Biomater.* **5** (2009) 2258.
[39] H.-M. Kim, T. Kokubo, S. Fujibayashi, S. Nishiguchi, and T. Nakamura, *J. Biomed. Mater. Res.* **52** (2000) 553.
[40]D. K. Pattanayak, K. Doi, H. Takadama, T. Nakamura, and T. Kokubo, *Mater. Sci. Eng. C* **29** (2009) 1974.
[41]K. Nishio, M. Neo, H. Akiyama, S. Nishiguchi, H.-M. Kim, T. Kokubo, and T. Nakamura, *J. Biomed. Mater. Res.* **52** (2000) 652.
[42]J. Isaac, S. Loty, A. Hamdan, T. Kokubo, H.-M. Kim, A. Berdal, and J. M. Sutier, *J. Biomed. Mater. Res.* **89A** (2009) 585.
[43]W.-Q. Yan, T. Nakamura, K. Kawanabe, S. Nishiguchi, M. Oka, and T. Kokubo, *Biomaterials.* **18** (1997) 1185.
[44]S. Nishiguchi, T. Nakamura, M. Kobayashi, H.-M. Kim, F. Miyaji, and T. Kokubo, *Biomaterials.* **20** (1999) 491.
[45]S. Nishiguchi, T. Nakamura, M. Kobayashi, H.-M. Kim, F. Miyaji, and T. Kokubo, *J. Biomed. Mater. Res. (Appl. Biomater.)* **48** (1999) 689.
[46]S. Nishiguchi, H. Kato, H. Fujita, M. Oka, H.-M. Kim, T. Kokubo, and T. Nakamura, *Biomaterials.* **22** (2001) 2525.
[47]S. Nishiguchi, S. Fujibayashi, H.-M. Kim, T. Kokubo, and T. Nakamura, *J. Biomed. Mater. Res.* **67A** (2003) 26.
[48]T. Kokubo, H.-M. Kim, M. Kawashita, and T. Nakamura, *J. Mater. Sci.: Mater. Med.* **15** (2004) 99.[49]S. Nishiguchi, H. Kato, M. Neo, M. Oka, H.-M. Kim, T. Kokubo, and T. Nakamura, *J. Biomed. Mater. Res.* **54** (2001) 198.
[50] K. Kawanabe, K. Ise, K. Goto, H. Akiyama, T. Nakamura, A. Kaneuji, T. Sugimori, and T. Mtsumoto, *J. Biomed. Mater. Res.* **90B** (2009) 476. 2009.
[51]T. Kokubo, *Mater. Sci. Eng. C* **25** (2005) 97.
[52]T. Hanawa, Y. Kamimura, S. Yamamoto, T. Kohgo, A. Amemiya, M. Ukai, K.

Murakami, K. Asaoka, *J. Biomed. Mater. Res.* **36** (1997) 131.
[53]D. A. Armitage, R. Mihoc, T. J. Tate, D. S. McPhail, R. Chater, J. A. Hobkirk, L. Shinawi, and F. H. Jones, *Appl. Surf. Sci.* **253** (2007) 4085.
[54]S. H. Nayab, F. H. Jones, and I. Olsen, *J. Biomed. Mater. Res.* **83A** (2007) 296.
[55]Y.-T. Sul, *Biomaterials.* **24** (2003) 3893.
[56]V. Fröjd, V. Franke-Stenport, L. Meirelles, and A. Wennerberg, *Int. J. Oral Maxillofac. Surg.* **37** (2008) 561.
[57]J. Wu, Z.-M. Liu, X.-H. Zhao, Y. Gao, J. Hu, and B. Gao, *J. Biomed. Mater. Res.* **92B** (2010) 298.
[58]W.-H. Song, H. S. Ryu, and S.-H. Hong, *J. Am. Ceram. Soc.* **88** (2005) 2642.
[59]M. Nakagawa, L. Zhang, K. Udoh, S. Matsuya, and K. Ishikawa, *J. Mater. Sci.: Mater. Med.* **16** (2005) 985.
[60]J.-W. Park, K.-B Park, and J.-Y. Suh, *Biomaterials.* **28** (2007) 3306.
[61]M. Ueda, M. Ikeda, and M. Ogawa, Mater. Sci. Eng. C **29** (2009) 994.
[62]X.-B. Chen, Y.-C. Li, J. D. Plessis, P. D. Hodgson, and C. Wen, *Acta Biomater.* **5** (2009) 1808.
[63]J.-W. Park, Y.-J. Kim, J.-H. Jang, T.-G. Kwon, Y.-C. Bae, and J.-Y. Suh, *Acta Biomater.* **6** (2010) 1661.
[64]X. Sun, and Y. Li, *Chem. Eur. J.* **9** (2003) 2229.
[65]T. Kokubo, and S. Yamaguchi, *Materials.* **3** (2010) 48.
[66]A. Rakngarm, Y. Miyashita, and Y. Mutoh, *J. Mater. Sci.: Mater. Med.* **19** (2008) 1953.
[67]T. Kizuki, H. Takadama, M. Matsushita, T. Nakamura, and T. Kokubo, *Acta Biomater.* **6** (2010) 2836.
[68]S. Yamaguchi, H. Takadama, T. Matsushita, T. Nakamura, and T. Kokubo, *J. Mater. Sci.: Mater. Med.* **21** (2010) 439.
[69]S. Yamaguchi, T. Kizuki, H. Takadama, T. Matsushita, T. Nakamura and T. Kokubo, in *The 31st annual meeting of the Japanese Society for Biomaterials*, 2009, p. 247.
[70]S. Yamaguchi, T. Kizuki, H. Takadama, T. Matsushita, T. Kokubo, A. Fukuda, M. Takemoto, S. Fujibayashi, and T. Nakamura, *Orthopedic Ceraic Implants,* in press.
[71]M. Niinomi, *Biomaterials* **24** (2003) 2673.
[72]T. Saito, T. Furuta, J.-H. Hwang, S. kuramoto, K. Nishino, N. Suzuki, R. Chen, A. Yamada, K. Ito, Y. Seno, T. Nonaka, H. ikehata, N. Nagasako, C. Iwamoto, Y. Ikuhara, and T. Sakuma, *Science* **300** (2003) 464.
[73]T. Kokubo, D. K. Pattanayak, S. Yamaguchi, H. Takadama, M. Matsushita, T. Kawai, M. Takemoto, S. Fujibayashi, and T. Nakamura, *J. R. Soc. Interface.* online pubsihed, May 5, 2010.
[74]Y.-T. Sul, C. B. Johansson, Y. Jeong, A. Wennerberg, and T. Albrektsson, *Clin. Oral. Impl. Res.* **13** (2002) 252.
[75]X. X. Wang, S. Hayakawa, K. Tsuru, and A. Osaka, *Biomaterials* **23** (2002) 1353.
[76]X. X. Wang, W. Yan, S. Hayakawa, K. Tsuru, and A. Osaka, *Biomaterials* **24** (2003) 4631.
[77]J.-M. Wu, S. Hayakawa, K. Tsuru and A. Osaka,*J. Am. Ceram. Soc.* **87** (2004) 1635.
[78]B. Yang, M. Uchida, H.-M. Kim, X. Zhang, and T. Kokubo, *Biomaterials* **25** (2004) 1003.
[79]X. Cui, H.-M.Kim, M. Kawashita, L. Wang, T. Xiong, T. Kokubo, and T. Nakamura, *Dent. Mater.* **25** (2009) 80.

[80]X. Lu, Z. Zhao, and Y. Leng, *Mater. Sci. Eng. C* **27** (2007) 700.

[81]W. Zhou, X. Zhong, X. Wu, L. Yuan, Q. Shu, Y. Xia, and K. Ostrikov, *J. Biomed. Mater. Res.* **81A** (2007) 453.

[82]W. Dong, T. Zhang, J. Epstein, L. Cooney, H. Wang, and Y. Li, *Chem. Mater.* **19** (2007) 4454.

[83]X. Lu, Y. Wang, X. Yang, Q. Zhang, Z. Zhao, L.-T. Weng, and Y. Leng, *J. Biomed. Mater. Res.* **84A** (2008) 523.

[84]S. Oh, C. Daraio, L.-H. Chen, T. R. Pisanic, R. R. Fiñones, S. Jin, *J. Biomed. Mater. Res.* **78A** (2006) 97.

[85]K. C. Popat, L. Leoni, C. A. Grimes, T. A. Desai, *Biomaterials* **28** (2007) 3188.

[86]L. M. Bjursten, L. Rasmusson, S. Oh, G. C. Smith, K. S. Brammer, S. Jin, *J. Biomed. Mater. Res.* **92A** (2010) 1218.

[87]A. Sugino, C. Ohtsuki, K. Tsuru, S. Hayakawa, T. Nakano, Y. Okazaki, and A. Osaka, *Acta Biomater.* **5** (2009) 298.

[88]M. Karthega, and N. Rajendran, *Appl. Surf. Sci.* **256** (2010) 2176.

[89]M. Uchida, H.-M. Kim, T. Kokubo, S. Fujibayashi, and T. Nakamura, *J. Biomed. Mater. Res.* **63** (2002) 522.

[90]D. K. Pattanayak, T. Kawai, T. Matsushita, H. Takadama, T. Nakamura, and T. Kokubo, *J. Mater. Sci.: Mater. Med.* **20** (2009) 2401.

[91]D. K. Pattanayak, T. Matsushita, H. Takadama, T. Nakamura and T. Kokubo, in *The 31st annual meeting of the Japanese Society for Biomaterials*, 2009, p. 130.

[92]S. Fujibayashi, T. Nakamura, S. Nishiguchi, J. Tamura, M. Uchida, H.-M. Kim, and T. Kokubo, *J. Biomed. Mater. Res.* **56** (2001) 562.

[93]B. Liang, S. Fujibayashi, M. Neo, J. Tamura, H.-M. Kim, M. Uchida, T. Kokubo, and T. Nakamura, *Biomaterials* **24** (2003) 4959.

[94]S. Fujibayashi, H.-M. Kim, M. Neo, M. Uchida, T. Kokubo, and T. Nakamura, *Biomaterials* **24** (2003) 3445.

[95]M. Takemoto, S. Fujibayashi, M. Neo, J. Suzuki, T. Kokubo, and T. Nakamura, *Biomaterials* **26** (2005) 6014.

[96]B. Otsuki, M. Takemoto, S. Fujibayashi, M. Neo, T. Kokubo, and T. Nakamura, *Biomaterials* **27** (2006) 5892.

[97]K. Tanaka, M. Takemoto, S. Fujibayashi, K. Kawanabe, T. Matsushita, T. Kokubo, and T. Nakamura, *Key Eng. Mater.* **396-398** (2009) 353. [98]S. Fujibayashi, M. Neo, H.-M. Kim, T. Kokubo, and T. Nakamura, *Biomaterials* **25** (2004) 443.

[99]M. Takemoto, S. Fujibayashi, M. Neo, J. Suzuki, T. Matsushita, T. Kokubo, and T. Nakamura, *Biomaterials* **27** (2006) 443. 2682.

[100]J. D. De Brujin, K. Shankar, H. Yuan and P. Habibovic, in *Bioceramics and their clinical applications,* Ed. by T. Kokubo, Woodhead Publishing, Cambridge, 2008, p. 199.

[101]M. Takemoto, S. Fujibayashi, M. Neo, K. So, N. Akiyama, T. Matsushita, T. Kokubo, and T. Nakamura, *J. Neurosurg. Spine* **7** (2007) 435.

[102]X. Zhao, X. Liu, and C. Ding, *J. Biomed. Mater. Res.* **75A** (2005) 888.

[103]X. Zhao, X. Liu, J. You, Z. Chen, and C. Ding, *Surf. Coat. Technol.* **202** (2008) 3221.

[104]T. Kokubo, T. Ueda, M. Kawashita, Y. Ikuhara, G. G. Takaoka, and T. Nakamura, *J. Mater. Sci.: Mater. Med.* **19** (2008) 695.

[105]T. Saito, M. Takemoto, S. Fujibayashi, M. Neo, T. Kokubo, D. Honjyo, T. Hiraide, and T. Nakamura, in *The 22nd European Conference on Biomaterials,* 2009.

[106]M. Uchida, H.-M. Kim, T. Kokubo, K. Tanaka, and T. Nakamura, *J Ceram Soc Japan* **110** (2002) 710.

[107]M. Uchida, H.-M. Kim, F. Miyaji, T. Kokubo, and T. Nakamura, *Biomaterials* **23** (2002) 313.
[108]X.-B.Chen, A. Nouri, Y.-C. Li, J. Lin, P. D. Hodgson, and C. Wen, *Biotecnol Bioeng* **101** (2008) 378.
[109]Y. Hang, Y. Yan, C. Lu, Y. Zhang, and K. Xu, *J. Biomed. Mater. Res.* **88A** (2009) 117.
[110]X. J. Wang, Y. C. Li, J. G. Lin, T. Yamada, P. D. Hodgson, and C. E. Wen, *Acta Biomater.* **4** (2008) 1530.
[111]T. Miyazaki, H.-M. Kim, F. Miyaji, T. Kokubo, H. Kato, and T. Nakamura, *J. Biomed. Mater. Res.* **50** (2000) 35.
[112]T. Miyazaki, H.-M. Kim, T. Kokubo, F. Miyaji, H. Kato, and T. Nakamura, *J. Mater. Sci.: Mater. Med.* **12** (2001) 683.
[113]T. Miyazaki, H.-M. Kim, T. Kokubo, C. Ohtsuki, H. Kato, and T. Nakamura, *J. Mater. Sci.: Mater. Med.* **13** (2002) 651.
[114]D. K. Pattanayak, T. Matsushita, H. Takadama, T. Nakamura and T. Kokubo, in *The 29th annual meeting of the Japanese Society for Biomaterials*, 2007, p. 213.
[115]H. Kato, T. Nakamura, S. Nishiguchi, Y. Matsusue, M. Kobayashi, T. Miyazaki, H.-M. Kim, and T. Kokubo, *J. Biomed. Mater. Res. (Appl. Biomater.)* **53** (2000) 28.

Tadashi Kokubo

Tadashi Kokubo is a Professor at the Department of Biomedical Sciences, College of Life and Health Sciences, Chubu University, Japan. Before joining Chubu University in 2003, he was with the Graduate School of Engineering, Kyoto University. Among several inventions of him, he has developed a high-strength bioactive glass-ceramic A-W for bone substitute, bioactive titanium metal and its alloys for artificial joints, a simulated body fluid (SBF) for evaluating bioactivity of a material *in vitro*. He has won several prestigious awards and honors, including Academic Awards from the Ceramic Society of Japan and Japanese Society for Biomaterials, the George Winter Award from The European Society for Biomaterials, and the Stookey Lecture of Discovery Award from the American Ceramic Society.

Seiji Yamaguchi

Seiji Yamaguchi is a researcher at the College of Life and Health Sciences, Chubu University, Japan. He received a doctorate in Energy Science from Kyoto University, Japan, in 2008. His field of expertise is inorganic solid-state chemistry, with focus on preparation of bioactive Ti-Nb-Ta-Zr alloy by chemical treatment.

Index

N. Eliaz (ed.), *Applications of Electrochemistry and Nanotechnology in Biology and Medicine I*, Modern Aspects of Electrochemistry 52,
DOI 10.1007/978-1-4614-0347-0,

GPSR Compliance

The European Union's (EU) General Product Safety Regulation (GPSR) is a set of rules that requires consumer products to be safe and our obligations to ensure this.

If you have any concerns about our products, you can contact us on ProductSafety@springernature.com

In case Publisher is established outside the EU, the EU authorized representative is:

Springer Nature Customer Service Center GmbH
Europaplatz 3
69115 Heidelberg, Germany

Batch number: 10761521

Printed by Printforce, the Netherlands